HOLT SCIENCE & TECHNOLOGY

Human Body Systems and Health

Annotated Teacher's Edition

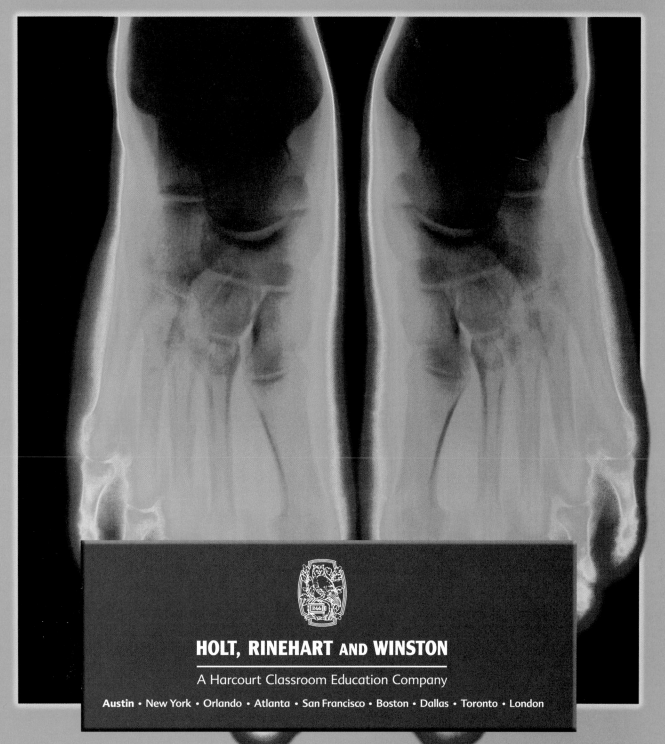

HOLT, RINEHART AND WINSTON

A Harcourt Classroom Education Company

Austin • New York • Orlando • Atlanta • San Francisco • Boston • Dallas • Toronto • London

Acknowledgments

Chapter Writers

Katy Z. Allen
Science Writer and Former Biology Teacher
Wayland, Massachusetts

Linda Ruth Berg, Ph.D.
Adjunct Professor–Natural Sciences
St. Petersburg Junior College
St. Petersburg, Florida

Jennie Dusheck
Science Writer
Santa Cruz, California

Mark F. Taylor, Ph.D.
Associate Professor of Biology
Baylor University
Waco, Texas

Lab Writers

Diana Scheidle Bartos
Science Consultant and Educator
Diana Scheidle Bartos, L.L.C.
Lakewood, Colorado

Carl Benson
General Science Teacher
Plains High School
Plains, Montana

Charlotte Blassingame
Technology Coordinator
White Station Middle School
Memphis, Tennessee

Marsha Carver
Science Teacher and Dept. Chair
McLean County High School
Calhoun, Kentucky

Kenneth E. Creese
Science Teacher
White Mountain Junior High School
Rock Springs, Wyoming

Linda Culp
Science Teacher and Dept. Chair
Thorndale High School
Thorndale, Texas

James Deaver
Science Teacher and Dept. Chair
West Point High School
West Point, Nebraska

Frank McKinney, Ph.D.
Professor of Geology
Appalachian State University
Boone, North Carolina

Alyson Mike
Science Teacher
East Valley Middle School
East Helena, Montana

C. Ford Morishita
Biology Teacher
Clackamas High School
Milwaukie, Oregon

Patricia D. Morrell, Ph.D.
Assistant Professor, School of Education
University of Portland
Portland, Oregon

Hilary C. Olson, Ph.D.
Research Associate
Institute for Geophysics
The University of Texas
Austin, Texas

James B. Pulley
Science Editor and Former Science Teacher
Liberty High School
Liberty, Missouri

Denice Lee Sandefur
Science Chairperson
Nucla High School
Nucla, Colorado

Patti Soderberg
Science Writer
The BioQUEST Curriculum Consortium
Beloit College
Beloit, Wisconsin

Phillip Vavala
Science Teacher and Dept. Chair
Salesianum School
Wilmington, Delaware

Albert C. Wartski
Biology Teacher
Chapel Hill High School
Chapel Hill, North Carolina

Lynn Marie Wartski
Science Writer and Former Science Teacher
Hillsborough, North Carolina

Ivora D. Washington
Science Teacher and Dept. Chair
Hyattsville Middle School
Washington, D.C.

Academic Reviewers

Renato J. Aguilera, Ph.D.
Associate Professor
Department of Molecular, Cell, and Developmental Biology
University of California
Los Angeles, California

David M. Armstrong, Ph.D.
Professor of Biology
Department of E.P.O. Biology
University of Colorado
Boulder, Colorado

Alissa Arp, Ph.D.
Director and Professor of Environmental Studies
Romberg Tiburon Center
San Francisco State University
Tiburon, California

Russell M. Brengelman
Professor of Physics
Morehead State University
Morehead, Kentucky

John A. Brockhaus, Ph.D.
Director of Mapping, Charting, and Geodesy Program
Department of Geography and Environmental Engineering
United States Military Academy
West Point, New York

Linda K. Butler, Ph.D.
Lecturer of Biological Sciences
The University of Texas
Austin, Texas

Barry Chernoff, Ph.D.
Associate Curator
Division of Fishes
The Field Museum of Natural History
Chicago, Illinois

Donna Greenwood Crenshaw, Ph.D.
Instructor
Department of Biology
Duke University
Durham, North Carolina

Hugh Crenshaw, Ph.D.
Assistant Professor of Zoology
Duke University
Durham, North Carolina

Joe W. Crim, Ph.D.
Professor of Biology
University of Georgia
Athens, Georgia

Peter Demmin, Ed.D.
Former Science Teacher and Chair
Amherst Central High School
Amherst, New York

Joseph L. Graves, Jr., Ph.D.
Associate Professor of Evolutionary Biology
Arizona State University West
Phoenix, Arizona

William B. Guggino, Ph.D.
Professor of Physiology and Pediatrics
The Johns Hopkins University School of Medicine
Baltimore, Maryland

David Haig, Ph.D.
Assistant Professor of Biology
Department of Organismic and Evolutionary Biology
Harvard University
Cambridge, Massachusetts

Roy W. Hann, Jr., Ph.D.
Professor of Civil Engineering
Texas A&M University
College Station, Texas

Copyright © 2002 by Holt, Rinehart and Winston

All rights reserved. No part of this publication may be reproduced or transmitted in any form or by any means, electronic or mechanical, including photocopy, recording, or any information storage and retrieval system, without permission in writing from the publisher.

Requests for permission to make copies of any part of the work should be mailed to the following address: Permissions Department, Holt, Rinehart and Winston, 10801 N. MoPac Expressway, Austin, Texas 78759.

For permission to reprint copyrighted material, grateful acknowledgment is made to the following sources:

sciLINKS is owned and provided by the National Science Teachers Association. All rights reserved.

The name of the **Smithsonian Institution** and the sunburst logo are registered trademarks of the Smithsonian Institution. The copyright in the Smithsonian Web site and Smithsonian Web site pages are owned by the Smithsonian Institution. All other material owned and provided by Holt, Rinehart and Winston under copyright appearing above.

Copyright © 2000 **CNN** and **CNNfyi.com** are trademarks of Cable News Network LP, LLLP, a Time Warner Company. All rights reserved. Copyright © 2000 Turner Learning logos are trademarks of Turner Learning, Inc., a Time Warner Company. All rights reserved.

Printed in the United States of America
ISBN 0-03-064781-9
1 2 3 4 5 6 7 048 05 04 03 02 01 00

Acknowledgments (cont.)

John E. Hoover, Ph.D.
Associate Professor of Biology
Millersville University
Millersville, Pennsylvania

Joan E. N. Hudson, Ph.D.
Associate Professor of Biological Sciences
Sam Houston State University
Huntsville, Texas

Laurie Jackson-Grusby, Ph.D.
Research Scientist and Doctoral Associate
Whitehead Institute for Biomedical Research
Massachusetts Institute of Technology
Cambridge, Massachusetts

George M. Langford, Ph.D.
Professor of Biological Sciences
Dartmouth College
Hanover, New Hampshire

Melanie C. Lewis, Ph.D.
Professor of Biology, Retired
Southwest Texas State University
San Marcos, Texas

V. Patteson Lombardi, Ph.D.
Research Assistant Professor of Biology
Department of Biology
University of Oregon
Eugene, Oregon

Glen Longley, Ph.D.
Professor of Biology and Director of the Edwards Aquifer Research Center
Southwest Texas State University
San Marcos, Texas

William F. McComas, Ph.D.
Director of the Center to Advance Science Education
University of Southern California
Los Angeles, California

LaMoine L. Motz, Ph.D.
Coordinator of Science Education
Oakland County Schools
Waterford, Michigan

Nancy Parker, Ph.D.
Associate Professor of Biology
Southern Illinois University
Edwardsville, Illinois

Barron S. Rector, Ph.D.
Associate Professor and Extension Range Specialist
Texas Agricultural Extension Service
Texas A&M University
College Station, Texas

Peter Sheridan, Ph.D.
Professor of Chemistry
Colgate University
Hamilton, New York

Miles R. Silman, Ph.D.
Assistant Professor of Biology
Wake Forest University
Winston-Salem, North Carolina

Neil Simister, Ph.D.
Associate Professor of Biology
Department of Life Sciences
Brandeis University
Waltham, Massachusetts

Lee Smith, Ph.D.
Curriculum Writer
MDL Information Systems, Inc.
San Leandro, California

Robert G. Steen, Ph.D.
Manager, Rat Genome Project
Whitehead Institute—Center for Genome Research
Massachusetts Institute of Technology
Cambridge, Massachusetts

Martin VanDyke, Ph.D.
Professor of Chemistry, Emeritus
Front Range Community College
Westminister, Colorado

E. Peter Volpe, Ph.D.
Professor of Medical Genetics
Mercer University School of Medicine
Macon, Georgia

Harold K. Voris, Ph.D.
Curator and Head
Division of Amphibians and Reptiles
The Field Museum of Natural History
Chicago, Illinois

Mollie Walton
Biology Instructor
El Paso Community College
El Paso, Texas

Peter Wetherwax, Ph.D.
Professor of Biology
University of Oregon
Eugene, Oregon

Mary K. Wicksten, Ph.D.
Professor of Biology
Texas A&M University
College Station, Texas

R. Stimson Wilcox, Ph.D.
Associate Professor of Biology
Department of Biological Sciences
Binghamton University
Binghamton, New York

Conrad M. Zapanta, Ph.D.
Research Engineer
Sulzer Carbomedics, Inc.
Austin, Texas

Safety Reviewer

Jack Gerlovich, Ph.D.
Associate Professor
School of Education
Drake University
Des Moines, Iowa

Teacher Reviewers

Barry L. Bishop
Science Teacher and Dept. Chair
San Rafael Junior High School
Ferron, Utah

Carol A. Bornhorst
Science Teacher and Dept. Chair
Bonita Vista Middle School
Chula Vista, California

Paul Boyle
Science Teacher
Perry Heights Middle School
Evansville, Indiana

Yvonne Brannum
Science Teacher and Dept. Chair
Hine Junior High School
Washington, D.C.

Gladys Cherniak
Science Teacher
St. Paul's Episcopal School
Mobile, Alabama

James Chin
Science Teacher
Frank A. Day Middle School
Newtonville, Massachusetts

Kenneth Creese
Science Teacher
White Mountain Junior High School
Rock Springs, Wyoming

Linda A. Culp
Science Teacher and Dept. Chair
Thorndale High School
Thorndale, Texas

Georgiann Delgadillo
Science Teacher
East Valley Continuous Curriculum School
Spokane, Washington

Alonda Droege
Biology Teacher
Evergreen High School
Seattle, Washington

Michael J. DuPré
Curriculum Specialist
Rush Henrietta Junior-Senior High School
Henrietta, New York

Rebecca Ferguson
Science Teacher
North Ridge Middle School
North Richland Hills, Texas

Susan Gorman
Science Teacher
North Ridge Middle School
North Richland Hills, Texas

Gary Habeeb
Science Mentor
Sierra-Plumas Joint Unified School District
Downieville, California

Karma Houston-Hughes
Science Mentor
Kyrene Middle School
Tempe, Arizona

Roberta Jacobowitz
Science Teacher
C. W. Otto Middle School
Lansing, Michigan

Kerry A. Johnson
Science Teacher
Isbell Middle School
Santa Paula, California

M. R. Penny Kisiah
Science Teacher and Dept. Chair
Fairview Middle School
Tallahassee, Florida

Kathy LaRoe
Science Teacher
East Valley Middle School
East Helena, Montana

Jane M. Lemons
Science Teacher
Western Rockingham Middle School
Madison, North Carolina

Scott Mandel, Ph.D.
Director and Educational Consultant
Teachers Helping Teachers
Los Angeles, California

Thomas Manerchia
Former Biology and Life Science Teacher
Archmere Academy
Claymont, Delaware

Maurine O. Marchani
Science Teacher and Dept. Chair
Raymond Park Middle School
Indianapolis, Indiana

Jason P. Marsh
Biology Teacher
Montevideo High School and Montevideo Country School
Montevideo, Minnesota

Edith C. McAlanis
Science Teacher and Dept. Chair
Socorro Middle School
El Paso, Texas

Kevin McCurdy, Ph.D.
Science Teacher
Elmwood Junior High School
Rogers, Arkansas

Kathy McKee
Science Teacher
Hoyt Middle School
Des Moines, Iowa

Acknowledgments continue on page 211.

D Human Body Systems and Health

CHAPTER 1	Chapter Interleaf 1A
	Body Organization and Structure 2
	1 Body Organization 4
	2 The Skeletal System 8
	3 The Muscular System 12
	4 The Integumentary System 16
CHAPTER 2	Chapter Interleaf 27A
	Circulation and Respiration 28
	1 The Cardiovascular System 30
	2 The Lymphatic System 38
	3 The Respiratory System 40
CHAPTER 3	Chapter Interleaf 51A
	The Digestive and Urinary Systems 52
	1 The Digestive System 54
	2 The Urinary System 62
CHAPTER 4	Chapter Interleaf 73A
	Communication and Control 74
	1 The Nervous System 76
	2 Responding to the Environment 83
	3 The Endocrine System 88
CHAPTER 5	Chapter Interleaf 99A
	Reproduction and Development 100
	1 Animal Reproduction 102
	2 Human Reproduction 106
	3 Growth and Development 110
CHAPTER 6	Chapter Interleaf 123A
	Body Defenses and Disease 124
	1 Disease 126
	2 Your Body's Defenses 130
	3 Challenges to the Immune System 135
CHAPTER 7	Chapter Interleaf 145A
	Staying Healthy 146
	1 What We Put into Our Bodies 148
	2 Risks of Alcohol and Other Drugs 155
	3 Healthy Habits 161

Skills Development

Process Skills

QuickLabs

Pickled Bones	9
Power in Pairs	13
Why Do People Snore?	43
Break It Up!	55
Knee Jerks	82
Where's the Dot?	85
Life Grows On	115
It's Only Skin Deep	131
A Healthy Diet	154

Chapter Labs

Muscles at Work	20
Build a Lung	44
Carbon Dioxide Breath	45
As the Stomach Churns	66
You've Gotta Lotta Nerve	92
It's a Comfy, Safe World!	116
Antibodies to the Rescue	138
To Diet or Not to Diet	166
Keep It Clean	167

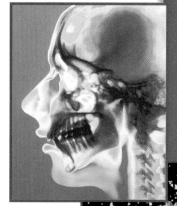

Skills Development (continued)

Research and Critical Thinking Skills

Apply

Blood Delivery	36
Beverage Ban	64
Fight or Flight?	89
Two Types of Twins	108
Cold Calamity	129
Stress SOS	163

Feature Articles

Science, Technology, and Society
- Engineered Skin 26
- Light on Lenses 98
- Technology in Its Infant Stages 123
- Bacteria at Your Service 172

Eureka!
- Hairy Oil Spills 27
- Pathway to a Cure 99

Weird Science
- Catching a Light Sneeze 50

Health Watch
- Goats to the Rescue 51
- A Voiceless Companion 73
- Frogs in the Medicine Cabinet? 145
- Meatless Munching 173

Across the Sciences
- Quench Your Thirst! 72
- Acne 122

Careers
- Naturopathic Physician 144

Connections

Chemistry Connection
- Don't Sweat the Small Stuff 14
- Breathing in the Mountains 41
- Confusing Chemicals 109

Environment Connection
- Drinkable Water 60

Physics Connection
- Too High to Hear 86
- Steamy Sterilization 128

Oceanography Connection
- Eating Seaweed 150

Mathematics

- Runner's Time 15
- The Beat Goes On 37
- How Much Water? 64
- Time to Travel 77
- Chromo-Combos 103
- Counting Eggs 107
- Epidemic! 129
- What Percentage? 153
- Deadly Averages 158

LabBook 174

Appendix 194
    Concept Mapping 186
    SI Measurement 187
    Temperature Scales 188
    Measuring Skills 189
    Scientific Method 190
    Making Charts and Graphs 193
    Periodic Table of the Elements 196
    The Six Kingdoms 198
    Using the Microscope 200

Glossary 202

Index 207

Self-Check Answers 212

Program Scope and Sequence

Selecting the right books for your course is easy. Just review the topics presented in each book to determine the best match to your district curriculum.

	A MICROORGANISMS, FUNGI, AND PLANTS	**B** ANIMALS
CHAPTER 1	**It's Alive!! Or, Is It?** ❏ Characteristics of living things ❏ Homeostasis ❏ Heredity and DNA ❏ Producers, consumers, and decomposers ❏ Biomolecules	**Animals and Behavior** ❏ Characteristics of animals ❏ Classification of animals ❏ Animal behavior ❏ Hibernation and estivation ❏ The biological clock ❏ Animal communication ❏ Living in groups
CHAPTER 2	**Bacteria and Viruses** ❏ Binary fission ❏ Characteristics of bacteria ❏ Nitrogen-fixing bacteria ❏ Antibiotics ❏ Pathogenic bacteria ❏ Characteristics of viruses ❏ Lytic cycle	**Invertebrates** ❏ General characteristics of invertebrates ❏ Types of symmetry ❏ Characteristics of sponges, cnidarians, arthropods, and echinoderms ❏ Flatworms versus roundworms ❏ Types of circulatory systems
CHAPTER 3	**Protists and Fungi** ❏ Characteristics of protists ❏ Types of algae ❏ Types of protozoa ❏ Protist reproduction ❏ Characteristics of fungi and lichens	**Fishes, Amphibians, and Reptiles** ❏ Characteristics of vertebrates ❏ Structure and kinds of fishes ❏ Development of lungs ❏ Structure and kinds of amphibians and reptiles ❏ Function of the amniotic egg
CHAPTER 4	**Introduction to Plants** ❏ Characteristics of plants and seeds ❏ Reproduction and classification ❏ Angiosperms versus gymnosperms ❏ Monocots versus dicots ❏ Structure and functions of roots, stems, leaves, and flowers	**Birds and Mammals** ❏ Structure and kinds of birds ❏ Types of feathers ❏ Adaptations for flight ❏ Structure and kinds of mammals ❏ Function of the placenta
CHAPTER 5	**Plant Processes** ❏ Pollination and fertilization ❏ Dormancy ❏ Photosynthesis ❏ Plant tropisms ❏ Seasonal responses of plants	
CHAPTER 6		
CHAPTER 7		

Scope and Sequence

Life Science

C — CELLS, HEREDITY, & CLASSIFICATION

Cells: The Basic Units of Life
- Cells, tissues, and organs
- Populations, communities, and ecosystems
- Cell theory
- Surface-to-volume ratio
- Prokaryotic versus eukaryotic cells
- Cell organelles

The Cell in Action
- Diffusion and osmosis
- Passive versus active transport
- Endocytosis versus exocytosis
- Photosynthesis
- Cellular respiration and fermentation
- Cell cycle

Heredity
- Dominant versus recessive traits
- Genes and alleles
- Genotype, phenotype, the Punnett square and probability
- Meiosis
- Determination of sex

Genes and Gene Technology
- Structure of DNA
- Protein synthesis
- Mutations
- Heredity disorders and genetic counseling

The Evolution of Living Things
- Adaptations and species
- Evidence for evolution
- Darwin's work and natural selection
- Formation of new species

The History of Life on Earth
- Geologic time scale and extinctions
- Plate tectonics
- Human evolution

Classification
- Levels of classification
- Cladistic diagrams
- Dichotomous keys
- Characteristics of the six kingdoms

D — HUMAN BODY SYSTEMS & HEALTH

Body Organization and Structure
- Homeostasis
- Types of tissue
- Organ systems
- Structure and function of the skeletal system, muscular system, and integumentary system

Circulation and Respiration
- Structure and function of the cardiovascular system, lymphatic system, and respiratory system
- Respiratory disorders

The Digestive and Urinary Systems
- Structure and function of the digestive system
- Structure and function of the urinary system

Communication and Control
- Structure and function of the nervous system and endocrine system
- The senses
- Structure and function of the eye and ear

Reproduction and Development
- Asexual versus sexual reproduction
- Internal versus external fertilization
- Structure and function of the human male and female reproductive systems
- Fertilization, placental development, and embryo growth
- Stages of human life

Body Defenses and Disease
- Types of diseases
- Vaccines and immunity
- Structure and function of the immune system
- Autoimmune diseases, cancer, and AIDS

Staying Healthy
- Nutrition and reading food labels
- Alcohol and drug effects on the body
- Hygiene, exercise, and first aid

E — ENVIRONMENTAL SCIENCE

Interactions of Living Things
- Biotic versus abiotic parts of the environment
- Producers, consumers, and decomposers
- Food chains and food webs
- Factors limiting population growth
- Predator-prey relationships
- Symbiosis and coevolution

Cycles in Nature
- Water cycle
- Carbon cycle
- Nitrogen cycle
- Ecological succession

The Earth's Ecosystems
- Kinds of land and water biomes
- Marine ecosystems
- Freshwater ecosystems

Environmental Problems and Solutions
- Types of pollutants
- Types of resources
- Conservation practices
- Species protection

Energy Resources
- Types of resources
- Energy resources and pollution
- Alternative energy resources

Scope and Sequence (continued)

	F INSIDE THE RESTLESS EARTH	**G** EARTH'S CHANGING SURFACE
CHAPTER 1	**Minerals of the Earth's Crust** ❏ Mineral composition and structure ❏ Types of minerals ❏ Mineral identification ❏ Mineral formation and mining	**Maps as Models of the Earth** ❏ Structure of a map ❏ Cardinal directions ❏ Latitude, longitude, and the equator ❏ Magnetic declination and true north ❏ Types of projections ❏ Aerial photographs ❏ Remote sensing ❏ Topographic maps
CHAPTER 2	**Rocks: Mineral Mixtures** ❏ Rock cycle and types of rocks ❏ Rock classification ❏ Characteristics of igneous, sedimentary, and metamorphic rocks	**Weathering and Soil Formation** ❏ Types of weathering ❏ Factors affecting the rate of weathering ❏ Composition of soil ❏ Soil conservation and erosion prevention
CHAPTER 3	**The Rock and Fossil Record** ❏ Uniformitarianism versus catastrophism ❏ Superposition ❏ The geologic column and unconformities ❏ Absolute dating and radiometric dating ❏ Characteristics and types of fossils ❏ Geologic time scale	**Agents of Erosion and Deposition** ❏ Shoreline erosion and deposition ❏ Wind erosion and deposition ❏ Erosion and deposition by ice ❏ Gravity's effect on erosion and deposition
CHAPTER 4	**Plate Tectonics** ❏ Structure of the Earth ❏ Continental drifts and sea floor spreading ❏ Plate tectonics theory ❏ Types of boundaries ❏ Types of crust deformities	
CHAPTER 5	**Earthquakes** ❏ Seismology ❏ Features of earthquakes ❏ P and S waves ❏ Gap hypothesis ❏ Earthquake safety	
CHAPTER 6	**Volcanoes** ❏ Types of volcanoes and eruptions ❏ Types of lava and pyroclastic material ❏ Craters versus calderas ❏ Sites and conditions for volcano formation ❏ Predicting eruptions	

Earth Science

H WATER ON EARTH	**I** WEATHER AND CLIMATE	**J** ASTRONOMY
The Flow of Fresh Water ❏ Water cycle ❏ River systems ❏ Stream erosion ❏ Life cycle of rivers ❏ Deposition ❏ Aquifers, springs, and wells ❏ Ground water ❏ Water treatment and pollution	**The Atmosphere** ❏ Structure of the atmosphere ❏ Air pressure ❏ Radiation, convection, and conduction ❏ Greenhouse effect and global warming ❏ Characteristics of winds ❏ Types of winds ❏ Air pollution	**Observing the Sky** ❏ Astronomy ❏ Keeping time ❏ Mapping the stars ❏ Scales of the universe ❏ Types of telescope ❏ Radioastronomy
Exploring the Oceans ❏ Properties and characteristics of the oceans ❏ Features of the ocean floor ❏ Ocean ecology ❏ Ocean resources and pollution	**Understanding Weather** ❏ Water cycle ❏ Humidity ❏ Types of clouds ❏ Types of precipitation ❏ Air masses and fronts ❏ Storms, tornadoes, and hurricanes ❏ Weather forecasting ❏ Weather maps	**Formation of the Solar System** ❏ Birth of the solar system ❏ Planetary motion ❏ Newton's Law of Universal Gravitation ❏ Structure of the sun ❏ Fusion ❏ Earth's structure and atmosphere
The Movement of Ocean Water ❏ Types of currents ❏ Characteristics of waves ❏ Types of ocean waves ❏ Tides	**Climate** ❏ Weather versus climate ❏ Seasons and latitude ❏ Prevailing winds ❏ Earth's biomes ❏ Earth's climate zones ❏ Ice ages ❏ Global warming ❏ Greenhouse effect	**A Family of Planets** ❏ Properties and characteristics of the planets ❏ Properties and characteristics of moons ❏ Comets, asteroids, and meteoroids
		The Universe Beyond ❏ Composition of stars ❏ Classification of stars ❏ Star brightness, distance, and motions ❏ H-R diagram ❏ Life cycle of stars ❏ Types of galaxies ❏ Theories on the formation of the universe
		Exploring Space ❏ Rocketry and artificial satellites ❏ Types of Earth orbit ❏ Space probes and space exploration

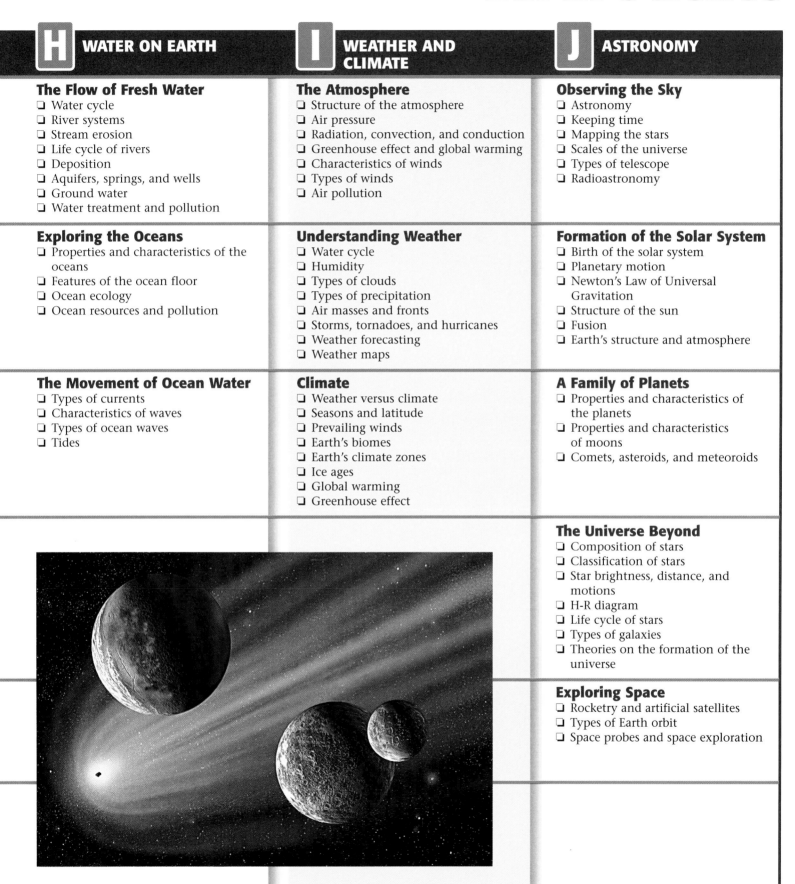

Scope and Sequence (continued)

	K INTRODUCTION TO MATTER	**L** INTERACTIONS OF MATTER
CHAPTER 1	**The Properties of Matter** ❏ Definition of matter ❏ Mass and weight ❏ Physical and chemical properties ❏ Physical and chemical change ❏ Density	**Chemical Bonding** ❏ Types of chemical bonds ❏ Valence electrons ❏ Ions versus molecules ❏ Crystal lattice
CHAPTER 2	**States of Matter** ❏ States of matter and their properties ❏ Boyle's and Charles's laws ❏ Changes of state	**Chemical Reactions** ❏ Writing chemical formulas and equations ❏ Law of conservation of mass ❏ Types of reactions ❏ Endothermic versus exothermic reactions ❏ Law of conservation of energy ❏ Activation energy ❏ Catalysts and inhibitors
CHAPTER 3	**Elements, Compounds, and Mixtures** ❏ Elements and compounds ❏ Metals, nonmetals, and metalloids (semiconductors) ❏ Properties of mixtures ❏ Properties of solutions, suspensions, and colloids	**Chemical Compounds** ❏ Ionic versus covalent compounds ❏ Acids, bases, and salts ❏ pH ❏ Organic compounds ❏ Biomolecules
CHAPTER 4	**Introduction to Atoms** ❏ Atomic theory ❏ Atomic model and structure ❏ Isotopes ❏ Atomic mass and mass number	**Atomic Energy** ❏ Properties of radioactive substances ❏ Types of decay ❏ Half-life ❏ Fission, fusion, and chain reactions
CHAPTER 5	**The Periodic Table** ❏ Structure of the periodic table ❏ Periodic law ❏ Properties of alkali metals, alkaline-earth metals, halogens, and noble gases	
CHAPTER 6		

Scope and Sequence

Physical Science

M — FORCES, MOTION, AND ENERGY

Matter in Motion
- Speed, velocity, and acceleration
- Measuring force
- Friction
- Mass versus weight

Forces in Motion
- Terminal velocity and free fall
- Projectile motion
- Inertia
- Momentum

Forces in Fluids
- Properties in fluids
- Atmospheric pressure
- Density
- Pascal's principle
- Buoyant force
- Archimedes' principle
- Bernoulli's principle

Work and Machines
- Measuring work
- Measuring power
- Types of machines
- Mechanical advantage
- Mechanical efficiency

Energy and Energy Resources
- Forms of energy
- Energy conversions
- Law of conservation of energy
- Energy resources

Heat and Heat Technology
- Heat versus temperature
- Thermal expansion
- Absolute zero
- Conduction, convection, radiation
- Conductors versus insulators
- Specific heat capacity
- Changes of state
- Heat engines
- Thermal pollution

N — ELECTRICITY AND MAGNETISM

Introduction to Electricity
- Law of electric charges
- Conduction versus induction
- Static electricity
- Potential difference
- Cells, batteries, and photocells
- Thermocouples
- Voltage, current, and resistance
- Electric power
- Types of circuits

Electromagnetism
- Properties of magnets
- Magnetic force
- Electromagnetism
- Solenoids and electric motors
- Electromagnetic induction
- Generators and transformers

Electronic Technology
- Properties of semiconductors
- Integrated circuits
- Diodes and transistors
- Analog versus digital signals
- Microprocessors
- Features of computers

O — SOUND AND LIGHT

The Energy of Waves
- Properties of waves
- Types of waves
- Reflection and refraction
- Diffraction and interference
- Standing waves and resonance

The Nature of Sound
- Properties of sound waves
- Structure of the human ear
- Pitch and the Doppler effect
- Infrasonic versus ultrasonic sound
- Sound reflection and echolocation
- Sound barrier
- Interference, resonance, diffraction, and standing waves
- Sound quality of instruments

The Nature of Light
- Electromagnetic waves
- Electromagnetic spectrum
- Law of reflection
- Absorption and scattering
- Reflection and refraction
- Diffraction and interference

Light and Our World
- Luminosity
- Types of lighting
- Types of mirrors and lenses
- Focal point
- Structure of the human eye
- Lasers and holograms

Scope and Sequence

Components Listing

Effective planning starts with all the resources you need in an easy-to-use package for each short course.

Directed Reading Worksheets Help students develop and practice fundamental reading comprehension skills and provide a comprehensive review tool for students to use when studying for an exam.

Study Guide Vocabulary & Notes Worksheets and Chapter Review Worksheets are reproductions of the Chapter Highlights and Chapter Review sections that follow each chapter in the textbook.

Science Puzzlers, Twisters & Teasers Use vocabulary and concepts from each chapter of the Pupil's Editions as elements of rebuses, anagrams, logic puzzles, daffy definitions, riddle poems, word jumbles, and other types of puzzles.

Reinforcement and Vocabulary Review Worksheets Approach a chapter topic from a different angle with an emphasis on different learning modalities to help students that are frustrated by traditional methods.

Critical Thinking & Problem Solving Worksheets Develop the following skills: distinguishing fact from opinion, predicting consequences, analyzing information, and drawing conclusions. Problem Solving Worksheets develop a step-by-step process of problem analysis including gathering information, asking critical questions, identifying alternatives, and making comparisons.

Math Skills for Science Worksheets Each activity gives a brief introduction to a relevant math skill, a step-by-step explanation of the math process, one or more example problems, and a variety of practice problems.

Science Skills Worksheets Help your students focus specifically on skills such as measuring, graphing, using logic, understanding statistics, organizing research papers, and critical thinking options.

LAB ACTIVITIES

Datasheets for Labs These worksheets are the labs found in the *Holt Science & Technology* textbook. Charts, tables, and graphs are included to make data collection and analysis easier, and space is provided to write observations and conclusions.

Whiz-Bang Demonstrations Discovery or Making Models experiences label each demo as one in which students discover an answer or use a scientific model.

Calculator-Based Labs Give students the opportunity to use graphing-calculator probes and sensors to collect data using a TI graphing calculator, Vernier sensors, and a TI CBL 2™ or Vernier Lab Pro interface.

EcoLabs and Field Activities Focus on educational outdoor projects, such as wildlife observation, nature surveys, or natural history.

Inquiry Labs Use the scientific method to help students find their own path in solving a real-world problem.

Long-Term Projects and Research Ideas Provide students with the opportunity to go beyond library and Internet resources to explore science topics.

ASSESSMENT

Chapter Tests Each four-page chapter test consists of a variety of item types including Multiple Choice, Using Vocabulary, Short Answer, Critical Thinking, Math in Science, Interpreting Graphics, and Concept Mapping.

Performance-Based Assessments Evaluate students' abilities to solve problems using the tools, equipment, and techniques of science. Rubrics included for each assessment make it easy to evaluate student performance.

TEACHER RESOURCES

Lesson Plans Integrate all of the great resources in the *Holt Science & Technology* program into your daily teaching. Each lesson plan includes a correlation of the lesson activities to the National Science Education Standards.

Teaching Transparencies Each transparency is correlated to a particular lesson in the Chapter Organizer.

Concept Mapping Transparencies, Worksheets, and Answer Key

Give students an opportunity to complete their own concept maps to study the concepts within each chapter and form logical connections. Student worksheets contain a blank concept map with linking phrases and a list of terms to be used by the student to complete the map.

TECHNOLOGY RESOURCES

One-Stop Planner CD-ROM

Finding the right resources is easy with the One-Stop Planner CD-ROM. You can view and print any resource with just the click of a mouse. Customize the suggested lesson plans to match your daily or weekly calendar and your district's requirements. Powerful test generator software allows you to create customized assessments using a databank of items.

The One-Stop Planner for each level includes the following:

- All materials from the Teaching Resources
- Bellringer Transparency Masters
- Block Scheduling Tools
- Standards Correlations
- Lab Inventory Checklist
- Safety Information
- Science Fair Guide
- Parent Involvement Tools
- Spanish Audio Scripts
- Spanish Glossary
- Assessment Item Listing
- Assessment Checklists and Rubrics
- Test Generator

 sciLINKS

*sci*LINKS numbers throughout the text take you and your students to some of the best on-line resources available. Sites are constantly reviewed and updated by the National Science Teachers Association. Special "teacher only" sites are available to you once you register with the service.

 go.hrw.com

To access Holt, Rinehart and Winston Web resources, use the home page codes for each level found on page 1 of the Pupil's Editions. The codes shown on the Chapter Organizers for each chapter in the Annotated Teacher's Edition take you to chapter-specific resources.

 Smithsonian Institution

Find lesson plans, activities, interviews, virtual exhibits, and just general information on a wide variety of topics relevant to middle school science.

CNNfyi.com

Find the latest in late-breaking science news for students. Featured news stories are supported with lesson plans and activities.

 Presents Science in the News Video Library

Bring relevant science news stories into the classroom. Each video comes with a Teacher's Guide and set of Critical Thinking Worksheets that develop listening and media analysis skills. Tapes in the series include:

- Eye on the Environment
- Multicultural Connections
- Scientists in Action
- Science, Technology & Society

 Guided Reading Audio CD Program

Students can listen to a direct read of each chapter and follow along in the text. Use the program as a content bridge for struggling readers and students for whom English is not their native language.

 Interactive Explorations CD-ROM

Turn a computer into a virtual laboratory. Students act as lab assistants helping Dr. Crystal Labcoat solve real-world problems. Activities develop students' inquiry, analysis, and decision-making skills.

Interactive Science Encyclopedia CD-ROM

Give your students access to more than 3,000 cross-referenced scientific definitions, in-depth articles, science fair project ideas, activities, and more.

ADDITIONAL COMPONENTS

Holt Anthology of Science Fiction

Science Fiction features in the Pupil's Edition preview the stories found in the anthology. Each story begins with a Reading Prep guide and closes with Think About It questions.

Professional Reference for Teachers

Articles written by leading educators help you learn more about the National Science Education Standards, block scheduling, classroom management techniques, and more. A bibliography of professional references is included.

Holt Science Posters

Seven wall posters highlight interesting topics, such as the Physics of Sports, or useful reference material, such as the Scientific Method.

 Holt Science Skills Workshop: Reading in the Content Area

Use a variety of in-depth skills exercises to help students learn to read science materials strategically.

Key	
	These materials are blackline masters.
	All titles shown in green are found in the *Teaching Resources* booklets for each course.

Science & Math Skills Worksheets

The *Holt Science and Technology* program helps you meet the needs of a wide variety of students, regardless of their skill level. The following pages provide examples of the worksheets available to improve your students' science and math skills, whether they already have a strong science and math background or are weak in these areas. Samples of assessment checklists and rubrics are also provided.

In addition to the skills worksheets represented here, *Holt Science and Technology* provides a variety of worksheets that are correlated directly with each chapter of the program. Representations of these worksheets are found at the beginning of each chapter in this Annotated Teacher's Edition. Specific worksheets related to each chapter are listed in the Chapter Organizer. Worksheets and transparencies are found in the softcover *Teaching Resources* for each course.

Many worksheets are also available on the HRW Web site. The address is **go.hrw.com.**

Science Skills Worksheets: Thinking Skills

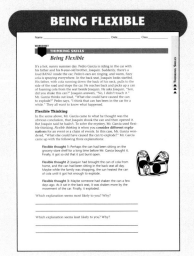
BEING FLEXIBLE

USING YOUR SENSES

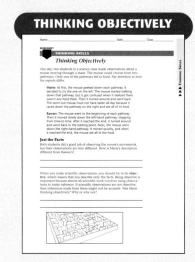
THINKING OBJECTIVELY

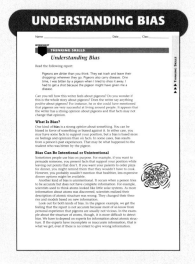
UNDERSTANDING BIAS

USING LOGIC

BOOSTING YOUR MEMORY

IMPROVING YOUR STUDY HABITS

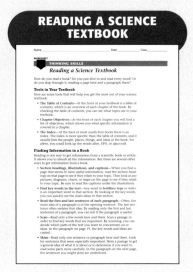
READING A SCIENCE TEXTBOOK

Science Skills Worksheets: Experimenting Skills

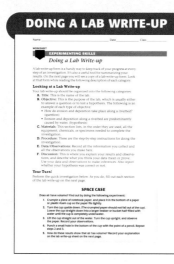

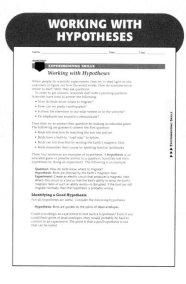

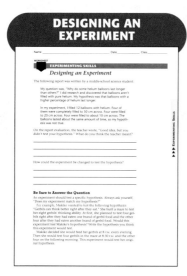

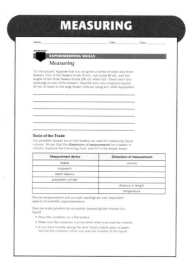

Science Skills Worksheets: Researching Skills

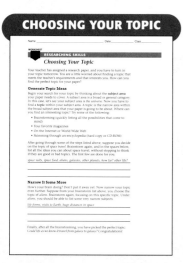
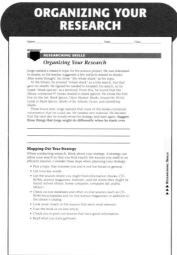

Science & Math Skills Worksheets

Science & Math Skills Worksheets (continued)

Science Skills Worksheets: Researching Skills (continued)

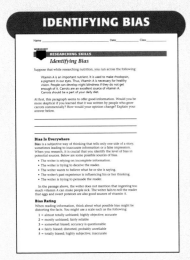
IDENTIFYING BIAS

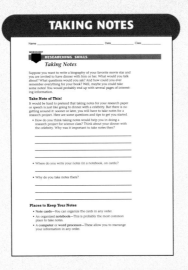
TAKING NOTES

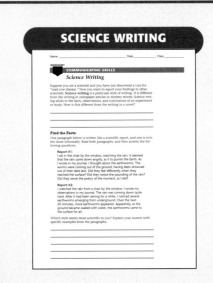
SCIENCE WRITING

Science Skills Worksheets: Communicating Skills

SCIENCE DRAWING

USING MODELS TO COMMUNICATE

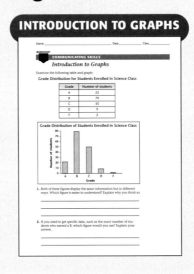
INTRODUCTION TO GRAPHS

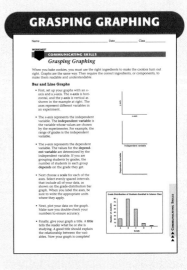
GRASPING GRAPHING

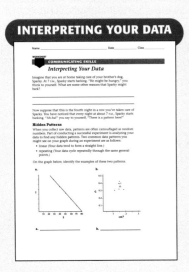
INTERPRETING YOUR DATA

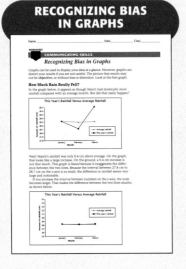
RECOGNIZING BIAS IN GRAPHS

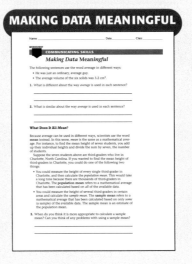
MAKING DATA MEANINGFUL

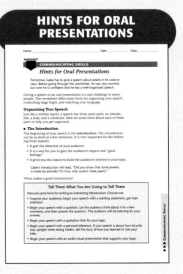
HINTS FOR ORAL PRESENTATIONS

Science & Math Skills Worksheets

Math Skills for Science

ADDITION AND SUBTRACTION

Worksheet preview: Addition Review and Subtraction Review pages with sample problems, procedures, and practice exercises ("Add It Up!" and "Take It Away!").

MULTIPLICATION

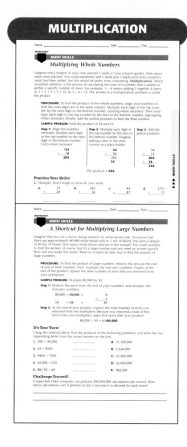

Worksheets: Multiplying Whole Numbers and A Shortcut for Multiplying Large Numbers.

DIVISION

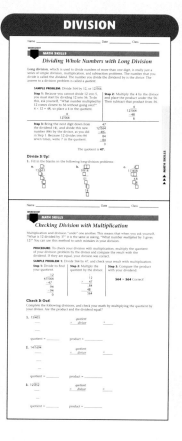

Worksheets: Dividing Whole Numbers with Long Division and Checking Division with Multiplication.

AVERAGES

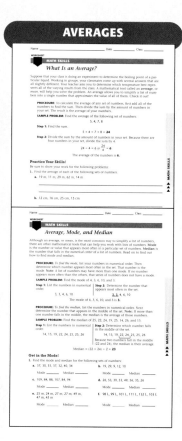

Worksheets: What is an Average? and Average, Mode, and Median.

POSITIVE AND NEGATIVE NUMBERS

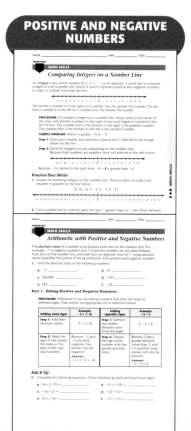

Worksheets: Comparing Integers on a Number Line and Arithmetic with Positive and Negative Numbers.

FRACTIONS

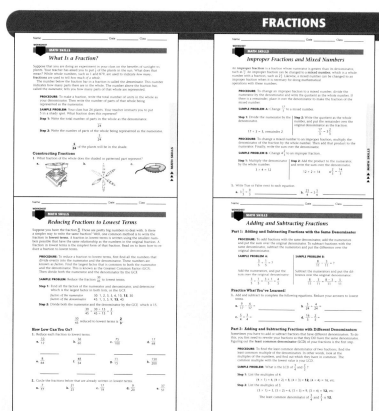

Worksheets: What Is a Fraction?, Improper Fractions and Mixed Numbers, Multiplying and Dividing Fractions, Reducing Fractions to Lowest Terms, and Adding and Subtracting Fractions.

Science & Math Skills Worksheets

Science & Math Skills Worksheets (continued)

Math Skills for Science (continued)

RATIOS AND PROPORTIONS
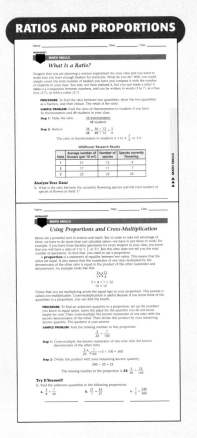

DECIMALS
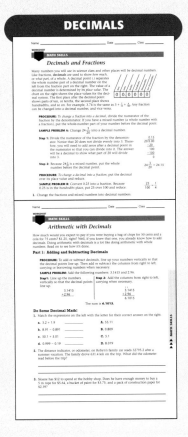

PERCENTAGES
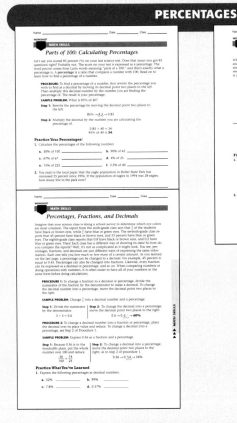

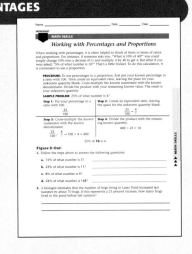

POWERS OF 10
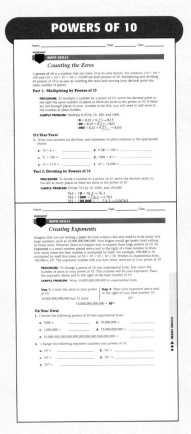

SCIENTIFIC NOTATION
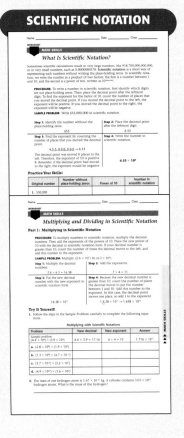

SI MEASUREMENT AND CONVERSION
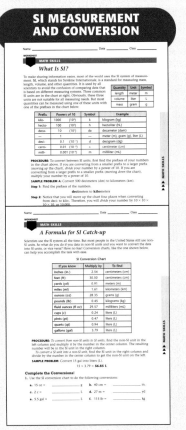

Math Skills for Science (continued)

GEOMETRY

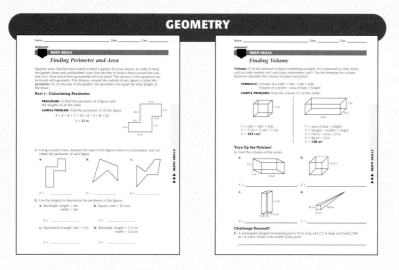

THE UNIT FACTOR AND DIMENSIONAL ANALYSIS

MATH IN SCIENCE: INTEGRATED SCIENCE

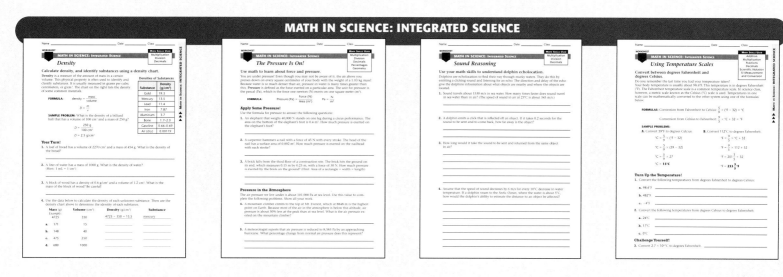

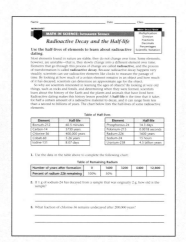

Science & Math Skills Worksheets

Science & Math Skills Worksheets (continued)

Math Skills for Science (continued)

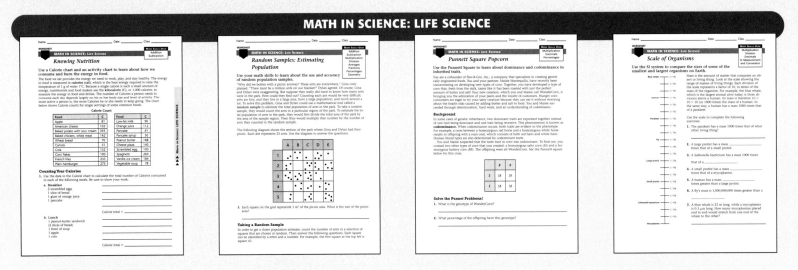

MATH IN SCIENCE: LIFE SCIENCE

Assessment Checklist & Rubrics

The following is just a sample of over 50 checklists and rubrics contained in this booklet.

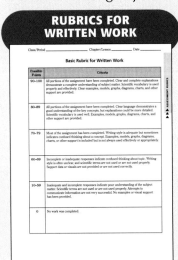

RUBRICS FOR WRITTEN WORK

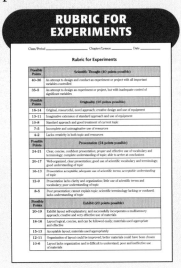

RUBRIC FOR EXPERIMENTS

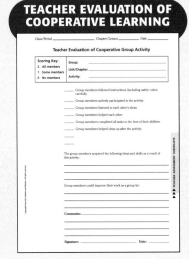

TEACHER EVALUATION OF COOPERATIVE LEARNING

TEACHER EVALUATION OF STUDENT PROGRESS

LIFE SCIENCE — NATIONAL SCIENCE EDUCATION STANDARDS CORRELATIONS

The following lists show the chapter correlation of **Holt Science and Technology: Human Body Systems and Health** with the *National Science Education Standards* (grades 5-8)

UNIFYING CONCEPTS AND PROCESSES

Standard	Chapter Correlation
Systems, order, and organization Code: UCP 1	Chapter 1 1.1, 1.3 Chapter 2 2.1, 2.2, 2.3 Chapter 3 3.1, 3.2 Chapter 4 4.1, 4.3 Chapter 5 5.2 Chapter 6 6.2 Chapter 7 7.1
Evidence, models, and explanation Code: UCP 2	Chapter 1 1.2, 1.3 Chapter 2 2.1, 2.3 Chapter 3 3.1, 3.2, 3.3 Chapter 4 4.1, 4.2 Chapter 5 5.3 Chapter 6 6.1, 6.2 Chapter 7 7.3
Change, constancy, and measurement Code: UCP 3	Chapter 1 1.1, 1.2, 1.3, 1.4 Chapter 2 2.1 Chapter 3 3.2 Chapter 4 4.3 Chapter 5 5.3 Chapter 6 6.2, 6.3
Evolution and equilibrium Code: UCP 4	Chapter 1 1.1, 1.3 Chapter 2 2.3 Chapter 3 3.2 Chapter 4 4.1, 4.2, 4.3 Chapter 6 6.2, 6.3 Chapter 7 7.1
Form and function Code: UCP 5	Chapter 1 1.2, 1.4 Chapter 2 2.1, 2.2, 2.3 Chapter 3 3.1, 3.2 Chapter 4 4.1, 4.2 Chapter 5 5.1, 5.2, 5.3 Chapter 7 7.2

SCIENCE & TECHNOLOGY

Standard	Chapter Correlation
Abilities of technological design Code: ST 1	Chapter 4 4.3
Understandings about science and technology Code: ST 2	Chapter 5 5.3 Chapter 6 6.1

SCIENCE AS INQUIRY

Standard	Chapter Correlation
Abilities necessary to do scientific inquiry Code: SAI 1	Chapter 1 1.2, 1.3 Chapter 2 2.3 Chapter 3 3.1, 3.2 Chapter 4 4.1, 4.2 Chapter 5 5.1, 5.2, 5.3 Chapter 6 6.1, 6.2 Chapter 7 7.1, 7.2, 7.3
Understandings about scientific inquiry Code: SAI 2	Chapter 1 1.3, 1.4 Chapter 2 2.3 Chapter 4 4.1 Chapter 6 6.1

HISTORY AND NATURE OF SCIENCE

Standard	Chapter Correlation
Science as a human endeavor Code: HNS 1	Chapter 3 3.1, 3.2 Chapter 6 6.1
History of science Code: HNS 3	Chapter 3 3.1 Chapter 6 6.1, 6.2

SCIENCE IN PERSONAL AND SOCIAL PERSPECTIVES

Standard	Chapter Correlation
Personal health Code: SPSP 1	Chapter 1 1.2, 1.3, 1.4 Chapter 2 2.1, 2.3 Chapter 3 3.1 Chapter 4 4.1, 4.2 Chapter 5 5.2, 5.3 Chapter 6 6.1 Chapter 7 7.1, 7.2, 7.3
Natural hazards Code: SPSP 3	Chapter 7 7.3
Risks and benefits Code: SPSP 4	Chapter 1 1.2 Chapter 2 2.1, 2.3 Chapter 5 5.2 Chapter 7 7.1, 7.2, 7.3
Science and technology in society Code: SPSP 5	Chapter 2 2.1, 2.3 Chapter 4 4.1, 4.2, 4.3 Chapter 5 5.3 Chapter 6 6.1 Chapter 7 7.2, 7.3

Standards Correlations **T23**

LIFE SCIENCE
NATIONAL SCIENCE EDUCATION CONTENT STANDARDS

STRUCTURE AND FUNCTION IN LIVING SYSTEMS

Standard	Chapter Correlation	
Living systems at all levels of organization demonstrate the complementary nature of structure and function. Important levels of organization for structure and function include cells, organs, tissues, organ systems, whole organisms, and ecosystems. Code: LS 1a	Chapter 1 Chapter 2 Chapter 3 Chapter 4 Chapter 5	1.1 2.1, 2.2, 2.3 3.1, 3.2 4.1, 4.2 5.2, 5.3
All organisms are composed of cells—the fundamental unit of life. Most organisms are single cells; other organisms, including humans, are multicellular. Code: LS 1b	Chapter 5 Chapter 6	5.3 6.2
Cells carry on the many functions needed to sustain life. They grow and divide, thereby producing more cells. This requires that they take in nutrients, which they use to provide energy for the work that cells do and to make the materials that a cell or an organism needs. Code: LS 1c	Chapter 1 Chapter 2 Chapter 3	1.4 2.1, 2.3 3.2
Specialized cells perform specialized functions in multicellular organisms. Groups of specialized cells co-operate to form a tissue, such as a muscle. Different tissues are in turn grouped together and form larger functional units, called organs. Each type of cell, tissue, and organ has a distinct structure and set of functions that serve the organism as a whole. Code: LS 1d	Chapter 1 Chapter 2 Chapter 3 Chapter 4 Chapter 5 Chapter 6	1.1, 1.2, 1.3, 1.4 2.1, 2.2, 2.3 3.1, 3.2 4.1, 4.2 5.2, 5.3 6.2, 6.3
The human organism has systems for digestion, respiration, reproduction, circulation, excretion, movement, control and coordination, and protection from disease. These systems interact with one another. Code: LS 1e	Chapter 1 Chapter 2 Chapter 3 Chapter 4 Chapter 6	1.1, 1.2, 1.3, 1.4 2.1, 2.2, 2.3 3.1, 3.2 4.1, 4.2, 4.3 6.2
Disease is the breakdown in structures or functions of an organism. Some diseases are the result of intrinsic failures of the system. Others are the result of damage by infection by other organisms. Code: LS 1f	Chapter 1 Chapter 2 Chapter 3 Chapter 4 Chapter 5 Chapter 6	1.4 2.1, 2.2, 2.3 3.1, 3.2 4.1, 4.3 5.2 6.1, 6.2, 6.3

REPRODUCTION AND HEREDITY

Standard	Chapter Correlation
Reproduction is a characteristic of all living systems; because no individual organism lives forever, reproduction is essential to the continuation of every species. Some organisms reproduce asexually. Others reproduce sexually. Code: LS 2a	**Chapter 5** 5.1
In many species, including humans, females produce eggs and males produce sperm. Plants also reproduce sexually—the egg and sperm are produced in the flowers of flowering plants. An egg and sperm unite to begin development of a new individual. The individual receives genetic information from its mother (via the egg) and its father (via the sperm). Sexually produced offspring never are identical to either of their parents. Code: LS 2b	**Chapter 5** 5.1, 5.2, 5.3
Hereditary information is contained in the genes, located in the chromosomes of each cell. Each gene carries a single unit of information. An inherited trait of an individual can be determined by one or by many genes, and a single gene can influence more than one trait. A human cell contains many thousands of different genes. Code: LS 2d	**Chapter 5** 5.1

REGULATION AND BEHAVIOR

Standard	Chapter Correlation
All organisms must be able to obtain and use resources, grow, reproduce, and maintain stable internal conditions while living in a constantly changing external environment. Code: LS 3a	**Chapter 1** 1.1, 1.4 **Chapter 3** 3.2 **Chapter 4** 4.1, 4.2, 4.3 **Chapter 6** 6.2
Regulation of an organism's internal environment involves sensing the internal environment and changing physiological activities to keep conditions within the range required to survive. Code: LS 3b	**Chapter 1** 1.4 **Chapter 2** 2.1 **Chapter 3** 3.1, 3.2 **Chapter 4** 4.1, 4.3 **Chapter 6** 6.2, 6.3
Behavior is one kind of response an organism can make to an internal or environmental stimulus. A behavioral response requires coordination and communication at many levels, including cells, organ systems, and whole organisms. Behavioral response is a set of actions determined in part by heredity and in part from experience. Code: LS 3c	**Chapter 4** 4.1, 4.3

Content Standards

Master Materials List

For added convenience, Science Kit® provides materials-ordering software on CD-ROM designed specifically for *Holt Science and Technology*. Using this software, you can order complete kits or individual items, quickly and efficiently.

Consumable Materials	Amount	Page
Agar, nutrient	30 mL	125, 167
Apple	1	131
Bag, plastic, sealable	2	53, 116
Bag, plastic, trash (small)	1	44
Balloon, small	1	44
Beef stew meat, 1 cm cubes	3	66
Bone, chicken	1	9
Bottle, soda, 2 L	1	44
Candy, hard	2 pieces	55
Clay, water-resistant, modeling	2 sticks	44
Cotton ball	30	116
Cup, plastic	2	55
Egg, chicken, soft-boiled	3–5	116
Flour	200 mL	53
Food coloring, various colors	1 box	131
Gloves, protective	1 pair	45, 66, 116, 125, 180, 184
Glue, white	1 bottle	138
Hydrochloric acid, 0.1 M	10 mL	66
Hydrogen peroxide, diluted	4 mL	180
Ice, in pieces	4–6	3
Liver, beef, 1 cm cubes	3	180
Marker, black, permanent	1	66, 178
Marker, fine-point, washable	1	92

Consumable Materials	Amount	Page
Marker, various colors	1 pack	166
Meat tenderizer with bromelain	1 bottle	66
Meat tenderizer with papain	1 bottle	66
Oil, cooking	100 mL	53
Oil, cooking	1 cup	116
Paper, construction, various colors	5 sheets	138
Paper, graphing	1 sheet	92, 178, 182
Paper, wax, 15 x 15 cm	1 sheet	43
Paper, white	1 sheet	166
Paper towel	1 roll	45
Pencil, assorted, colored	4	182
Pencil, wax	1	167
Phenol (carbolic acid) solution	4 drops	45
Pipe cleaner	3	138
Plastic wrap, clear, approx. 1 x 2 ft	1 sheet	131
Soap, liquid, antibacterial	1 bottle	125, 167
Sodium hydroxide	15 g	184
Straw, drinking	1	44, 45
Tape, masking	25 cm	66
Tape, transparent	1 roll	44, 125, 138, 167
Tissue, facial	1	92
Vinegar	2 cups	9

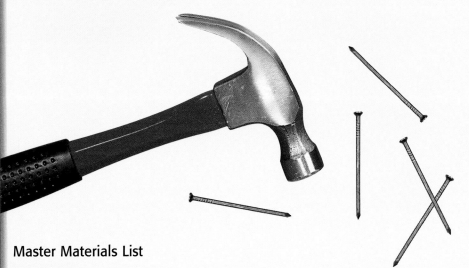

Nonconsumable Equipment	Amount	Page
Beaker, 250 mL	1	184
Cork, small	1	92
Eyedropper, plastic	1	45, 66, 92, 131, 184
Flask, Erlenmeyer, 150 mL	1	45
Graduated cylinder, 10 mL	1	180
Graduated cylinder, 25 mL	1	66
Graduated cylinder, 150 mL	1	45
Hammer	1	55
Incubator	1	167
Jar, wide-mouthed	1	9
Marble	2	125
Measuring spoons	1 set	66
Meterstick	1	75, 101
Mortar and pestle	1	180
Pan, square	1	3
Petri dish, plastic	2	125

Nonconsumable Equipment	Amount	Page
Petri dish, plastic	4	167
Pin, dissecting	1	92
Plate, small	1	180
Rubber band	2	44
Ruler, metric	1	44, 92, 115, 178
Scissors	1	131, 138
Scrub brush, new	1	167
Spatula	1	180
Stopwatch	1	20, 29, 45
Tape measure	1	101
Test tube	3	180
Test tube	4	66
Test-tube rack	1	66, 180
Thermometer	1	20
Towel	1	55
Tweezers	1	180

Master Materials List

Answers to Concept Mapping Questions

The following pages contain sample answers to all of the concept mapping questions that appear in the Chapter Reviews. Because there is more than one way to do a concept map, your students' answers may vary.

CHAPTER 1 Body Organization and Structure

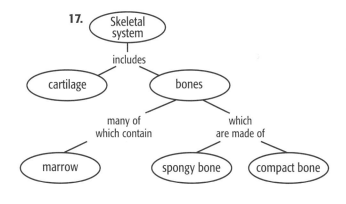

CHAPTER 2 Circulation and Respiration

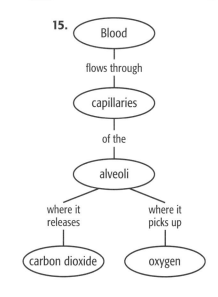

CHAPTER 3 The Digestive and Urinary Systems

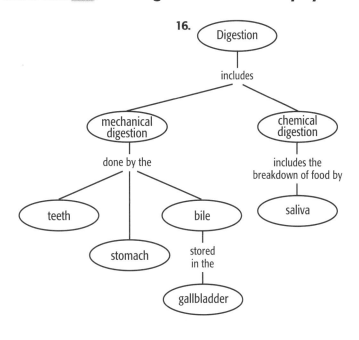

CHAPTER 4 Communication and Control

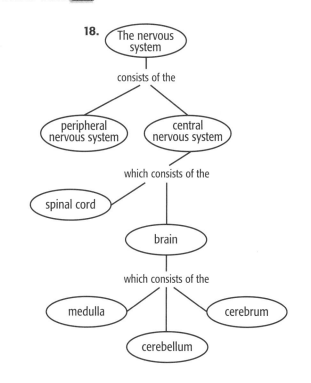

Concept Mapping Answers

CHAPTER 5 Reproduction and Development

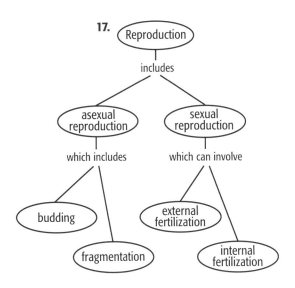

CHAPTER 6 Body Defenses and Disease

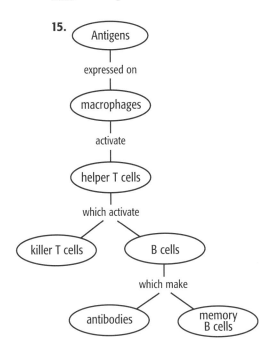

CHAPTER 7 Staying Healthy

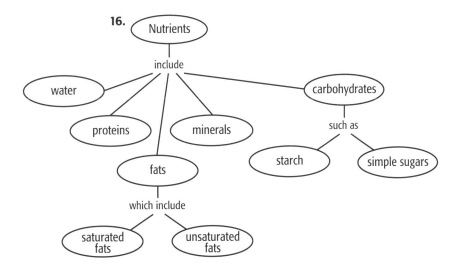

Concept Mapping Answers

To the Student

This book was created to make your science experience interesting, exciting, and fun!

Go for It!

Science is a process of discovery, a trek into the unknown. The skills you develop using *Holt Science & Technology*—such as observing, experimenting, and explaining observations and ideas—are the skills you will need for the future. There is a universe of exploration and discovery awaiting those who accept the challenges of science.

Science & Technology

You see the interaction between science and technology every day. Science makes technology possible. On the other hand, some of the products of technology, such as computers, are used to make further scientific discoveries. In fact, much of the scientific work that is done today has become so technically complicated and expensive that no one person can do it entirely alone. But make no mistake, the creative ideas for even the most highly technical and expensive scientific work still come from individuals.

Activities and Labs

The activities and labs in this book will allow you to make some basic but important scientific discoveries on your own. You can even do some exploring on your own at home! Here's your chance to use your imagination and curiosity as you investigate your world.

Keep a ScienceLog

In this book, you will be asked to keep a type of journal called a ScienceLog to record your thoughts, observations, experiments, and conclusions. As you develop your ScienceLog, you will see your own ideas taking shape over time. You'll have a written record of how your ideas have changed as you learn about and explore interesting topics in science.

Know "What You'll Do"

The "What You'll Do" list at the beginning of each section is your built-in guide to what you need to learn in each chapter. When you can answer the questions in the Section Review and Chapter Review, you know you are ready for a test.

Check Out the Internet

You will see this 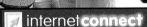 logo throughout the book. You'll be using sciLINKS as your gateway to the Internet. Once you log on to sciLINKS using your computer's Internet link, type in the sciLINKS address. When asked for the keyword code, type in the keyword for that topic. A wealth of resources is now at your disposal to help you learn more about that topic.

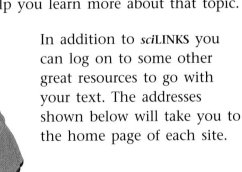

In addition to sciLINKS you can log on to some other great resources to go with your text. The addresses shown below will take you to the home page of each site.

internet connect

This textbook contains the following on-line resources to help you make the most of your science experience.

Visit **go.hrw.com** for extra help and study aids matched to your textbook. Just type in the keyword HG2 HOME.

Visit **www.scilinks.org** to find resources specific to topics in your textbook. Keywords appear throughout your book to take you further.

  Smithsonian Institution®
Internet Connections

Visit **www.si.edu/hrw** for specifically chosen on-line materials from one of our nation's premier science museums.

Visit **www.cnnfyi.com** for late-breaking news and current events stories selected just for you.

To the Student

Chapter Organizer

CHAPTER ORGANIZATION	TIME MINUTES	OBJECTIVES	LABS, INVESTIGATIONS, AND DEMONSTRATIONS
Chapter Opener pp. 2–3	45	National Standards: UCP 3, SAI 2, LS 3a	**Start-Up Activity,** Too Cold for Comfort, p. 3
Section 1 Body Organization	90	▶ Identify the major tissues found in the body. ▶ Compare an organ with an organ system. ▶ Describe a major function of each organ system. UCP 1, 3, LS 1a, 1d, 1e, 3a	
Section 2 The Skeletal System	90	▶ Identify the major organs of the skeletal system. ▶ Describe the functions of bones. ▶ Illustrate the internal structure of bones. ▶ Compare three types of joints. ▶ Discuss how bones function as levers. UCP 2–5, SAI 1, SPSP 1, 4, LS 1a, 1d, 1e	**QuickLab,** Pickled Bones, p. 9 **Demonstration,** Bone Dissection, p. 9 in ATE
Section 3 The Muscular System	90	▶ List the major parts of the muscular system. ▶ Describe the different types of muscle. ▶ Describe how skeletal muscles move bones. ▶ Compare aerobic exercise with resistance exercise. ▶ Give an example of a muscle injury. UCP 1, 4, SPSP 1, LS 1d, 1e; Labs UCP 2, 3, SAI 1, 2	**QuickLab,** Power in Pairs, p. 13 **Design Your Own,** Muscles at Work, p. 20 **Datasheets for LabBook,** Muscles at Work **Inquiry Labs,** On a Wing and a Layer
Section 4 The Integumentary System	90	▶ Describe the major functions of the integumentary system. ▶ List the major parts of the skin, and discuss their functions. ▶ Describe the structure and function of hair and nails. ▶ Describe some common types of damage that can affect skin. UCP 5, SPSP 1, LS 1c–1f, 3a, 3b; Labs UCP 3, SAI 2	**Skill Builder,** Seeing Is Believing, p. 178 **Datasheets for LabBook,** Seeing Is Believing **Long-Term Projects & Research Ideas,** Mapping the Human Body

See page **T23** *for a complete correlation of this book with the*

NATIONAL SCIENCE EDUCATION STANDARDS.

TECHNOLOGY RESOURCES

 Guided Reading Audio CD English or Spanish, Chapter 1

 One-Stop Planner CD-ROM with Test Generator

 CNN Scientists in Action, Segments 8 and 27

Science, Technology & Society, Manufactured Body Parts, Segment 27

Multicultural Connections, African-American Burial Ground, Segment 5

Chapter 1 • Body Organization and Structure

CLASSROOM WORKSHEETS, TRANSPARENCIES, AND RESOURCES	SCIENCE INTEGRATION AND CONNECTIONS	REVIEW AND ASSESSMENT
Directed Reading Worksheet **Science Puzzlers, Twisters & Teasers**		
Directed Reading Worksheet, Section 1 **Transparency 78,** Organ Systems	**Cross-Disciplinary Focus,** p. 5 in ATE	**Section Review,** p. 7 **Quiz,** p. 7 in ATE **Alternative Assessment,** p. 7 in ATE
Directed Reading Worksheet, Section 2 **Transparency 79,** What's in a Bone? **Transparency 231,** Machines Change the Size or Direction (or Both) of a Force **Math Skills for Science Worksheet,** Mechanical Advantage **Reinforcement Worksheet,** The Hipbone's Connected to the… **Critical Thinking Worksheet,** The Tissue Engineering Debate	**Multicultural Connection,** p. 9 in ATE **Connect to Physical Science,** p. 10 in ATE	**Section Review,** p. 11 **Quiz,** p. 11 in ATE **Alternative Assessment,** p. 11 in ATE
Teaching Transparency 80, Types of Muscle **Directed Reading Worksheet,** Section 3 **Math Skills for Science Worksheet,** The Unit Factor and Dimensional Analysis **Reinforcement Worksheet,** Muscle Map	**Chemistry Connection,** p. 14 **Math and More,** p. 14 in ATE **MathBreak,** Runner's Time, p. 15	**Self-Check,** p. 14 **Homework,** p. 14 in ATE **Section Review,** p. 15 **Quiz,** p. 15 in ATE **Alternative Assessment,** p. 15 in ATE
Teaching Transparency 81, The Skin **Directed Reading Worksheet,** Section 4	**Real-World Connection,** p. 18 in ATE **Science, Technology, and Society:** Engineered Skin, p. 26 **Eureka!** Hairy Oil Spills, p. 27	**Self-Check,** p. 17 **Section Review,** p. 19 **Quiz,** p. 19 in ATE **Alternative Assessment,** p. 19 in ATE

 Holt, Rinehart and Winston On-line Resources
go.hrw.com

For worksheets and other teaching aids related to this chapter, visit the HRW Web site and type in the keyword: **HSTBD1**

 National Science Teachers Association
www.scilinks.org

Encourage students to use the *sci*LINKS numbers listed in the internet connect boxes to access information and resources on the **NSTA** Web site.

END-OF-CHAPTER REVIEW AND ASSESSMENT

Chapter Review in Study Guide
Vocabulary and Notes in Study Guide
Chapter Tests with Performance-Based Assessment, Chapter 1 Test
Chapter Tests with Performance-Based Assessment, Performance-Based Assessment 1
Concept Mapping Transparency 22

Chapter Resources & Worksheets

Visual Resources

TEACHING TRANSPARENCIES

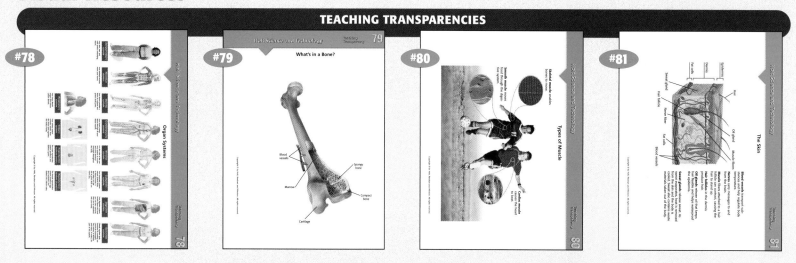

TEACHING TRANSPARENCIES

CONCEPT MAPPING TRANSPARENCY

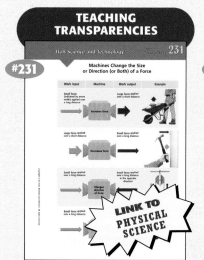

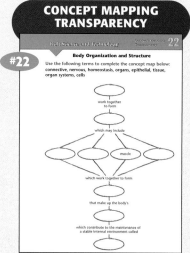

Meeting Individual Needs

DIRECTED READING

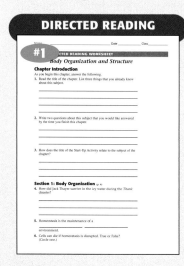

REINFORCEMENT & VOCABULARY REVIEW

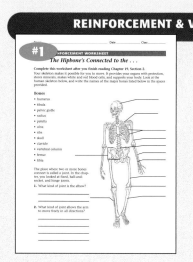

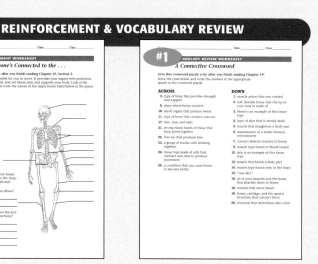

SCIENCE PUZZLERS, TWISTERS & TEASERS

Chapter 1 • Body Organization and Structure

Review & Assessment

STUDY GUIDE

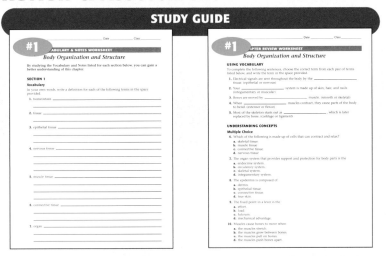

CHAPTER TESTS WITH PERFORMANCE-BASED ASSESSMENT

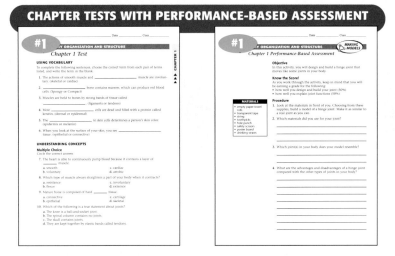

Lab Worksheets

INQUIRY LABS

LONG-TERM PROJECTS & RESEARCH IDEAS

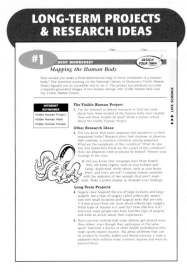

DATASHEETS FOR LABBOOK

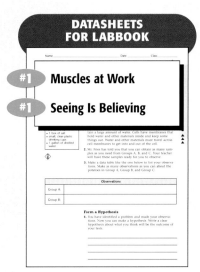

Applications & Extensions

CRITICAL THINKING & PROBLEM SOLVING

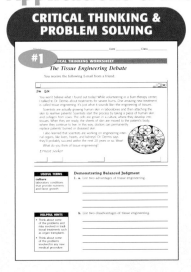

MULTICULTURAL CONNECTIONS

SCIENCE TECHNOLOGY

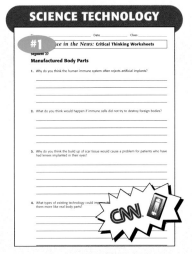

SCIENTISTS IN ACTION

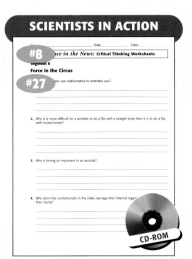

Chapter 1 • Chapter Resources & Worksheets

Chapter Background

SECTION 1

Body Organization

▶ **Tissues**

Tissues differ from each other in a number of ways: shape and size of cells, amount and kind of material between the cells, and the special functions they perform to maintain proper functioning of the body.

- The cells that make up epithelial tissue have little space between them; most are welded to adjacent cells, creating a barrier to movement of materials between them; they form continuous sheets of tissue.

- Connective tissue is the most abundant tissue in the body, and it displays the most variety in form and type. All connective tissue, however, can be classified into one of four types: dense connective tissue—cartilage and bone; loose connective tissue—beneath the skin and around nerves, blood vessels, and body organs; liquid connective tissue—blood and lymph; and fat (adipose) tissue, where the body stores energy as droplets of fat.

- Although bone is considerably harder than other body tissues, it accounts for only about 14 percent of a person's total body weight.

SECTION 2

The Skeletal System

▶ **The Human Skeleton**

The skeleton provides more than just support for soft tissue. It is key to the regulation of body minerals. It also produces both red and white blood cells. There are officially 206 bones in the adult human, but extra bones, particularly those in the hands and feet, increase that number. The number of bones in children varies with age. This is because the skeleton forms from over 800 centers of ossification. All of the bony elements are generally not completely united to form an adult skeleton until a person reaches his or her mid-20s. The clavicles are usually the last bones to complete fusion.

- The skeletons of male and female humans are slightly different. The most pronounced differences are in the pelvis. That is because a female's pelvis is adapted for childbearing and thus has a larger pelvic inlet. Women who are malnourished during their childhood typically do not develop the wider pelvis. This can make childbirth dangerous or even fatal for them.

- It is possible to determine an individual's age by looking at the skeleton alone. A younger individual's dentition and bone fusion patterns are indicative of his or her age. In adults, age determination is much more difficult because one must rely solely on signs of skeletal deterioration.

▶ **Bones**

Each bone is surrounded by a strong fibrous covering called a periosteum. Articular surfaces are also covered in cartilage.

- Bones are made of three types of cells: osteoblasts, osteocytes, and osteoclasts. Osteoblasts are bone-producing cells. Osteocytes are bone-maintaining cells. Osteoclasts are bone-destroying cells.

- For its weight, bone is five times stronger than steel.

▶ **Joints**

Doctors typically classify joints by structure rather than movement. The three types of joint structures are called fibrous, cartilaginous, and synovial. Fibrous joints are immovable joints (such as those in the skull) in which a fibrous tissue or a hyaline cartilage connects the bones. Cartilaginous joints are slightly moveable joints (such as those in the rib cage) in which cartilage connects the

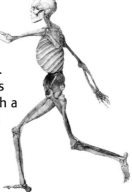

Chapter 1 • Body Organization and Structure

bones. Synovial joints are freely moving joints (such as the knee) in which slick synovial membranes cover the cartilage and ligaments connecting the bone.

Section 3

The Muscular System

▶ The Muscular System
There are more than 600 skeletal muscles in the human body. To simplify the study of these muscles, they are organized into the following groups: muscles of the head and the neck, muscles of the trunk, muscles of the upper limbs (arms and hands), and muscles of the lower limbs (legs and feet).

▶ Types of Muscle Cells
When observed through a microscope, the three types of muscles are clearly identifiable. Cells of smooth muscles have a long tapered shape, no clearly defined striations, and a large central nucleus. Skeletal muscle cells are characterized by distinct light- and dark-colored bands across the long tapered cell; each cell has multiple nuclei because several cells merge and the cell membranes become indistinct. The cells of cardiac muscle have one or more nuclei and have an irregular, branched shape.

Section 4

The Integumentary System

▶ The Skin
One square inch of the skin can hold as many as 650 sweat glands, 20 blood vessels, and more than 1,000 nerve endings.

- Each person has a unique series of ridges and indentations on the tips of his or her fingers called fingerprints. No two people have the same fingerprints. Fingerprints help the fingers to grip slippery surfaces. Toes also have a unique pattern on their tips, so along with fingerprints, we all have toe prints as well.

IS THAT A FACT!

◆ More than three-quarters of the dust in some homes is made up of dead skin cells!

▶ Hair and Nails
Only mammals have true hair; all mammals have hair somewhere on their bodies.

- The body's most visible signs of aging occur in the integumentary system. Skin becomes thin, dry, wrinkled, and less supple. Dark-colored age spots may develop. Hair turns gray or white and may begin to fall out; hair follicles decrease in number. Sweat glands become less active, causing older people to be less tolerant and adaptable to extremely hot weather.

- Hair that is kept short grows an average of 2 cm per month. Growth slows to about 1 cm per month when the hair reaches about 30 cm long. Fingernails grow about 2 cm each year. The fastest-growing nail is on the middle finger. Toenails grow three to four times more slowly than fingernails.

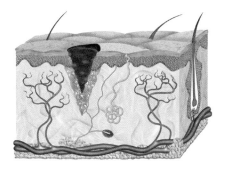

For background information about teaching strategies and issues, refer to the *Professional Reference for Teachers*.

Body Organization and Structure

Pre-Reading Questions

Students may not know the answers to these questions before reading the chapter, so accept any reasonable response.

Suggested Answers
1. Organs are made of tissues, and tissues are made of cells.
2. Skin provides protection from the elements; muscles provide movement; and bones provide structure and protection.

Body Organization and Structure

Sections

1. Body Organization 4
   Internet Connect 7
2. The Skeletal System 8
   QuickLab 9
3. The Muscular System 12
   QuickLab 13
   Chemistry Connection . 14
   MathBreak 15
   Internet Connect 15
4. The Integumentary System 16
   Internet Connect 19

Chapter Lab 20
Chapter Review 24
Feature Articles 26, 27
LabBook 178–179

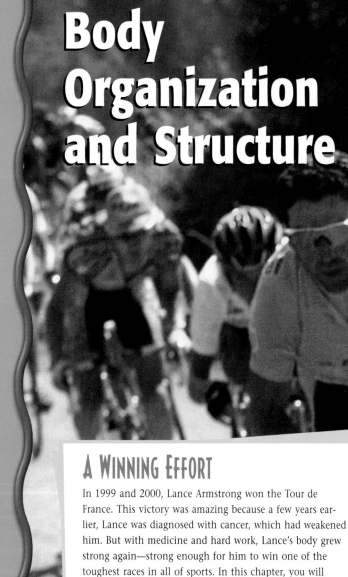

1. What is the relationship between cells, tissues, and organs?
2. How do your skin, muscles, and bones help to keep you well?

A Winning Effort

In 1999 and 2000, Lance Armstrong won the Tour de France. This victory was amazing because a few years earlier, Lance was diagnosed with cancer, which had weakened him. But with medicine and hard work, Lance's body grew strong again—strong enough for him to win one of the toughest races in all of sports. In this chapter, you will learn more about how each of the organ systems of the human body works to keep you healthy.

---

internet connect

HRW On-line Resources
go.hrw.com
For worksheets and other teaching aids, visit the HRW Web site and type in the keyword: **HSTBD1**

www.scilinks.com
Use the *sci*LINKS numbers at the end of each chapter for additional resources on the **NSTA** Web site.

Smithsonian Institution
www.si.edu/hrw
Visit the Smithsonian Institution Web site for related on-line resources.

www.cnnfyi.com
Visit the CNN Web site for current events coverage and classroom resources.

START-UP Activity

TOO COLD FOR COMFORT

Did you know that your nervous system sends you messages about your body's cells? For example, the pain you feel when someone steps on your toe is a message that you should move your toe to safety. Try this exercise to watch your nervous system in action.

Procedure

1. Hold a **few pieces of ice** in one hand. Allow the melting water to drip into a **dish**. Hold the ice until the cold is uncomfortable. Then release the ice into the dish. What message did you receive from your nervous system?

2. Look at the hand that held the ice, and then look at your other hand. What changes in your skin do you see? How quickly does the cold hand return to normal?

Analysis

3. What organ systems do you think were involved in restoring your hand to normal?

4. Think of a time when your nervous system sent you a message, such as an uncomfortable feeling of heat, cold, or pain. How did your body react? Which organ systems do you think were involved in the reaction?

START-UP Activity

TOO COLD FOR COMFORT

MATERIALS
For Each Group:
• ice
• waterproof dish

Safety Caution

Students should clean up the water that results from the melting ice.

Answers to START-UP Activity

1. Answers will vary, but students should feel some discomfort intense enough to make them want to drop the ice.

2. Students might notice redness when their hand becomes very cold. Individuals will experience warming at different rates.

3. Answers will vary, but students might suggest the cardiovascular system. Point out to students that the redness is an increased blood supply to the cold area. The blood brings warmth to the hand, helping to restore it to normal.

4. Answers will vary, but students may know that the nervous system detected the discomfort and sent signals to other body systems to respond.

SECTION 1

Focus

Body Organization

This section introduces the basic organization of the human body. Students identify the four major tissues of the body and describe how the body's tissues are organized into organs. Students also learn that the body's organs are arranged by function into 11 organ systems. Students will be able to describe the major functions of each body system.

Bellringer

Write the names of the body systems below and their functions on the board or an overhead projector. Scramble the columns so that systems and functions do not match. Ask students to copy both columns in their ScienceLog and draw a line between the body system and its correct function.

- Respiratory system—absorbs oxygen
- Muscular system—moves bones
- Digestive system—breaks down food
- Circulatory system—pumps blood
- Endocrine system—regulates body functions

Directed Reading Worksheet Section 1

TOPIC: Tissues and Organs
GO TO: www.scilinks.org
sciLINKS NUMBER: HSTL530

SECTION 1
READING WARM-UP

Terms to Learn

homeostasis
tissue
epithelial tissue
nervous tissue
muscle tissue
connective tissue
organ

What You'll Do

- Identify the major tissues found in the body.
- Compare an organ with an organ system.
- Describe a major function of each organ system.

Figure 1 Your body has four types of tissue, and each type has a special function in your body.

Body Organization

Your body has an amazing ability to survive, even in the face of harsh conditions. How does a person stay alive even though the environment around him or her is so cold? A short answer is that the body did not allow its internal conditions to change enough to stop the cells from working properly. The maintenance of a stable internal environment is called **homeostasis** (HOH mee OH STAY sis). If homeostasis is disrupted, cells suffer and sometimes die.

Four Types of Tissue

Making sure your internal environment remains stable enough to support healthy cells is not an easy task. Many different "jobs" must be done to maintain homeostasis. Fortunately, not every cell has to do all those jobs because the cells are organized into different teams. Just as each member of a soccer team has a special role in the game, each cell in your body has a specific job in maintaining homeostasis. A group of similar cells working together forms a **tissue.** Your body contains four main types of tissue—epithelial tissue, connective tissue, muscle tissue, and nervous tissue, as shown in **Figure 1.**

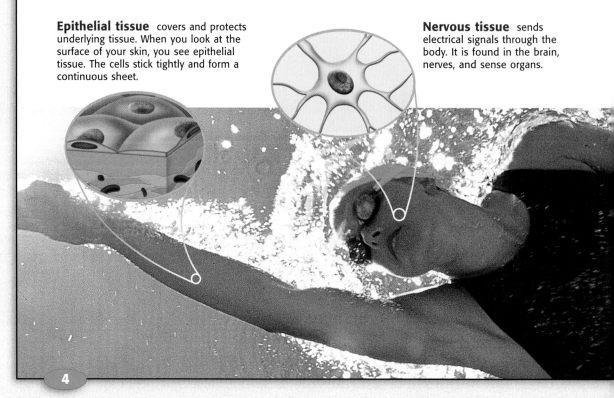

Epithelial tissue covers and protects underlying tissue. When you look at the surface of your skin, you see epithelial tissue. The cells stick tightly and form a continuous sheet.

Nervous tissue sends electrical signals through the body. It is found in the brain, nerves, and sense organs.

IS THAT A FACT!

The Pompeii worm, *Alvinella pompejana*, can survive a temperature difference of 60°C between its head and its tail! Scientists theorize that a coating of furry bacteria living on the worm's back allow the worm to endure such extreme temperature differences.

4 Chapter 1 • Body Organization and Structure

Tissues Form Organs

Two or more tissues working together form an **organ**. One type of tissue alone cannot do all the things that several types working together can do. Your stomach, as shown in **Figure 2**, uses several different types of tissue to carry out digestion.

Organs Form Systems

Your stomach does much to help you digest your food, but it doesn't do it all. It works together with other organs, such as the small intestine and large intestine, to digest your food. Organs working together make up an *organ system*. The failure of any part can affect the entire system. Your body has 11 major organ systems, which are illustrated on the next two pages. Are there any that you have not heard of before?

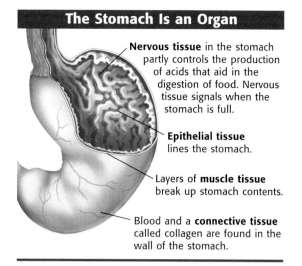

The Stomach Is an Organ

Nervous tissue in the stomach partly controls the production of acids that aid in the digestion of food. Nervous tissue signals when the stomach is full.

Epithelial tissue lines the stomach.

Layers of **muscle tissue** break up stomach contents.

Blood and a **connective tissue** called collagen are found in the wall of the stomach.

Figure 2 *The four types of tissue work together so that the stomach can carry out digestion.*

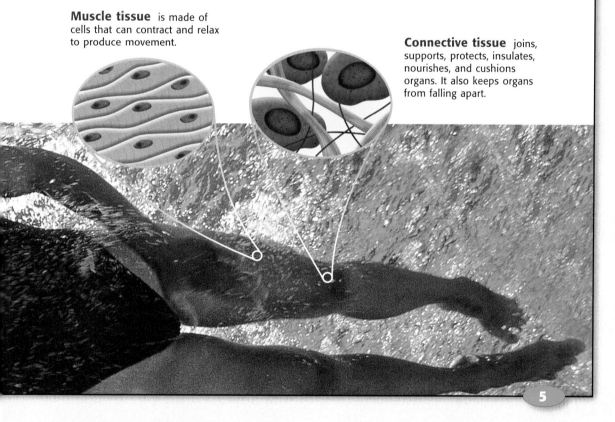

Muscle tissue is made of cells that can contract and relax to produce movement.

Connective tissue joins, supports, protects, insulates, nourishes, and cushions organs. It also keeps organs from falling apart.

1) Motivate

DISCUSSION

Homeostasis Before students read this section, ask them to think about what they've already learned about endotherms and ectotherms and how each adjusts to outside temperatures. How does the idea of homeostasis fit into what they already know?

2) Teach

READING STRATEGY

Prediction Guide Before students read the pages that describe the body's tissues and organs, ask whether the following statements are true or false. Students will discover the answers as they read Section 1.

- Homeostasis is the body's ability to maintain a stable internal environment. (true)
- The human body has four main types of tissues. (true)
- An organ is a group of tissues that work together. (true)

USING THE FIGURE

Refer to **Figure 2** as you review with students the four types of body tissue. Encourage students to consider what role each tissue might play in other organs, such as the heart. **Sheltered English**

CROSS-DISCIPLINARY FOCUS

History The first transplant of a human heart was performed in Cape Town, South Africa, on December 3, 1967, by Dr. Christiaan Barnard and a team of 30 physicians. In an operation that lasted 5 hours, Barnard removed a heart from the body of a 25-year-old woman and transplanted it into a 55-year-old man named Louis Washkansky. Washkansky lived for 18 days following the transplant before he died of pneumonia. Have interested students conduct research to find out when other organs such as kidneys and livers were first transplanted.

2 Teach, continued

DEBATE

Transplant Ethics There are more than 53,000 people in the United States waiting for organ transplants. There are approximately 20,000 organ transplants performed every year in the United States. The average cost of an organ transplant is $120,000 for an American citizen. Encourage students to research the topic, and have them debate the ethical issues surrounding transplants. Some suggested topics:

- Should transplants happen at all?
- Who should get a transplant?
- Should a child receive a transplant before an older person?

MEETING INDIVIDUAL NEEDS

Learners Having Difficulty
To help students understand and identify the 11 major organ systems of the body, have them make a table with the following headings: Name of organ system, Function(s), and Main organs. Have students use the table to organize the information presented on these pages. Encourage students to continue to fill in their tables as they read the remainder of this book.
Sheltered English

Teaching Transparency 78
"Organ Systems"

TOPIC: Body Systems
GO TO: www.scilinks.org
sciLINKS NUMBER: HSTL535

Organ Systems

Integumentary system
Your skin, hair, and nails protect underlying tissue.

Muscular system
Your skeletal muscles move your bones.

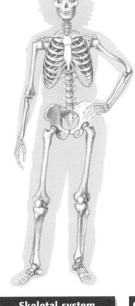
Skeletal system
Your bones provide a frame to support and protect body parts.

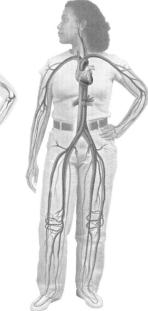
Cardiovascular system
Your heart pumps blood through all your blood vessels.

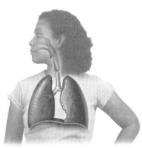
Respiratory system
Your lungs absorb oxygen and release carbon dioxide.

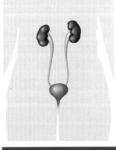
Urinary system
Your urinary system removes wastes from the blood and regulates the body's fluids.

Reproductive system (male)
The male reproductive system produces and delivers sperm.

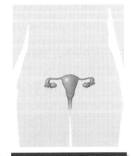
Reproductive system (female)
The female reproductive system produces eggs and nourishes and shelters the unborn baby.

IS THAT A FACT!

Can you learn about the human body by cutting open animals instead of human bodies? In ancient Rome, physician Galen (129–199) did just that! He wrote more than 500 treatises on the human body and, because of the Roman belief that it was wrong to cut open a human corpse, Galen studied animals and the wounds of fallen gladiators. He never saw inside the human body!

Chapter 1 • Body Organization and Structure

Organ Systems

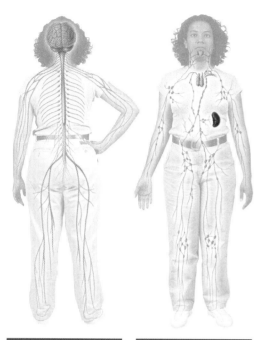

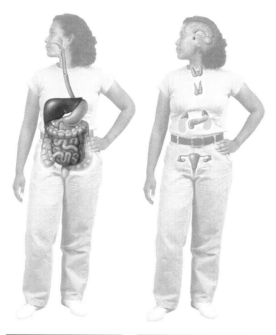

Nervous system
It is the role of the nervous system to receive and send electrical messages throughout the body.

Lymphatic system
Your lymphatic system returns leaked fluids to blood vessels. It also helps you get rid of germs that can harm you.

Digestive system
Your digestive system breaks down the food you eat into nutrients that can be absorbed into your body.

Endocrine system
Glands regulate body functions by sending out chemical messengers. The ovaries, in females, and testes, in males, are part of this system.

SECTION REVIEW

1. Explain the relationship between cells, tissues, organs, and organ systems.
2. Compare the four kinds of tissue found in the human body.
3. **Using Graphics** Make a chart that lists the major organ systems and their functions.
4. **Relating Concepts** Describe a time when homeostasis was disrupted in your body. Which body systems do you think were affected?

internetconnect
sciLINKS NSTA
TOPIC: Tissues and Organs, Body Systems
GO TO: www.scilinks.org
sciLINKS NUMBER: HSTL530, HSTL535

---

3) Extend

GROUP ACTIVITY
Divide the class into 11 equal groups. Assign each group one of the body's organ systems. Have students use library references or the Internet to prepare an oral report for the class that identifies the main organs of each system and their functions. Encourage students to include a labeled diagram with their reports. In each group, assign two students as researchers and one or two students as poster artists and report writers.

4) Close

Quiz
Ask students whether these statements are true or false. Have them correct the false statements to make them true.

1. Homeostasis is the maintenance of a stable internal environment. (true)
2. Epithelial tissue sends electrical signals throughout the body. (False; this is the function of nervous tissue.)
3. Blood is a type of connective tissue. (true)
4. The lymphatic system returns leaked fluids to blood vessels. (true)

ALTERNATIVE ASSESSMENT

Writing In their ScienceLog, have students describe in as much detail as possible three of the organ systems covered in this chapter. Have them describe the functions and primary organs of each system and include drawings of each system.

▼ Answers to Section Review

1. The body is organized into cells, tissues, organs, and organ systems. Cells that have similar functions make up tissue; two or more types of tissue work together to form an organ; organs work together to form a system.
2. Nervous tissue sends electrical signals or messages; epithelial tissue covers and protects underlying tissue; muscle tissue is made of cells that can contract and relax to produce movement; connective tissue joins, supports, protects, insulates, nourishes, cushions, and keeps organs from falling apart.
3. The chart should reflect the material found on the last two pages of this section.
4. Answers will vary.

Section 1 • Body Organization

SECTION 2

Focus

The Skeletal System

This section introduces the major organs of the skeletal system and describes the function of bones. The section also illustrates the major internal structure of bones and compares the three major types of joints. The section concludes with a discussion of how bones and joints form levers.

 **Bellringer**

Have students write in their ScienceLog five problems they would have if they lacked bones. (They would have no defined structure, mineral storage, protection, red blood cells, or mobility.)

1 Motivate

ACTIVITY

Locating Bones Review with students that the skeletal system supports the body and protects delicate body parts. Ask students to name the main organ of the skeletal system. (bones)

Encourage students to press the skin in various parts of their body, to feel their bones. Ask students to describe any parts of their body where they cannot feel their bones. (Answers will vary but should include the abdomen.)

As you point to various parts of the body, ask students what organ the bones protect. The skull, for example, protects the brain, and the ribs protect the heart and lungs.
Sheltered English

 **SECTION 2** READING WARM-UP

Terms to Learn

skeletal system   cartilage
compact bone   joint
spongy bone   ligament

What You'll Do

◆ Identify the major organs of the skeletal system.
◆ Describe the functions of bones.
◆ Illustrate the internal structure of bones.
◆ Compare three types of joints.
◆ Discuss how bones function as levers.

The Skeletal System

When you hear the word *skeleton,* you may think of the remains of something that has died. But your skeleton is not dead; it is very much alive. Your bones are not dry and brittle. They are just as alive and active as the muscles that are attached to them. Bones, cartilage, and the special structures that connect them make up your **skeletal system.**

The Burden of Being a Bone

Bones do a lot more than just hold you up. Your bones perform several important functions inside your body. The names of some of your bones are identified in **Figure 3.**

Protection Your heart and lungs are shielded by your ribs, your spinal cord is protected by your vertebrae, and your brain is protected by your skull.

Storage Bones store minerals that help the nerves and muscles function properly. Your arm and leg bones also store fat that can be used for energy.

Movement Skeletal muscles pull on the bones to produce movement. Without bones, you would not be able to sit, stand, walk, or run.

Blood Cell Formation Some of your bones are filled with a special material that makes blood cells.

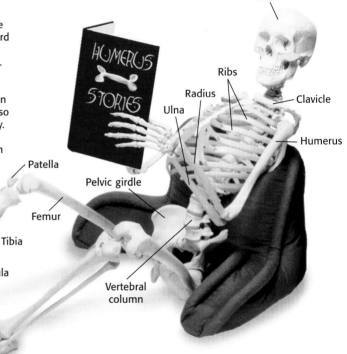

Figure 3 The adult human skeleton has approximately 206 bones. Several major bones are identified in this skeleton.

 Directed Reading Worksheet Section 2

 **SCIENCE HUMOR**

Q: Why didn't the skeleton cross the road?

A: It didn't have the guts.

What's in a Bone?

A bone may seem lifeless, but it is a living organ made of several different tissues. Bone is composed of connective tissue and minerals that are deposited by living cells called *osteoblasts*.

If you look inside a bone, you will notice there are two different kinds of bone tissue. If the tissue does not have any visible open spaces, it is called **compact bone.** Bone tissue that has many open spaces is called **spongy bone.** Spongy bone provides most of the strength and support for a bone. It acts like the trusses of a bridge.

Down to the Marrow Bones contain a soft tissue called *marrow*. Red marrow, sometimes found in spongy bone, produces red blood cells. Yellow marrow, found in the central cavity of long bones, stores fat. Tiny canals within the compact bone contain small blood vessels. **Figure 4** shows a cross section of a femur.

QuickLab

Pickled Bones

This activity lets you see how a bone changes when it is exposed to an acid, such as vinegar. Place a **clean chicken bone** in a **jar of vinegar.** After 1 week, remove the bone and rinse it with water. Make a list of changes that you can see or feel. How has the bone's strength changed? What did the vinegar remove?

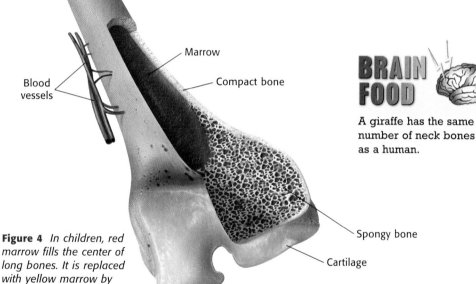

Figure 4 *In children, red marrow fills the center of long bones. It is replaced with yellow marrow by adulthood.*

Labels: Blood vessels, Marrow, Compact bone, Spongy bone, Cartilage

BRAIN FOOD

A giraffe has the same number of neck bones as a human.

2) Teach

QuickLab

MATERIALS

FOR EACH STUDENT:
- clean chicken bone
- jar of vinegar

Safety Caution: Remind students to review all safety cautions and icons before beginning this lab activity.

Answer to QuickLab

The bone becomes more flexible, because the hard minerals in the leg bone dissolve in the vinegar.

DEMONSTRATION

Bone Dissection Ask a local butcher to cut a long bone from a pig, cow, or sheep in half lengthwise to expose the bone's internal structure. If a bone is not available from a butcher, obtain a preserved long bone from a biological supply house. Point out to students the differences in structure and location of spongy bone and compact bone. Also point out any cartilage that might remain on the bone. Ask students what makes the bone relatively lightweight for its size. (Students should infer that the air spaces in the spongy bone make the bone relatively lightweight for its size.)
Sheltered English

 Teaching Transparency 79
"What's in a Bone?"

Multicultural CONNECTION

Writing On November 2 each year, people across Mexico celebrate *El Día de los Muertos* (the Day of the Dead). They take food and gifts to the graves of their loved ones and set up elaborate shrines and altars in their homes. These shrines are often decorated with figures of skeletons depicting the person's job when he or she was alive, such as that of a doctor or a teacher. Foods eaten during this holiday are *pan del muerto* (bread of the dead) and candies shaped like skulls. Have interested students write a research paper on this holiday to share with the class.

Section 2 • The Skeletal System

2) Teach, continued

DISCUSSION

Joint Review After students read about the different kinds of joints, ask them the following questions. You may want to have students move their own bodies before they answer each question.

- What kind of joint do you use when you bend your elbow? (hinge joint) your knee? (hinge joint)
- What kind of joint moves when you swing your arm back and forth? (ball-and-socket joint) Name another location in your body where this type of joint is located. (the hips) **Sheltered English**

CONNECT TO PHYSICAL SCIENCE

Students may be surprised to learn that their arms and legs are "machines." Our limbs are levers, and levers are the simplest kind of machine. They allow us to apply, increase, and change the direction of force. Have students think of other levers they may use or know about. (Examples include a seesaw, the claw end of a hammer, a crowbar, and a garlic press.)

Use Teaching Transparency 231 to help students understand how levers make work easier.

Teaching Transparency 231 "Machines Change the Size or Direction (or Both) of a Force" LINK TO PHYSICAL SCIENCE

Growing Bones

Did you know that most of your skeleton used to be soft and rubbery? Most bones start out as a soft, flexible tissue called **cartilage**. When you were born, you had little true bone. But as you grew, the cartilage was replaced by bone. During childhood, growth plates of cartilage remain in most bones, providing a place for those bones to continue to grow.

Feel the end of your nose, or bend the top of your ear. As shown in **Figure 5**, some areas, like these, never become bone. The flexible material beneath your skin in these areas is cartilage.

What's the Point of a Joint?

The place where two or more bones connect is called a **joint**. Your joints have special designs that allow your body to move when your muscles contract. Some joints allow a lot of movement, while other joints are fixed, which means they allow little or no movement. For example, the joints in the skull are fixed. Joints that have a wide range of movement tend to be more susceptible to injury than those that are less flexible. Some examples of movable joints are shown in **Figure 6**.

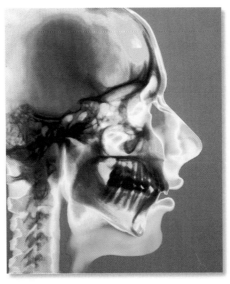

Figure 5 *The skull and neck bones in this computer-colored X ray are shown mostly in blue.*

Figure 6 *Joints are shaped according to their function in the body.*

Sliding joint
Sliding joints allow bones in the hand to glide over one another, giving some flexibility to the area.

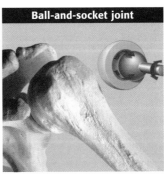
Ball-and-socket joint
Like a joystick on a computer game, the shoulder enables your arm to move freely in all directions.

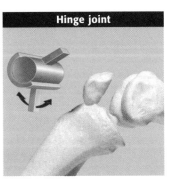
Hinge joint
Like a hinge on a door, the knee enables you to flex and extend your lower leg.

Math Skills Worksheet "Mechanical Advantage"

IS THAT A FACT!

The only bone in the human body that is not connected to another bone is the hyoid bone. The hyoid bone is found in the throat above the larynx. This bone is easily broken when a person is strangled, and it is often an important piece of evidence in suspected strangulation deaths.

Bone to Bone Joints are kept together with strong elastic bands of connective tissue called **ligaments.** If a ligament is stretched too far, it becomes strained. A strained ligament will usually heal with time, but a torn ligament will not. A torn ligament must be repaired surgically. Cartilage helps cushion the area where two bones meet. If cartilage wears away, the joint becomes arthritic.

Can Levers Lessen Your Load?

You may not think of your limbs as being machines, but they are. The action of a muscle pulling on a bone often works like a type of simple machine called a *lever*. A lever is a rigid bar that moves on a fixed point known as a *fulcrum*. Any force applied to the lever is called the *effort*. A force that resists the motion of the lever, such as the downward force exerted by a weight on the bar, is called the *load* or the *resistance*. **Figure 7** shows how three types of levers are used in the human body.

Figure 7 *There are three classes of levers, based on the location of the fulcrum, the load, and the effort.*

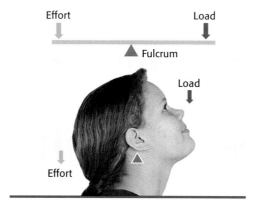

First-class lever
The fulcrum lies between the load and the effort.

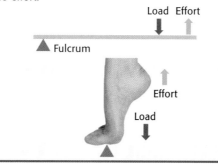

Second-class lever
The load lies between the fulcrum and the effort.

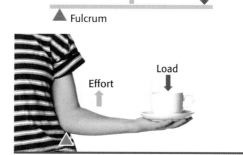

Third-class lever
The effort lies between the fulcrum and the load.

SECTION REVIEW

1. Describe four important functions of bones.
2. Draw a bone, and label the inside and outside structures. Use colored pencils to color and label spongy bone, blood vessels, marrow cavity, compact bone, and cartilage.
3. List three hinge joints in your body.
4. **Interpreting Models** Study the models of levers pictured in Figure 7. Use a small box (load), a ruler (bar), and a pencil (fulcrum) to create models of each type of lever.

3) Extend

RESEARCH

Have students work in pairs or groups of three to do library research on the different kinds of bone fractures. Encourage students to make posters or models showing each type of fracture, including a caption or label on how each type of fracture is treated.

4) Close

Quiz

1. What is the difference between compact bone and spongy bone? (Compact bone has no visible, open spaces. Spongy bone has many visible spaces and contains a tissue called marrow.)
2. Where in the body are ball-and-socket joints found? (hip and shoulder) hinge joints? (knee and elbow) fixed joints? (skull)

ALTERNATIVE ASSESSMENT

Writing Have students write an essay about bones. Essays should address what bones do, how they are specialized, and how they are joined.

▼ **Answers to Section Review**

1. Bones support your body, store and release minerals, and enable your muscles to move the body; some bones make blood cells.
2. Illustrations will vary but should reflect the structures shown in **Figure 4.**
3. examples include: knee, knuckles, toes, elbow, jaw
4. Models will vary but should reflect the information on levers described in **Figure 7.**

Reinforcement Worksheet
"The Hipbone's Connected to the ..."

Critical Thinking Worksheet
"The Tissue Engineering Debate"

Section 2 • The Skeletal System **11**

SECTION 3

Focus

The Muscular System

This section introduces students to the major parts of the muscular system and describes the different types of muscle. This section also describes how skeletal muscles move bones. Students compare aerobic exercise to resistance exercise. The section concludes with a discussion of typical muscle injuries and how to prevent them.

Bellringer

On the board or an overhead projector, write the following:

In your ScienceLog, list at least five parts of your body that you use to drink a glass of water. (Sample answer: fingers, hands, arm, lips, tongue)

When the students are done, remind them that all the parts needed, including the eyes they used to see the glass, are controlled by muscles.

1) Motivate

ACTIVITY

Poster Project Have students draw an outline of a human body. Have them add smooth muscles, skeletal muscles, and cardiac muscles to their drawings. Point out that certain kinds of muscles are located in certain parts of the body. Ask students to identify the function of each type of muscle. Make sure that students understand that all tissue movement in the body is caused by muscle movement.
Sheltered English

SECTION 3
READING WARM-UP

Terms to Learn

muscular system   skeletal muscle
smooth muscle   tendon
cardiac muscle

What You'll Do

◆ List the major parts of the muscular system.
◆ Describe the different types of muscle.
◆ Describe how skeletal muscles move bones.
◆ Compare aerobic exercise with resistance exercise.
◆ Give an example of a muscle injury.

The Muscular System

Have you ever tried to be perfectly still for just 1 minute? Try as you might, you just can't do it. Somewhere in your body, certain muscles are always working. For example, muscles continuously push blood through your blood vessels. A muscle makes you breathe. And muscles hold you upright. If all your muscles rested at the same time, you would collapse. Your muscles are made of muscle tissue and connective tissue. Muscles that attach to bones and the connective tissue that attaches them make up the **muscular system**.

Types of Muscle

There are three types of muscle tissue that make up the muscles in your body. **Smooth muscle** is found in the digestive tract and the blood vessels. **Cardiac muscle** is a special type of muscle found only in your heart. **Skeletal muscles** are attached to your bones for movement, and they help protect your inner organs. The three types of muscles are shown in **Figure 8**.

Muscle action can be voluntary or involuntary. Muscle action that is under your control is *voluntary*. Muscle action that is not under your control is *involuntary*. The actions of smooth muscle and cardiac muscle are involuntary. The actions of skeletal muscles can be both voluntary and involuntary. For example, you can blink your eyes any time you want to, but your eyes will also blink automatically if you do not think about it.

Figure 8 Your body has smooth muscle, cardiac muscle, and skeletal muscle.

Skeletal muscle enables bones to move.

Smooth muscle moves food through the digestive system.

Cardiac muscle causes the heart to beat.

Teaching Transparency 80
"Types of Muscle"

IS THAT A FACT!

Horses can sleep standing up. Their legs can support their weight on their bones and tendons without the use of any muscles at all. When they fall asleep and their muscles relax, their leg bones lock in place underneath them, holding them upright for the duration of their nap.

12 Chapter 1 • Body Organization and Structure

Making Your Move

Skeletal muscles produce hundreds of different voluntary movements. This is demonstrated by a ballet dancer, a swimmer, or even someone making a funny face, as shown in **Figure 9.** When you want to make a movement, you cause electrical signals to travel from the brain to the skeletal muscle cells. The muscle cells respond to these signals by contracting or getting shorter.

Muscles to Bones Strands of tough connective tissue called **tendons** connect your skeletal muscles to your bones. When a muscle gets shorter, a pulling action occurs, bringing the bones closer to each other. For example, the biceps muscle, shown in **Figure 10,** is attached by tendons to a bone in your shoulder and to another bone in your forearm. When the biceps contracts, your arm bends.

Figure 9 *It takes an average of 13 muscles to smile and an average of 43 muscles to frown.*

Working in Pairs Your skeletal muscles work in pairs to cause smooth, controlled movements. Many basic movements are the result of muscle pairs that cause bending and straightening. If a muscle bends part of your body, then that muscle is called a *flexor*. If the muscle straightens part of your body, then it is called an *extensor*. The flexor muscle of the arm is the biceps. The extensor muscle of the arm is the triceps. Discover some of your own flexor and extensor muscles by doing the QuickLab at right.

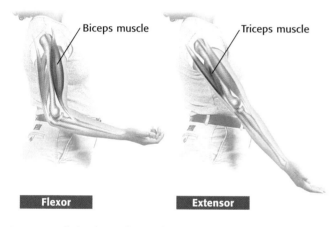

Figure 10 *Skeletal muscles, such as the biceps and triceps muscles, work in pairs. When the biceps muscle contracts, the elbow bends. When the triceps muscle contracts, the elbow straightens.*

QuickLab

Power in Pairs

1. While sitting in a chair, place one of your hands palm up under the edge of a **table.** Apply gentle upward pressure.
2. With your free hand, feel the front and back of your upper arm.
3. Next place your hand palm down on top of the table. Apply pressure downward.
4. Again with your free hand, feel the front and back of your upper arm.
5. What did you notice when you were pressing up? when you were pressing down?

TRY at HOME

SCIENTISTS AT ODDS

In the early part of this century, tens of thousands of people, mostly children, were stricken with polio, a viral disease that paralyzes muscles. Often, polio would leave its victims unable to walk or move. One Australian nurse, Sister Elizabeth Kenny, treated patients using flexible hot wraps and exercise instead of hard, immobilizing casts. Using her treatments, patients avoided paralysis. Largely because Kenny was self-taught and had no formal medical education, the doctors and hospital administrators of the time fought against her practices. Eventually, she and her successes became well known, and her contributions became the beginning of the field of physical therapy. Have students research and report to the class Sister Kenny's story and the opposition she faced in Australia and the United States.

2) Teach

ACTIVITY

Muscle Contraction Ask a student volunteer to stand in a doorway with his or her arms and hands relaxed with the palms turned inward. Ask the student to raise his or her hands against the door frame (backs of the hands on the frame) and press steadily against the frame for about 30 to 40 seconds. Then ask the student to relax and step away from the door. Have the rest of the class observe what happens to the student's arms. (The student's arms should rise slowly without obvious effort by the student.)

Explain to students that the arms rise because the muscles that were pushing against the door frame are still shortened, or contracted. **Sheltered English**

Answer to QuickLab

5. Students should feel their biceps tighten when they press up and their triceps tighten when they press down.

Directed Reading Worksheet Section 3

 internet**connect**

TOPIC: The Muscular System
GO TO: www.scilinks.org
*sci*LINKS NUMBER: HSTL540

Section 3 • The Muscular System   **13**

3) Extend

MATH and MORE

Explain to students that a regular program of aerobic exercise enables the heart to work more efficiently, pumping the same volume of blood in fewer beats. Ask students to imagine that their heart normally beats 75 times per minute. After a 3-month program of aerobic exercise, their heart now beats 65 times per minute. What is the difference in heart rate per minute? (75 beats − 65 beats = 10 beats)

How many fewer times will the heart beat in 1 hour? (10 beats/minute × 60 minutes/hour = 600 beats/hour)

How many fewer times will the heart beat in 1 day? (600 beats/hour × 24 hours/day = 14,400 beats/day)

Math Skills Worksheet "The Unit Factor and Dimensional Analysis"

GOING FURTHER

Concept Mapping Divide the class into cooperative groups of three or four students. Have each group create a concept map for the ideas in this lesson. They should connect at least 10 terms and link them with apt phrases. Have students record their finished concept maps in their ScienceLog. **Sheltered English**

Answers to Self-Check

Curl-ups use flexor muscles; push-ups use extensor muscles.

Chemistry CONNECTION

Body chemistry is very important for healthy muscle functioning. If there is a chemical imbalance in a muscle due to excessive sweating, poor diet, tension, or illness, spasms or cramping may occur. Sodium, calcium, and potassium—three chemicals called *electrolytes*—must be in proper balance to avoid cramps and spasms. Relaxation and massage usually help the muscle restore its chemical balance.

Figure 12 *Aerobic exercise is a great way to have fun while strengthening your heart.*

Use It or Lose It

When someone breaks an arm and has to wear a cast, the muscles surrounding the injured bone change. That's because these muscles are not exercised, and they become smaller and weaker. On the other hand, exercised muscles are stronger and larger. Certain exercises can give muscles more endurance. This means they're able to work longer without getting tired. Strong muscles benefit other systems in your body too. When a muscle contracts, blood vessels in that muscle get squeezed. This helps push blood along, increasing blood flow without demanding more work from the heart.

Resistance Exercises To develop the size and strength of your skeletal muscles, resistance exercises are the most effective form of exercise. Resistance exercises require muscles to overcome the resistance (weight) of another object. Some resistance exercises, like the bent knee curl-up shown in **Figure 11**, require you to overcome your own weight.

Figure 11 *Resistance exercises are tough, but they can really help you build strong muscles.*

Aerobic Exercise Steady, moderate-intensity activity, such as jogging, cycling, skating, swimming laps, or walking, is called aerobic exercise. Aerobic exercise increases the size and strength of your skeletal muscles somewhat, but mostly it strengthens the heart while increasing the endurance of your skeletal muscles. Many people, like the girl in **Figure 12**, enjoy doing aerobic exercise.

✓ Self-Check

Which kind of skeletal muscle do you use to perform a curl-up? Which kind do you use to do a push-up? (See page 212 to check your answers.)

Homework

Researching Injuries For one month, have students read the sports section in the local newspaper or look for articles in sports magazines about injuries to the muscular and skeletal systems sustained by athletes. Have them research and write about these injuries in their ScienceLog. Ask students to identify and count the types of injuries—such as sprained ankles, torn or pulled muscles, bruised ribs, and torn or damaged ligaments. Have students compile their information on bar graphs in which they record the kinds of injuries on the *x*-axis and the number and frequency of injuries on the *y*-axis.

Muscle Injury

Any exercise program should be started gradually so that the muscles gain strength and endurance without injury. Muscles should also be warmed up gradually to reduce the risk of injury. However, as shown in **Figure 13**, the muscular system can experience damage. A muscle strain, commonly called a pulled muscle, is the overstretching or even tearing of a muscle. Muscle strain often occurs because the muscle has not been properly conditioned for the work it is doing.

Tendons, as well as muscles, can get injured from overuse. A damaged tendon can become hot or inflamed as your body tries to repair it. This painful condition is called tendinitis, and an extended period of rest is often required for the tendon to heal.

Figure 13 A pulled hamstring is a tear or strain of one of the muscles or tendons on the back of the thigh.

The Dangers of Anabolic Steroids Some people try to make their muscles larger and stronger by taking hormones called *anabolic steroids*. Anabolic steroids are powerful chemicals that resemble testosterone, a male sex hormone. Using anabolic steroids not only gives athletes an unfair advantage in competition but also puts the user at risk for serious long-term health problems. The use of anabolic steroids threatens the heart, liver, and kidneys, and it can cause high blood pressure. If taken before the skeleton is mature, anabolic steroids can cause the bones to stop growing.

SECTION REVIEW

1. List three types of muscle tissue, and describe their functions in the body.
2. Compare aerobic exercise with resistance exercise, and give two examples of each.
3. **Applying Concepts** Describe the muscle action required to pick up a book. Make a sketch that illustrates the muscle action.

MATH BREAK

Runner's Time

Jan, who has been a runner for several years, has decided to enter a race. She now runs 5 km in 30 minutes. She would like to decrease her time by 15 percent before the race. What will her time be when she meets her goal?

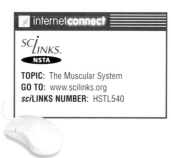

internet connect

SC_{LINKS}
NSTA

TOPIC: The Muscular System
GO TO: www.scilinks.org
*sci*LINKS NUMBER: HSTL540

4) Close

Quiz

1. What is the difference between voluntary muscle action and involuntary muscle action? Give an example of each. (Voluntary muscle action, such as lifting your arm, is action that you can control. Involuntary muscle action, such as the beating of your heart, is not under your control.)

2. What kind of muscle bends part of your body? (flexor) What kind of muscle straightens part of your body? (extensor)

3. What is the danger of using anabolic steroids? (These powerful drugs can damage the heart, liver, and kidneys and can cause baldness. They can also cause the bones to stop growing.)

ALTERNATIVE ASSESSMENT

Writing Divide the class into cooperative groups of four. Have groups develop crossword puzzles using the vocabulary terms in this section and the definitions of the terms as the clues. Have groups of students exchange puzzles with other groups.

Answers to MATHBREAK

She will run 5 km in 25.5 minutes.

Reinforcement Worksheet
"Muscle Map"

▼ Answers to Section Review

1. Smooth muscle helps move materials through the digestive tract and blood vessels; cardiac muscle causes the heart to beat; and skeletal muscle enables bones to move.

2. Resistance exercises, such as curl-ups and push-ups, build up the size and strength of skeletal muscles; they usually involve overcoming weight. Aerobic exercise, such as steady jogging, walking, or swimming, strengthens the heart while increasing endurance.

3. Illustrations and descriptions will vary but should show that the biceps muscle shortens to bring the forearm upward, lifting the book.

SECTION 4

Focus

The Integumentary System

This section introduces students to the major functions of the integumentary system; it also describes the major parts of the skin and discusses their functions. Students learn about the structure and functions of hair and nails and about common injuries to the skin.

Bellringer

Write the following questions on the board or an overhead projector, and ask students to write their answer in their ScienceLog:

When do you see dogs panting? Why do you think they pant? (Dogs cannot sweat, and panting cools them down on hot days or after strenuous activity. A dog pants in order to regulate its body temperature.)

1) Motivate

DISCUSSION

Homeostasis Relay the following story to students:

Dr. Charles Blagden was secretary of the Royal Society of London more than 200 years ago. He wanted to test how mammals regulate their body temperature. He spent 45 minutes in a room with a few friends, a dog, and a steak. The temperature in the room measured 126°C (260°F). Ask students what they think happened to the man, the dog, and the steak after the 45 minutes were up. (Everyone and the dog emerged from the room unharmed, but the steak was cooking! Mammals can regulate their temperature.)

16   Chapter 1 • Body Organization and Structure

SECTION 4 READING WARM-UP

Terms to Learn

integumentary system
sweat glands
melanin
epidermis
dermis
hair follicle

What You'll Do

- Describe the major functions of the integumentary system.
- List the major parts of the skin, and discuss their functions.
- Describe the structure and function of hair and nails.
- Describe some common types of damage that can affect skin.

The Integumentary System

Here's a quiz for you. What part of your body has to be partly dead to keep you alive? Here are some clues: it comes in a variety of colors, it is the largest organ in the body, and it protects you from the outside world. Oh, and guess what—it is showing right now. Did you guess your skin? If you did, you guessed correctly.

Your skin, hair, and nails make up your **integumentary** (in TEG yoo MEN tuhr ee) **system**. (*Integument* means "covering.") Like all organ systems, the integumentary system helps your body maintain a healthy internal environment.

The Skin: More than Just a "Coat"

Why do you need skin? Here are four good reasons:

- Skin protects you by keeping moisture in your body and foreign particles out of your body.

- Skin keeps you "in touch" with the outside world. The nerve endings in your skin allow you to feel what's around you.

- Skin helps regulate your body's temperature. For example, small organs in the skin called **sweat glands** produce sweat, a salty liquid that flows to the surface of the skin. As sweat evaporates, the skin cools.

- Skin helps get rid of wastes. Several types of waste chemicals can leave the bloodstream and be removed in sweat.

What Determines Skin Color? A darkening chemical in skin called **melanin** determines skin color, as shown in **Figure 14**. If a lot of melanin is present, the skin is very dark. If only a little melanin is produced, the skin is very light. Melanin in the upper layer of the skin absorbs much of the harmful radiation from the sun, reducing DNA damage that can lead to cancer. However, *all* skin is vulnerable to cancer and therefore should be protected from sun exposure whenever possible.

Figure 14 *Variety in skin color is caused by the pigment melanin. The amount of melanin varies from person to person.*

IS THAT A FACT!

In an average adult, the skin has a surface area of about 2 m² and weighs about 4 kg. The skin on the human body varies in thickness from about 5 mm on the soles of the feet to about 0.5 mm on the eyelids.

A Tale of Two Layers

As you already know, the skin is the largest organ of your body. In fact, the skin of an adult covers an area of about 2 m^2! However, there's a lot more to skin than meets the eye. The skin has two main layers: the dermis and the epidermis. The **epidermis** is the thinner layer of the two. It's what you see when you look at your skin. (*Epi* means "on top of.") The deeper, thicker layer is known as the **dermis**.

Epidermis The epidermis is composed of a type of epithelial tissue. Even though the epidermis has many layers of cells, it is only as thick as two sheets of notebook paper over most of the body. It is thicker in the palms of your hands and the soles of your feet. Most epidermal cells are dead and are filled with a protein called keratin, which helps make the skin tough.

Dermis The dermis lies underneath the epidermis. It is mostly connective tissue, and it contains many fibers made of a protein called collagen. The fibers provide strength and allow skin to bend without tearing. The dermis also contains a variety of small structures, as shown in **Figure 15**.

Self-Check

To what system do the skin's blood vessels belong? *(See page 212 to check your answer.)*

Answer to Self-Check

Blood vessels belong to the cardiovascular system.

ACTIVITY

Measuring Temperature

MATERIALS

FOR EACH GROUP:
- 2 Celsius thermometers
- cotton balls
- water at room temperature
- fan
- clock or watch

Divide class into groups of three to five students. Have students wrap the bulb of each thermometer with a cotton ball and then wet one of the cotton balls. Have students record the beginning temperature of each thermometer and then hold both thermometers in front of a fan. Students should record the temperature of each thermometer every minute for 5 minutes. How do the temperatures of the thermometers differ? Why? (The thermometer with the wet cotton has a lower temperature. Evaporation lowers the temperature.) How does this relate to what happens when your body sweats? (As sweat evaporates from the body, the skin becomes cooler.)

Figure 15 *Beneath the surface, your skin is a complex organ made of blood vessels, nerves, glands, and muscles.*

Blood vessels transport substances and help regulate body temperature.

Nerves carry messages to and from the brain.

Muscle fibers attached to a hair follicle can contract, causing the hair to stand up.

Hair follicles in the dermis produce hair.

Oil glands release oil that keeps hair flexible and helps waterproof the epidermis.

Sweat glands release sweat. As sweat evaporates, heat is removed from the skin, and the body is cooled. Sweat also contains waste materials taken out of the body.

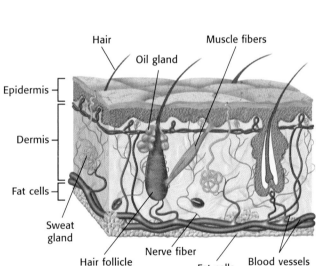

 Teaching Transparency 81 "The Skin"

 Directed Reading Worksheet Section 4

A hippo's sweat is a natural sunscreen. Hippopotamuses secrete a pink fluid onto the surface of their skin that helps protect them from the sun's harmful rays.

2) Teach, continued

Seeing Is Believing — PG 178

READING STRATEGY

Prediction Guide Before students read this page, ask them if they agree or disagree with the following statements. Students will discover the answers as they explore Section 4.

- Dark hair gets its color from melanin. (true)
- Hair is made up of both living and dead cells. (true)
- Nails grow longer as new cells form at the nail roots. (true)

REAL-WORLD CONNECTION

Protection from the Sun Skin cancer is the most common kind of cancer. More than 800,000 new cases of skin cancer are reported each year. You can take several preventive measures to reduce your risk of skin cancer. Avoid being in the sun from 10:00 A.M. to 3:00 P.M., when the sun's rays are most direct. Wear sunglasses, a wide-brimmed hat, and a long-sleeved shirt and long pants. Wear plenty of sunscreen with an SPF (sun protection factor) of 15 or higher.

internetconnect
TOPIC: Integumentary System
GO TO: www.scilinks.org
sciLINKS NUMBER: HSTL545

How fast do your fingernails grow? Find out on *page 178*.

Hair and Nails

A hair, shown in **Figure 16,** is formed at the bottom of a tiny sac called a **hair follicle.** The hair grows as new cells are added at the hair follicle and older cells get pushed upward. The only living cells in a hair are in the hair follicle, where the hair is produced.

Letting Your Hair Down Hairs protect skin from ultraviolet light and can help keep particles, such as dust and insects, out of your eyes and nose. Like skin, hair gets its color from the pigment melanin. Dark hair contains more melanin than blond hair. In most mammals, hair also helps regulate body temperature. A contraction of a tiny muscle attached to the hair follicle causes the follicle to bend. In humans, the bending follicle pushes up the epidermis to make a goose bump. If the follicle contains a hair, the hair "stands up." The lifted hairs function like a sweater to trap warm air around the body.

A Nail Tale Nails protect the tips of your fingers and toes so that they can remain soft and sensitive. This allows you to have a keen sense of touch. Nails form from *nail roots* under the skin at the base and sides of nails. As new cells form, the nail grows longer. The parts of a nail are shown in **Figure 17.**

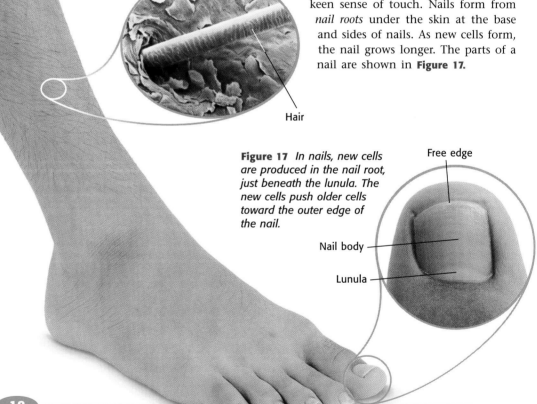

Figure 16 A hair is actually layers of dead, tightly packed, keratin-filled cells.

Figure 17 In nails, new cells are produced in the nail root, just beneath the lunula. The new cells push older cells toward the outer edge of the nail.

IS THAT A FACT!
Chemicals in human skin use the energy in ultraviolet light to make vitamin D.

SCIENCE HUMOR
Q: Where do sheep get their hair cut?
A: at the baa-baa shop

18 Chapter 1 • Body Organization and Structure

Living in Harm's Way

Skin is often damaged. The damage may be minor—a blister, an insect bite, or a small cut. Fortunately, your skin has an amazing ability to repair itself, as shown in **Figure 18.**

Figure 18 How Skin Heals

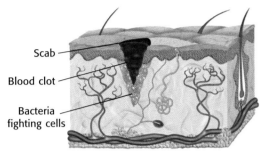
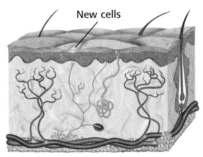

① When you get a cut, a blood clot forms to prevent bacteria from entering the wound. Bacteria-fighting cells then come to the area to kill bacteria.

② Damaged cells are replaced through cell division. Eventually, all that is left on the surface is a scar.

Other damage to the skin is very serious. Damage to the genetic material in skin cells can result in uncontrolled cell division, producing a mass of skin cells called a tumor. The term *cancer* is used to describe a tumor that invades other tissue. Darkened areas on the skin, such as moles, should be watched carefully for signs of cancer. **Figure 19** shows an example of a mole that has possibly become cancerous.

Your skin may also be affected by hormones that cause the oil glands in your skin to produce excess oil. This oil combines with dead skin cells and bacteria to clog hair follicles and cause infections. Proper cleansing and daily skin care can be helpful in decreasing the amount of infections.

Figure 19 *This mole has two halves that do not match, a characteristic that might indicate skin cancer.*

SECTION REVIEW

1. Why does skin color vary from person to person?
2. List six structures found in the dermis and the function of each one.
3. **Making Inferences** Why do you feel pain when you pull on your hair or nails but not when you cut them?

TOPIC: Integumentary System
GO TO: www.scilinks.org
sciLINKS NUMBER: HSTL545

3) Extend

DEBATE

Sun Safety Is sun tanning beneficial or dangerous? Lead students in a debate about the benefits of sunlight and the potential dangers and safety issues associated with getting a suntan.

4) Close

Quiz

1. What are five functions of the skin? (keeps moisture in and foreign particles out; provides information about the outside world; helps to regulate body temperature; and removes some wastes)
2. Describe the two layers of the skin. (The epidermis is the thinner outer layer of the skin and contains epithelial tissue and mostly dead cells. The dermis, the deeper, thicker layer of skin, is made up mostly of connective tissue.)
3. How are pimples formed? (Oil glands in the skin produce excess oil that combines with dead skin cells and bacteria to clog hair follicles and cause an infection. The result is a pimple.)

ALTERNATIVE ASSESSMENT

Have students make a colorful drawing in their ScienceLog of a cross section of skin. Have students make their drawings from memory and label the parts and describe the function of each: dermis, epidermis, fat cells, blood vessels, nerves, muscle, hair follicle, oil gland, sweat gland.
Sheltered English

▼ Answers to Section Review

1. The amount of melanin in the skin determines skin color.
2. nerves, sense; blood vessels, transport; muscle, motion; oil glands, waterproofing; sweat glands, heat regulation; fat cells, insulation
3. The hair and nails don't contain nerves, but the hair follicle and nail root do.

Design Your Own Lab

Muscles at Work
Teacher's Notes

Time Required

One 45-minute class period

Lab Ratings

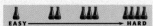

EASY → HARD

- TEACHER PREP 🧪
- STUDENT SET-UP 🧪
- CONCEPT LEVEL 🧪🧪
- CLEAN UP 🧪

Safety Caution

Remind students to review all safety cautions before beginning this lab activity.

A digital thermometer that measures temperature from the ear is recommended.

Because of the vigorous nature of the exercise, you may want to ask for volunteers to do the exercising. Also, you should be aware of any health concerns your students have.

 Datasheets for LabBook

Design Your Own Lab

USING SCIENTIFIC METHODS

Muscles at Work

Have you ever exercised outside on a cold fall day, wearing only a thin warm-up suit or shorts? How did you stay warm? The answer is that your muscle cells contracted. When contraction takes place, some energy is used to do work and the rest is converted to thermal energy. The thermal energy helps your body maintain a constant temperature in cold conditions. When you exercise strenuously on a hot summer day, your muscles can cause your body to become overheated. In this activity, your job is to find out how the release of energy can cause a change in your body temperature.

MATERIALS

- clock or watch with a second hand
- small hand-held thermometer
- other materials as approved by your teacher

Ask a Question

1. Form a group of four students. In your group, discuss what you already know about how muscle contractions can affect body temperature. During your discussion, ask several questions about how the release of energy can cause a change in body temperature. As a group, pick one of the questions that you think you can answer by performing an experiment.

Form a Hypothesis

2. Formulate a testable hypothesis to answer the question you chose. Write your hypothesis in your ScienceLog.

3. Plan an investigative procedure that includes the steps that are necessary to test your hypothesis. You will need to select the appropriate equipment to use during your experiment. Be sure to get your teacher's approval before you begin.

Conduct an Experiment

4. Assign tasks such as note taking, data recording, and timing to individuals in the group. What observations and data will you be recording? Design tables using a computer or graph paper to organize and examine the data you collect.

5. Perform your experiment as planned by your group. Be sure to record in your data tables all of the observations made during the experiment.

Analyze the Results

6 After you complete your experiment, review the data that you collected. Use a computer or graph paper to organize the data into graphs and charts.

7 Using your tables, charts, and graphs, can you make any inferences about how muscle contractions affect body temperature?

8 Do you recognize any patterns in your data? What trends can you predict? What might happen to your body temperature when you sleep?

Draw Conclusions

9 Was your hypothesis supported by your data? Communicate your conclusions in a written report. Describe how you could improve your experimental method.

Going Further
Why do humans shiver in the cold? Do all animals shiver? Find out why shivering is one of the first signs that your body is becoming too cold.

Design Your Own Lab

Answers
7–9. All answers will depend on the students' observations and their own hypotheses.

Going Further
In a process known as shivering thermogenesis, muscle tone is gradually increased. Shivering increases the workload of the muscles and elevates oxygen and energy consumption. The heat that is produced warms the deep vessels. Shivering can elevate body temperature effectively. It can increase the rate of heat generation by as much as 400 percent. Endothermic animals have the capacity to shiver. Shivering is an automatic response of the body to cold.

Kathy LaRoe
East Valley Middle School
East Helena, Montana

Chapter 1 • Design Your Own Lab

Chapter Highlights

VOCABULARY DEFINITIONS

SECTION 1

homeostasis the maintenance of a stable internal environment

tissue a group of similar cells that work together to perform a specific job in the body

epithelial tissue one of the four main types of tissue in the body; the tissue that covers and protects underlying tissue

nervous tissue one of the four main types of tissue in the body; the tissue that sends electrical signals through the body

muscle tissue one of the four main types of tissue in the body; contains cells that contract and relax to produce movement

connective tissue one of the four main types of tissue in the body; functions include support, protection, insulation, and nourishment

organ a combination of two or more tissues that work together to perform a specific function in the body

SECTION 2

skeletal system a collection of organs whose primary function is to support and protect the body; the organs in this system include bones, cartilage, ligaments, and tendons

compact bone the type of bone tissue that does not have open spaces; the tissue that gives a bone its strength

spongy bone the type of bone tissue that has many open spaces and contains marrow

cartilage a flexible, white tissue that gives support and protection but is not rigid like bone

joint the place where two or more bones connect

ligament a strong band of tissue that connects bones to bones

Chapter Highlights

SECTION 1

Vocabulary
homeostasis (p. 4)
tissue (p. 4)
epithelial tissue (p. 4)
nervous tissue (p. 4)
muscle tissue (p. 5)
connective tissue (p. 5)
organ (p. 5)

Section Notes

- Your body maintains a stable internal environment called homeostasis.
- Four types of tissues work to maintain homeostasis. Each tissue has a special job to do.
- Tissues work together to form organs.
- A group of organs working together for a common purpose is called an organ system.
- There are 11 major organ systems in the human body.

SECTION 2

Vocabulary
skeletal system (p. 8)
compact bone (p. 9)
spongy bone (p. 9)
cartilage (p. 10)
joint (p. 10)
ligament (p. 11)

Section Notes

- The skeletal system includes bones, cartilage, and ligaments.
- Bones support and protect the body, store minerals and fat, and produce blood cells.
- A typical bone contains marrow, spongy bone, compact bone, blood vessels, and cartilage.

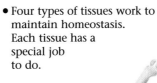

☑ Skills Check

Math Concepts

CALCULATING A PERCENTAGE In the MathBreak on page 15 you were asked to calculate a percentage of a number. To do this, first express the percentage as a decimal or a fraction. Then multiply it by the number. For example, 25 percent can be written as 0.25 or $25 \div 100$. To find 25 percent of 48, multiply by either 0.25 or $25 \div 100$.

$$0.25 \times 48 = 12$$
or
$$(25 \div 100) \times 48 = 12$$

Visual Understanding

MOVING WITH JOINTS Take another look at the three kinds of joints on page 10. Consider how your joints work when you throw a ball or walk up stairs. The hinge joint in your knee can move freely in only two directions. The ball-and-socket joint in your shoulder can move in many directions. The sliding joints in your hand allow bones to glide over one another.

Lab and Activity Highlights

Muscles at Work PG 20

Seeing Is Believing PG 178

 Datasheets for LabBook
(blackline masters for these labs)

SECTION 2

- A joint is where two bones meet. Some joints allow a lot of movement, and some allow little or no movement.
- Bones are attached to bones by connective tissue called ligaments.
- The action of muscle on bone and joints often works like a simple machine called a lever.

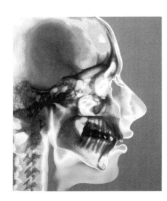

SECTION 3

Vocabulary
muscular system *(p. 12)*
smooth muscle *(p. 12)*
cardiac muscle *(p. 12)*
skeletal muscle *(p. 12)*
tendon *(p. 13)*

Section Notes
- Skeletal muscles and tendons make up the muscular system.
- You have three types of muscle: smooth, cardiac, and skeletal.
- Muscles are attached to bones by tendons.
- Exercise helps keep your muscular system healthy.

SECTION 4

Vocabulary
integumentary system *(p. 16)*
sweat glands *(p. 16)*
melanin *(p. 16)*
epidermis *(p. 17)*
dermis *(p. 17)*
hair follicle *(p. 18)*

Section Notes
- Your skin, hair, and nails make up your integumentary system.
- Your skin has two layers that contain a variety of small organs.
- Your hair and nails help protect your body.
- Skin can be damaged, but it has an amazing ability to repair itself.

Labs
Seeing Is Believing *(p. 178)*

VOCABULARY DEFINITIONS, continued

SECTION 3

muscular system a collection of organs whose primary function is movement; organs in this system include the muscles and the connective tissue that attaches them to bones

smooth muscle the type of muscle found in the blood vessels and the digestive tract

cardiac muscle the type of muscle found in the heart

skeletal muscle the type of muscle that moves the bones and helps protect the inner organs

tendon a tough connective tissue that connects skeletal muscles to bones

SECTION 4

integumentary system a collection of organs whose primary function is to help the body maintain a stable and healthy internal environment; the organs in this system include skin, hair, and nails

sweat glands small organs in the dermis layer of the skin that release sweat

melanin a darkening chemical in the skin that determines skin color

epidermis the outermost layer of the skin; also the outermost layer of cells covering roots, stems, leaves, and flower parts

dermis the layer of skin below the epidermis

hair follicle a small organ in the dermis layer of the skin that produces hair

---

internetconnect

 GO TO: go.hrw.com

Visit the **HRW** Web site for a variety of learning tools related to this chapter. Just type in the keyword:

KEYWORD: HSTBD1

  GO TO: www.scilinks.org

Visit the **National Science Teachers Association** on-line Web site for Internet resources related to this chapter. Just type in the *sci*LINKS number for more information about the topic:

TOPIC	*sci*LINKS NUMBER
Tissues and Organs	HSTL530
Body Systems	HSTL535
Skeletal System	HSTL537
The Muscular System	HSTL540
Integumentary System	HSTL545

Lab and Activity Highlights

LabBank

 Inquiry Labs, On a Wing and a Layer

Long-Term Projects & Research Ideas,
Mapping the Human Body

 Vocabulary Review Worksheet

 Blackline masters of these Chapter Highlights can be found in the **Study Guide.**

Chapter Review Answers

USING VOCABULARY

1. nervous
2. integumentary
3. skeletal
4. flexor
5. cartilage

UNDERSTANDING CONCEPTS

Multiple Choice

6. b
7. c
8. b
9. c
10. c
11. b

Short Answer

12. Epithelial tissue—covers and protects; muscle tissue—produces movement; nervous tissue—communicates by transfer of electrical signals or messages throughout the body; connective tissue—joins, supports, and transports other tissues. (Illustrations will vary but should be similar to those in **Figure 1**.)
13. Skin protects underlying tissue from dehydration, germs, and harmful radiation from the sun.
14. Muscle attached to the hair follicle contracts, pushing up the epidermis as it pulls on the hair follicle.
15. a. Skeletal muscle is attached to bone by tendons and causes bones to move; cardiac muscle does not attach to bone.
    b. Skeletal muscle can be voluntary or involuntary, but cardiac muscle is involuntary.
16. Bones in your skeleton depend on muscles to move them. Bones give muscles a frame on which to move, and bones supply muscles with calcium. Bones protect muscles that carry out vital functions. The ribs protect the heart and lungs, which in turn supply the bones with oxygenated blood.

Chapter Review

USING VOCABULARY

To complete the following sentences, choose the correct term from each pair of terms listed below:

1. Electrical signals are sent throughout the body by the __?__ tissue. (*epithelial* or *nervous*)
2. Your __?__ system is made up of skin, hair, and nails. (*integumentary* or *muscular*)
3. Bones are moved by __?__ muscle. (*smooth* or *skeletal*)
4. When __?__ muscles contract, they cause parts of the body to bend. (*extensor* or *flexor*)
5. Most of the skeleton starts out as __?__, which is later replaced by bone. (*cartilage* or *ligaments*)

UNDERSTANDING CONCEPTS

Multiple Choice

6. Which of the following is made up of cells that can contract and relax?
    a. skeletal tissue
    b. muscle tissue
    c. connective tissue
    d. nervous tissue

7. The organ system that provides support and protection for body parts is the
    a. endocrine system.
    b. circulatory system.
    c. skeletal system.
    d. respiratory system.

8. The epidermis is composed of
    a. dermis.
    b. epithelial tissue.
    c. connective tissue.
    d. true skin.

9. The fixed point in a lever is the
    a. effort.
    b. load.
    c. fulcrum.
    d. mechanical advantage.

10. Muscles cause bones to move when
    a. the muscles stretch.
    b. the muscles grow between bones.
    c. the muscles pull on bones.
    d. the muscles push bones apart.

11. Ligaments are the connective tissue that attaches
    a. bones to muscles.
    b. bones to other bones.
    c. muscles to other muscles.
    d. muscles to dermis.

Short Answer

12. Summarize the functions of the four types of tissues, and draw a sketch of each type.
13. How does the skin help protect the body?
14. What is a goose bump?
15. What are two ways skeletal muscle differs from cardiac muscle?
16. How do the functions of the skeletal system relate to the functions of the muscular system?

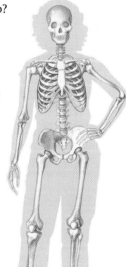

Concept Mapping

17. An answer to this exercise can be found at the front of this book.

 Concept Mapping Transparency 22

24 Chapter 1 • Body Organization and Structure

Concept Mapping

17. Use the following terms to create a concept map: bones, marrow, skeletal system, spongy bone, compact bone, cartilage.

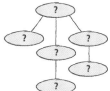

CRITICAL THINKING AND PROBLEM SOLVING

Write one or two sentences to answer the following questions:

18. Why do some muscles not work when a bone is broken?

19. Unlike human bones, some bird bones have air-filled cavities. What advantage does this give birds?

20. Compare the shapes of the bones of the human skull with the shapes of the bones of the human leg. Why is their shape important?

21. Compare the texture and sensitivity of the skin on your elbows with those of the skin on your fingertips. How can you explain the differences?

MATH IN SCIENCE

22. Your muscles make up about 40 percent of your overall mass. What is the muscle mass of a person whose total body mass is 60 kg?

23. The average person blinks 700 times an hour. How many times would the average person blink in a week if he or she were awake for 16 hours each day?

INTERPRETING GRAPHICS

Look at the picture below, and answer the questions that follow.

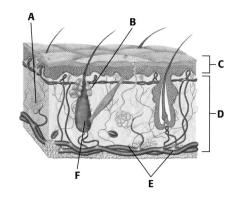

24. What is *D* called? What type of tissue is most abundant in this layer?

25. What is the name and function of *A*?

26. What is the name and function of *B*?

27. What part of the skin is made up of epithelial tissue that contains dead cells?

28. How does skin help regulate body temperature?

Take a minute to review your answers to the Pre-Reading Questions found at the bottom of page 2. Have your answers changed? If necessary, revise your answers based on what you have learned since you began this chapter.

CRITICAL THINKING AND PROBLEM SOLVING

18. A muscle needs a bone for leverage. If that bone is broken in two, the muscle can only move part of the bone, or none of it depending on where the break is in relation to where the muscle's tendon attaches to the bone.

19. Hollow bones are lighter, which makes flight possible.

20. The bones in the skull are shaped to form a strong helmet-like protection around the brain. The leg bone has a design that makes standing upright and moving around possible.

21. Students should notice that the skin on their elbows is hard and rough compared with the delicate, sensitive skin on their fingertips. The skin on an elbow has to be tough and insensitive because we rest our upper body on it.

MATH IN SCIENCE

22. Forty percent of 60 kg is 24 kg.

23. $700 \frac{\text{blinks}}{\text{hours}} \times 16 \frac{\text{hours}}{\text{day}} \times 7 \frac{\text{days}}{\text{week}} = 78{,}400 \frac{\text{blinks}}{\text{week}}$

INTERPRETING GRAPHICS

24. the dermis, connective tissue

25. Sweat glands produce sweat that helps regulate body temperature.

26. Oil glands produce oil, which waterproofs the skin.

27. C. Epidermis

28. Sample answer: Sweat glands produce sweat, which helps cool the body.

 Blackline masters of this Chapter Review can be found in the **Study Guide.**

SCIENCE, TECHNOLOGY, AND SOCIETY

Engineered Skin

Background

The engineered skin is far better than the scar tissue that would form without it. Scar tissue is weaker and more brittle than the skin it replaces. It does not stretch and grow, making a particularly difficult problem for children suffering from burns. Engineered skin also helps reduce the disfigurement associated with scarring.

One significant limitation of the engineered skin is that when it is new, it lacks sweat glands. Because patients cannot perspire in the areas that have skin grafts, patients with large grafts need to be cautious about overexercising and exposure to the sun.

Science, Technology, and Society

Engineered Skin

Your skin is more than just a well-fitting suit—it's your first line of defense against the outside world. Your skin keeps you safe from dehydration and infection, and the oil glands in your skin keep you waterproof. But what happens when a significant portion of skin is damaged?

More Skin Is the Answer

Sometimes doctors perform a skin graft, transferring some of a person's healthy skin to a damaged area of skin. This is because skin is really the best "bandage" for a wound. It protects the wound but still allows it to breathe. And unlike manufactured cloth or plastic bandages, skin can regenerate itself as it covers a wound. Sometimes, though, a person's skin is so severely damaged (as often occurs in burn victims) that the person doesn't have enough skin to spare.

Tissue Engineering

In the past few years, scientists have been studying tissue engineering to learn more about how the human body heals itself naturally. Using a small piece of young, healthy human skin and some collagen from cows, scientists can now engineer human skin. During the engineering process, cells form the dermal and epidermal layers of skin just as they would if they were still on the body. The living human skin that results can even heal itself if it is cut before it is used for a skin graft. Because it is living, the skin must be kept on a medium that provides it with nutrients until it is placed on a wound. Over time, the color of the grafted skin changes to match the color of the skin that surrounds it.

A Woven Dermis

Tissue engineers have also created another kind of skin, except this one has an unusual dermal

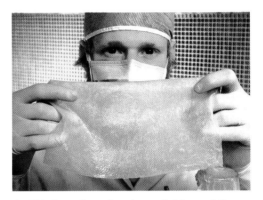

▲ *This is a piece of engineered skin used for grafting.*

and epidermal layer. In this skin, the dermis is made of woven collagen fibers. The wounded area digests these fibers and uses them as a guide to create a new dermis. The epidermal layer is a temporary layer of silicone. It shields the body from infection and protects against dehydration while new skin is being made.

After a new dermal layer forms under the protective silicone epidermis, the body is in better condition to accept a skin graft. Doctors can also graft a thinner portion of skin. A thinner graft is better for the body in the long run because it is easier to take from another part of the body. The new dermal layer also gives the body more time to strengthen on its own before the trauma of transplanting healthy skin to other areas.

On Your Own

▶ In the past, doctors have harvested skin from the bodies of people who, before they died, chose to be organ donors. What kinds of problems could arise if this harvested skin were used on burn victims?

Answer to On Your Own

Skin harvested from cadavers can be rejected, and it can also introduce viruses and disease. The bits of skin used to create engineered skin are from the foreskins of babies, and there seems to be a readily available supply. Additionally, youthful skin grows very quickly.

Eureka!

Hairy Oil Spills

Oil and water don't mix, right? Oil floats on the surface of water and is clearly visible to the naked eye. Proving this in your kitchen isn't difficult, nor is it dangerous. But what happens when the water is the ocean and the oil is crude oil? You have an environmental disaster that costs millions of dollars to clean up. The worst example in American waters was in 1989 when the *Exxon Valdez* oil tanker spilled nearly 42 million liters of crude oil into the waters of Prince William Sound on the Alaskan coast.

▲ *Phil McCrory among bags of discarded human hair.*

▲ *This otter was drenched with oil spilled from the* Exxon Valdez.

Backyard Testing

A Huntsville, Alabama, hairdresser asked a brilliant question when he saw an otter whose fur was drenched with oil from the *Valdez* spill. If the otter's fur soaked up all the oil, why wouldn't human hair do the same? The hairdresser, Phil McCrory, gathered hair from the floor of his salon and took it home to perform his own experiments. He stuffed 2.2 kg of hair into a pair of his wife's pantyhose and tied the ankles together to form a bagel-shaped bundle. After filling his son's wading pool with water, McCrory floated the bundle in the pool. Next, McCrory poured used motor oil into the center of the ring. When he pulled the ring closed, not a drop of oil remained in the water!

How Does Hair Do This?

What McCrory discovered was that hair **ad**sorbs oil instead of **ab**sorbing it. To adsorb means to collect a liquid or gas in layers on a surface. Because tiny cuticles cover every hair shaft like fish scales, the oil can bind to the surface of hair. Compare this process with the way a sponge works. A sponge completely absorbs a liquid. This means it is wet throughout, not just on the surface.

McCrory approached the National Aeronautics and Space Administration (NASA) with his discovery. In controlled tests performed by NASA, hair proved to be the fastest adsorber around. A little more than 1 kg of hair can adsorb over 3.5 L of oil in just 2 minutes!

It is estimated that within a week, 64 million kilograms of hair in reusable mesh pillows could have soaked up *all* of the oil spilled by the *Valdez*. Unfortunately, the $2 billion spent on the cleanup removed only about 12 percent of the spill. Did you ever think that the hair from your head could have a purpose beyond keeping your head warm?

Compare the Facts

▶ Research how McCrory's discovery compares with the methods currently used to clean up oil spills. Share your findings with the class.

Eureka!
Hairy Oil Spills

Background

Bioremediation is the use of biological processes and products to eliminate organic contaminants from soils. Hair is a biological waste product that doesn't degrade well in landfills. Using hair for the bioremediation of oil spills would reduce the amount of waste in landfills. Also, since the hair adsorbs the oil, wringing the hair out means the oil can be recovered and the hair can be used again. As a last resort, the oil-saturated mesh pillows can be burned as fuel in order to recover the value of the oil they contain.

Activity

Have a few students work in a group, and ask them to research the current state of Prince William Sound. Has the ecosystem fully recovered? Are there still problems? Ask these students to report their findings to the class.

Answer to Compare the Facts

Current methods of cleanup include skimmers, booms, and dispersing agents. Skimmers recover oil from the water's surface. There are three types of skimmers—weir, oleophilic, and suction. Dispersing agents contain surfactants, which break liquids such as oil into small drops. Booms are used for containment so that other areas are not contaminated. Booms also concentrate the oil into thicker layers on the water's surface, which facilitates recovery. One type of boom has natural oil-eating microbes within it. These microbes biodegrade the contaminants and then the boom biodegrades itself.

McCrory's method has the potential to save millions of dollars. Current remediation procedures cost about $10 per 3.8 L of oil, while his method may cost as little as $2.

Chapter Organizer

CHAPTER ORGANIZATION	TIME MINUTES	OBJECTIVES	LABS, INVESTIGATIONS, AND DEMONSTRATIONS
Chapter Opener pp. 28–29	45	National Standards: UCP 2, SAI 1, ST 2, SPSP 5, LS 1e, 3b	**Start-Up Activity,** Exercise and Your Heart, p. 29
Section 1 The Cardiovascular System	90	▶ Describe the functions of the cardiovascular system. ▶ Compare and contrast the three types of blood vessels. ▶ Describe the path that blood travels as it circulates through the body. ▶ Distinguish between blood types. UCP 1–3, 5, ST 2, SPSP 1, 4, 5, LS 1a, 1c–1f, 3b	**Whiz-Bang Demonstrations,** Get the Beat! Demo
Section 2 The Lymphatic System	90	▶ Discuss the functions of the lymphatic system. ▶ Identify the relationship between lymph and blood. ▶ Describe the organs of the lymphatic system. UCP 1, 5, LS 1a, 1c–1f	
Section 3 The Respiratory System	90	▶ Describe the flow of air through the respiratory system. ▶ Discuss the relationship between the respiratory system and the cardiovascular system. ▶ Identify respiratory disorders. UCP 1, 2, 4, 5, SPSP 1, 4, LS 1a, 1c–1f; Labs UCP 2, 5, SAI 1, 2, LS 1a	**QuickLab,** Why Do People Snore? p. 43 **Making Models,** Build a Lung, p. 44 **Datasheets for LabBook,** Build a Lung **Skill Builder,** Carbon Dioxide Breath, p. 45 **Datasheets for LabBook,** Carbon Dioxide Breath **EcoLabs & Field Activities,** There's Something in the Air **Whiz-Bang Demonstrations,** Take a Deep Breath **Long-Term Projects & Research Ideas,** Getting to the Heart

See page **T23** for a complete correlation of this book with the

NATIONAL SCIENCE EDUCATION STANDARDS.

TECHNOLOGY RESOURCES

 Guided Reading Audio CD
English or Spanish, Chapter 2

 One-Stop Planner CD-ROM with Test Generator

 CNN. Science, Technology & Society, Breakthrough Bandage, Segment 30
Modern Acupuncture, Segment 31

Chapter 2 • Circulation and Respiration

Chapter 2 • Circulation and Respiration

CLASSROOM WORKSHEETS, TRANSPARENCIES, AND RESOURCES	SCIENCE INTEGRATION AND CONNECTIONS	REVIEW AND ASSESSMENT
Directed Reading Worksheet **Science Puzzlers, Twisters & Teasers**		
Directed Reading Worksheet, Section 1 **Math Skills for Science Worksheet,** The Unit Factor and Dimensional Analysis **Transparency 82,** The Flow of Blood Through the Heart **Transparency 83,** The Flow of Blood Through the Body **Reinforcement Worksheet,** Matchmaker, Matchmaker **Reinforcement Worksheet,** Colors of the Heart **Critical Thinking Worksheet,** Doctor for a Day	**Math and More,** p. 32 in ATE **Apply,** p. 36 **MathBreak,** The Beat Goes On, p. 37 **Health Watch:** Goats to the Rescue, p. 51	**Homework,** pp. 32, 33 in ATE **Self-Check,** p. 34 **Section Review,** p. 35 **Section Review,** p. 37 **Quiz,** p. 37 in ATE **Alternative Assessment,** p. 37 in ATE
Directed Reading Worksheet, Section 2		**Self-Check,** p. 38 **Section Review,** p. 39 **Quiz,** p. 39 in ATE **Alternative Assessment,** p. 39 in ATE
Transparency 84, The Respiratory System **Directed Reading Worksheet,** Section 3 **Transparency 225,** Air Pressure and Breathing	**Chemistry Connection,** p. 41 **Multicultural Connection,** p. 41 in ATE **Connect to Physical Science,** p. 42 in ATE **Weird Science,** Catching a Light Sneeze, p. 50	**Section Review,** p. 43 **Quiz,** p. 43 in ATE **Alternative Assessment,** p. 43 in ATE

END-OF-CHAPTER REVIEW AND ASSESSMENT

Chapter Review in Study Guide
Vocabulary and Notes in Study Guide
Chapter Tests with Performance-Based Assessment, Chapter 2 Test
Chapter Tests with Performance-Based Assessment, Performance-Based Assessment 2
Concept Mapping Transparency 23

internet connect

 Holt, Rinehart and Winston On-line Resources
 go.hrw.com

For worksheets and other teaching aids related to this chapter, visit the HRW Web site and type in the keyword: **HSTBD2**

 National Science Teachers Association
 www.scilinks.org

Encourage students to use the *sci*LINKS numbers listed in the internet connect boxes to access information and resources on the **NSTA** Web site.

Chapter 2 • Chapter Organizer

Chapter Resources & Worksheets

Visual Resources

TEACHING TRANSPARENCIES

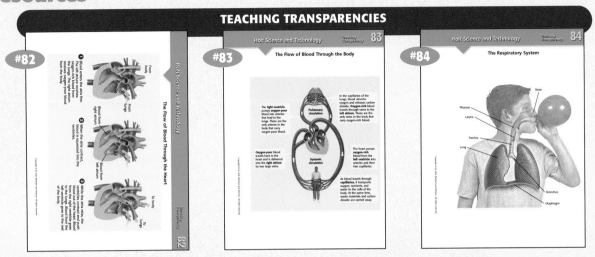

TEACHING TRANSPARENCIES

CONCEPT MAPPING TRANSPARENCY

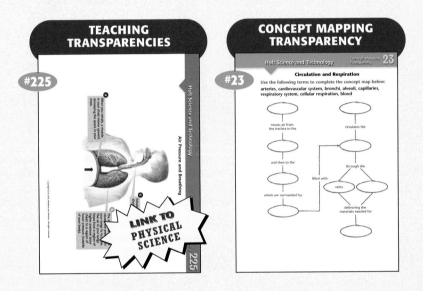

Meeting Individual Needs

DIRECTED READING

REINFORCEMENT & VOCABULARY REVIEW

SCIENCE PUZZLERS, TWISTERS & TEASERS

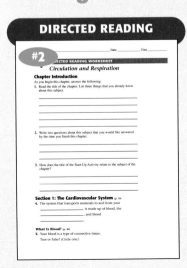
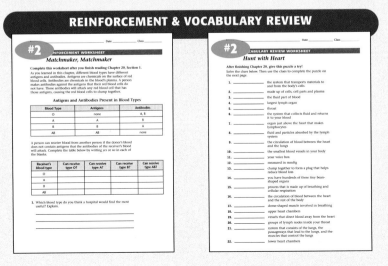
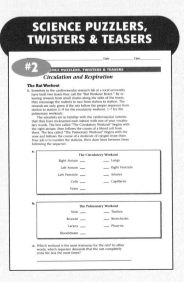

Chapter 2 • Circulation and Respiration

Chapter 2 • Circulation and Respiration

Review & Assessment

STUDY GUIDE

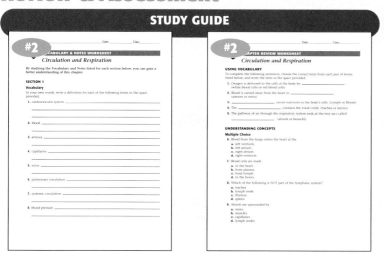

CHAPTER TESTS WITH PERFORMANCE-BASED ASSESSMENT

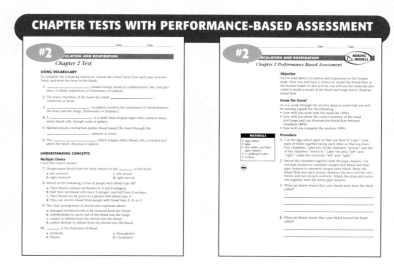

Lab Worksheets

ECOLABS & FIELD ACTIVITIES

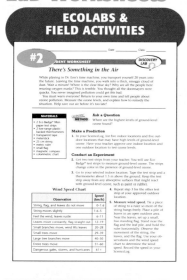

WHIZ-BANG DEMONSTRATIONS

LONG-TERM PROJECTS & RESEARCH IDEAS

DATASHEETS FOR LABBOOK

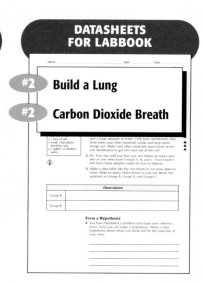

Applications & Extensions

CRITICAL THINKING & PROBLEM SOLVING

SCIENCE TECHNOLOGY

Chapter 2 • Chapter Resources & Worksheets

Chapter Background

SECTION 1

The Cardiovascular System

▶ The Flow of Blood
William Harvey (1578–1657) is credited with being the first European to discover the circulation of the blood through the body.

- Based on his dissections of animals, Harvey rightly concluded that the heart was a muscle that served to pump blood through the body. In contrast to conventional wisdom and theories of the famous physician Galen (A.D. 129–c. 201), Harvey also correctly maintained that arteries carry blood away from the heart and veins carry blood toward the heart. Harvey was ridiculed for his views by many other physicians.

▶ Atherosclerosis
Atherosclerosis is a disease of the arteries in which the inside layer of the arterial walls thickens with plaque. The plaque forms in areas of turbulent blood flow. As the walls of an artery thicken, the diameter of the vessel narrows, impeding blood flow.

- Cigarette smoking, high blood pressure, obesity, inactivity, high cholesterol level, and a family history of heart disease are all risk factors for atherosclerosis.

▶ Hypertension
Like atherosclerosis, people afflicted with hypertension may not exhibit symptoms of the disease for years. In fact, hypertension is often called "the silent killer." Although blood pressure varies within a wide range across the population, a person whose resting blood pressure is consistently at the high end of that range is said to have hypertension.

- Smoking, obesity, stress, excessive consumption of alcohol, and diabetes mellitus all exacerbate high blood pressure.

▶ Stroke
A stroke occurs when the brain is damaged due to an interruption in the blood flow or leaking of blood from the blood vessels. Atherosclerosis and hypertension are some of the causes of strokes.

▶ Heart Attack
A heart attack occurs when part of the heart muscle dies due to blood and oxygen deprivation. About 1 million people in the United States suffer a heart attack each year.

▶ Blood
If red blood cells (RBCs) have an Rh antigen (a particular kind of protein on the surface of the cell), the blood is positive (+). If an Rh antigen is not present, the blood is negative (−). People have one of the following blood types: A+, A−, B+, B−, AB+, AB−, O+, or O−.

- People make antibodies against the antigens that they do not have on their own RBCs. However, people who have Rh− blood do not always have antibodies against the Rh antigen. These antibodies are made only after the Rh− person has been exposed to blood that is Rh+. For example, type B− people make A antibodies that attack any blood cell with an A antigen on it. They may also make Rh antibodies that will attack any blood cell with an Rh antigen on it. This means that people with type B− blood can't be given A+, A−, B+, AB+, or AB− blood. Type O− blood can be given to anyone because its RBCs have no antigens on their surface that a patient's antibodies could attack. Because of this, a type O− person is said to be a *universal donor*. Type AB+ people are *universal recipients;* they can be given any type of blood because they do not make any antibodies against A, B, or Rh antigens.

Chapter 2 • Circulation and Respiration

SECTION 2

The Lymphatic System

▶ **Lymphocytes**
Lymphocytes are white blood cells that specialize in fighting pathogens. The two main kinds of lymphocytes are B cells and T cells.

- About 10 percent of lymphocytes are B cells. When confronted with foreign antigens, B cells produce antibodies that destroy the antigens. This process is called humoral immunity.

- About 90 percent of lymphocytes are T cells, which are formed in the bones and mature in the thymus. Killer T cells locate and attack cells that have foreign antigens on their surface. This type of immunity is called cell-mediated immunity.

- HIV infects and destroys lymphocytes called helper T cells. When a person's helper T cell count falls below 200 cells per cubic millimeter of blood, the person is diagnosed with AIDS.

IS THAT A FACT!

- The tonsils reach their largest size when a person is about seven years old. Then the tonsils begin to shrink.

SECTION 3

The Respiratory System

▶ **Control of Breathing**
Unless a person consciously holds his or her breath or changes the rate of his or her breathing, breathing is controlled automatically by breathing control centers in the base of the brain (in the medulla oblongata and in the pons).

- Hiccups are caused by a sudden jerky contraction of the diaphragm. When a person eats too much food, the full stomach may irritate the diaphragm muscle, causing it to contract jerkily. Generally, though, the cause of hiccups is unknown.

- Yawning is caused by a buildup of carbon dioxide in the lungs. The brain controls breathing by monitoring carbon dioxide, rather than oxygen, levels in the body. Thus, shallow breathing may result in the accumulation of carbon dioxide in the lungs, resulting in a signal from the brain to take a deep breath and exhale the extra carbon dioxide.

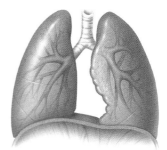

▶ **Smoking**
Tobacco smoking has been implicated in more than 90 percent of lung cancers among men. Among people who do not smoke, 3,000 cases of lung cancer are linked to secondhand cigarette smoke each year.

IS THAT A FACT!

- More than 3,000 adolescents in the United States start to smoke each day. The habit will eventually kill one-third of these children.

- Each year, parents who smoke at least 10 cigarettes a day cause 8,000 to 26,000 new cases of asthma in children. Moreover, each year between 200,000 and 1 million children who already have asthma have their condition worsened by parental secondhand smoke.

For background information about teaching strategies and issues, refer to the *Professional Reference for Teachers*.

CHAPTER 2

Circulation and Respiration

Pre-Reading Questions

Students may not know the answers to these questions before reading the chapter, so accept any reasonable response.

Suggested Answers

1. Blood is a type of connective tissue made of cell parts, plasma, and two types of cells. Blood carries nutrients and oxygen to all the cells of the body and carries carbon dioxide and wastes away. White blood cells fight infection, and platelets aid in clotting.

2. Inhaling brings oxygen into the lungs, where it is transferred to the blood and transported throughout the body. Exhaling allows carbon dioxide in the lungs to leave the body.

CHAPTER 2

Circulation and Respiration

Sections

1. The Cardiovascular
   System 30
      Apply 36
      MathBreak 37
      Internet Connect 37
2. The Lymphatic
   System 38
      Internet Connect 39
3. The Respiratory
   System 40
      Chemistry Connection . 41
      QuickLab 43
      Internet Connect 43

Chapter Labs 44, 45
Chapter Review 48
Feature Articles 50, 51

Pre-Reading Questions

1. What is blood, and what is its function in your body?
2. Why do you need to breathe?

Small but Mighty!

These donut-shaped objects are red blood cells like those that can be found throughout your body. Red blood cells are smaller than most other body cells. In fact, millions of them can be found in a single drop of blood. These cells may be small, but they perform a very important function. They are so important that your body makes about 200 billion new red blood cells every day. Why does your body need so many red blood cells? In this chapter, you will learn how these tiny cells enable all your body cells to carry out cellular respiration.

---

internet connect

HRW On-line Resources
go.hrw.com
For worksheets and other teaching aids, visit the HRW Web site and type in the keyword: **HSTBD2**

sciLINKS / NSTA
www.scilinks.com
Use the *sci*LINKS numbers at the end of each chapter for additional resources on the **NSTA** Web site.

Smithsonian Institution
www.si.edu/hrw
Visit the Smithsonian Institution Web site for related on-line resources.

CNNfyi.com
www.cnnfyi.com
Visit the CNN Web site for current events coverage and classroom resources.

START-UP Activity

EXERCISE AND YOUR HEART

Your heart pumps blood throughout your body. How does your heart respond to exercise? You can determine this reaction by measuring your pulse. You can take your pulse by placing your fingers on the inside of your wrist just below your thumb.

Procedure

1. Take your pulse while remaining still. Using a **watch with a second hand,** count the number of beats in 15 seconds. Then multiply this number by 4 to calculate the number of beats in 1 minute

2. Do jumping jacks or jog in place for 30 seconds. Then stop and calculate your heart rate again.

    Caution: Do not perform this exercise if you have difficulty breathing, have high blood pressure, or easily get dizzy.

3. Rest for 5 minutes, and then take your pulse again.

Analysis

4. How did exercise affect your heart rate? Why do you think this happened?

5. How does your heart rate affect the rate at which red blood cells travel throughout your body?

6. Why did your heart rate return to normal after you rested?

START-UP Activity

EXERCISE AND YOUR HEART

MATERIALS

For Each Group:
- watch with a second hand

Safety Caution

Have students bring in a signed permission slip for this activity. Any students who have a health problem that is worsened by exercise should be excused from the exercise portion of this activity. Instruct students who feel pain or become dizzy or tired to stop exercising immediately. Some students may feel embarrassed to exercise in front of their peers. You may want to invite a few volunteers to perform this part of the activity instead of having all students exercise.

Answers to START-UP Activity

4–5. Students should notice that their heart rate goes up when they are exercising. Explanations will vary. Students might suggest that while exercising, more blood is needed to deliver energy and oxygen to the muscles.

6. Once exercise is complete, less blood is needed, so the heart can slow down.

SECTION 1

Focus

The Cardiovascular System

This section introduces the structures and functions of the cardiovascular system. Students compare and contrast three different types of blood vessels and trace the path of blood through these vessels in the body. Students also learn to distinguish between the different blood types.

Bellringer

Ask students to list in their ScienceLog as many song titles, phrases, and slogans that contain the word *heart* as they can in 2–3 minutes. Ask for examples, and list them on the board. Ask for reasons why the word *heart* is the focus of so many songs and slogans.

1) Motivate

DISCUSSION

Invite students to describe a time when their skin was cut. Encourage students to describe not only what happened but also what their blood looked like. Make a list of the words students use to describe their blood. Based on students' experiences with blood and bleeding, lead a discussion about the structure and functions of blood. Then have students read these pages and compare their descriptions of blood with the one in the textbook. Ask:

Can you see individual blood cells when you bleed? Why or why not? How are red blood cells different from other cells in the body? (They lack a nucleus, organelles, and DNA, and they carry oxygen to other cells in the body.)

SECTION 1
READING WARM-UP

Terms to Learn

cardiovascular system
blood
arteries
capillaries
veins
pulmonary circulation
systemic circulation
blood pressure

What You'll Do

- Describe the functions of the cardiovascular system.
- Compare and contrast the three types of blood vessels.
- Describe the path that blood travels as it circulates through the body.
- Distinguish between blood types.

The Cardiovascular System

When you hear the word *heart*, what do you think of first? Many people think of romance. But the heart is much more than a symbol of love. It's the pump that drives your cardiovascular system. The **cardiovascular system** transports materials to and from your cells. The word *cardio* means "heart," and the word *vascular* means "vessel." The cardiovascular system, which is shown in **Figure 1**, is made up of three parts: blood, the heart, and blood vessels.

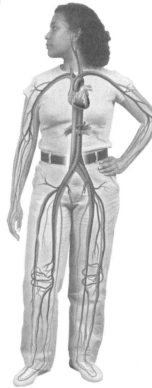

Figure 1 The Cardiovascular System

What Is Blood?

The human body contains about 5 L of blood. **Blood** is a connective tissue made up of two types of cells, cell parts, and plasma. *Plasma* is the fluid part of blood. It is a mixture of water, minerals, nutrients, sugars, proteins, and other substances. Red blood cells, white blood cells, and platelets float in the plasma.

Figure 2 Red blood cells deliver oxygen.

Red Blood Cells Red blood cells, or RBCs, are the most abundant cells in blood. RBCs, shown in **Figure 2**, supply your cells with oxygen. As you have learned, cells need oxygen to carry out cellular respiration. Each RBC contains a protein called *hemoglobin* (HEE moh GLOH bin). Hemoglobin, which gives RBCs their red color, clings to the oxygen you inhale. This allows RBCs to transport oxygen throughout the body. The shape of RBCs gives them a large amount of surface area for absorbing and releasing oxygen.

RBCs are made in the bone marrow. Before RBCs enter the bloodstream, they lose their nucleus and other organelles. Without a nucleus, which contains DNA, the RBCs cannot replace worn-out proteins. RBCs therefore can live only about 4 months.

---

internetconnect

TOPIC: The Cardiovascular System
GO TO: www.scilinks.org
sciLINKS NUMBER: HSTL555

IS THAT A FACT!

About half of the volume of blood is cells. The other half is plasma. Plasma consists of 95 percent water.

---

30 Chapter 2 • Circulation and Respiration

White Blood Cells Sometimes *pathogens*—bacteria, viruses, and other microscopic particles that can make you sick—are able to enter your body. When they do, they often encounter your white blood cells, or WBCs. WBCs, shown in **Figure 3,** help you stay healthy by destroying pathogens and helping to clean wounds.

WBCs fight pathogens in several ways. Some squeeze out of vessels and move around in tissues, searching for pathogens. When they find a pathogen, they engulf it. Other WBCs release chemicals called *antibodies,* which help destroy pathogens. WBCs also keep you healthy by engulfing body cells that have died or been damaged. WBCs are made in bone marrow. Some of them mature in lymphatic organs, which will be discussed later.

Platelets Drifting among the blood cells are tiny particles called platelets. *Platelets* are pieces of larger cells found in bone marrow. These larger cells remain in the bone marrow, but they pinch off fragments of themselves, which enter the blood. Although platelets last for only 5 to 10 days, they are an important part of blood. When you cut or scrape your skin, you bleed because blood vessels have been opened. As soon as bleeding occurs, platelets begin to clump together in the damaged area and form a plug that helps reduce blood loss, as shown in **Figure 4.** Platelets also release a variety of chemicals that react with proteins in the plasma and cause tiny fibers to form. The fibers create a blood clot.

Figure 3 *White blood cells defend the body against pathogens. These white blood cells have been colored yellow to make their shape easier to see.*

Figure 4 *Platelets release chemicals in damaged vessels and cause fibers to form. The fibers make a "net" that traps blood cells and stops bleeding.*

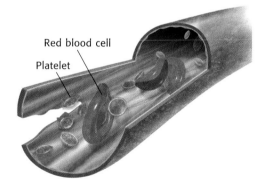
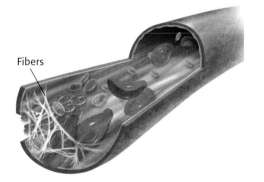

MISCONCEPTION ALERT

Students may think that deoxygenated blood is blue because it is often depicted that way in illustrations and because blood appears blue in vessels through the skin. The color of blood does change depending on the amount of oxygen it contains; however, human blood is not blue. It varies in color from scarlet to deep red.

Directed Reading Worksheet Section 1

2) Teach

READING STRATEGY

Activity Before students read the text on this page, have them read the headings aloud. Then ask students to formulate one question that they expect the text under each heading to answer. Have students write the questions in their ScienceLog.

USING THE FIGURE

Have students compare **Figure 2,** which shows red blood cells, with **Figure 3,** which shows white blood cells. Ask students to relate any differences in shape or size to the different functions of these two types of blood cells. Help students understand that the flat shape and small size of red blood cells enable them to carry oxygen through tiny capillaries to cells throughout the body. Help students link the larger size and sticky-looking surface of white bloods cells to their ability to fight pathogens. **Sheltered English**

MEETING INDIVIDUAL NEEDS

Learners Having Difficulty
To help students visualize the components of blood, have them make a model of blood. Provide students with a variety of materials, such as a resealable plastic sandwich bag, thick corn syrup, plastic-foam balls, and other craft materials. Then have students use the materials to make a model of the parts of blood. Encourage students to be creative in their representations of the parts of blood. **Sheltered English**

Section 1 • The Cardiovascular System

2) Teach, continued

MATH and MORE

Your heart beats about 100,800 times per day. With every beat, about 70 mL of blood is pumped out of your heart. In 1 hour, how much blood does your heart pump out? (294 L)

About how much blood does your heart pump out in a day? (7,056 L)

Help students visualize this amount by showing them a liter of water.

Math Skills Worksheet
"The Unit Factor and Dimensional Analysis"

Homework

Making Models Have students make a model of the human heart to show the path of blood through it. Models can be drawings or three-dimensional constructions. Have students present their completed model to the class. Provide yarn or a pen light so that students can demonstrate to the class the flow of blood through their model heart.
Sheltered English

Teaching Transparency 82
"The Flow of Blood Through the Heart"

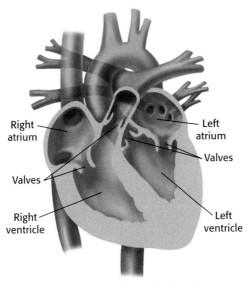

Figure 5 *The heart is a four-chambered organ that pumps blood through cardiovascular vessels. The vessels carrying oxygen-rich blood are shown in red. The vessels carrying oxygen-poor blood are shown in blue.*

Have a Heart

Your heart is a muscular organ about the size of your fist. It is found in the center of your chest cavity. The heart pumps oxygen-poor blood to the lungs and oxygen-rich blood to the body. Like the hearts of all mammals, your heart has a left side and a right side that are separated by a thick wall. As you can see in **Figure 5,** each side has an upper chamber and a lower chamber. Each upper chamber is called an *atrium* (plural, *atria*). Each lower chamber is called a *ventricle*.

Flaplike structures called *valves* are located between the atria and ventricles and also where large arteries are attached to the heart. As blood moves through the heart, the valves close and prevent blood from going backward. The lub-dub, lub-dub sound that a beating heart makes is caused by the closing of the valves. The flow of blood through the heart is shown in the diagram below.

The Flow of Blood Through the Heart

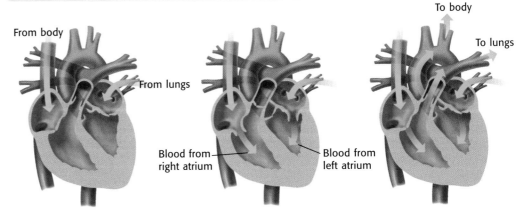

1 Blood enters the atria first. The left atrium receives oxygen-rich blood from the lungs. The right atrium receives oxygen-poor blood from the body.

2 When the atria contract, blood is squeezed into the ventricles.

3 While the atria relax, the ventricles contract and push blood out of the heart. Blood from the right ventricle goes to the lungs. Blood from the left ventricle goes to the rest of the body.

MISCONCEPTION ALERT

The words *left* and *right* as used to diagram the anatomy of the heart might confuse students. When students are looking at a picture of the heart, the left atrium appears on their right, and the right atrium appears on their left. Help students understand anatomical left and right by facing them and asking them to identify your left and right hands in relation to their own.

Blood Vessels

Blood travels throughout your body in blood vessels. A blood vessel is a hollow tube that transports blood. There are three types of blood vessels—arteries, capillaries, and veins. Their structures and their relationship to each other are shown in **Figure 6.**

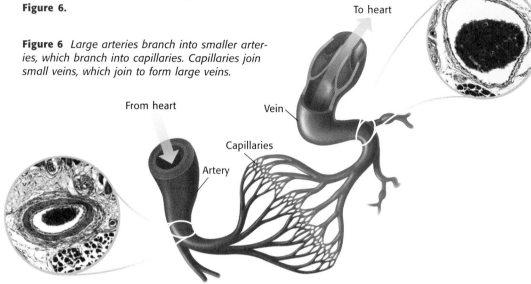

Figure 6 *Large arteries branch into smaller arteries, which branch into capillaries. Capillaries join small veins, which join to form large veins.*

Arteries **Arteries** are blood vessels that direct blood away from the heart. Arteries have thick elastic walls that contain a layer of smooth muscle. Each time the heart beats, blood is pumped out of the heart at high pressure. The thick walls of arteries have the strength to withstand this pressure. The rhythmic change in blood pressure is called a *pulse*.

Capillaries A strand of hair is about 10 times wider than a capillary. **Capillaries** are the smallest blood vessels in your body. Capillary walls are only one cell thick. As shown in **Figure 7,** capillaries are so narrow that blood cells must pass through them in single file. The simple structure of a capillary allows nutrients, oxygen, and many other kinds of substances to diffuse easily through capillary walls. No cell in the body is more than three or four cells away from a capillary.

Veins After leaving capillaries, the blood enters veins. **Veins** are blood vessels that direct the blood back to the heart. As blood travels through veins, valves keep the blood from flowing backward. When skeletal muscles contract, they squeeze nearby veins and help push blood toward the heart.

Figure 7 *These red blood cells are traveling through a capillary.*

If all the blood vessels in your body were strung together, the total length would be more than twice the circumference of the Earth.

WEIRD SCIENCE

Babies born with certain congenital heart defects are blue at birth. The condition, called cyanosis, can be caused by low levels of oxygen in the blood. The low levels of oxygen can be a result of defects in the anatomy of the heart. Deoxygenated blood is pumped to the body before it is pumped to the lungs. This defect can be repaired surgically.

2 Teach, continued

GROUP ACTIVITY
Circulation Relay

MATERIALS

FOR EACH CLASS:
- 5 inflated red balloons (or paper disks)
- 5 inflated blue balloons (or paper disks)
- diagram of The Flow of Blood Through the Body
- 8 stations (cardboard boxes, flags, and so on)

Divide the class into five teams for a relay race.

1. Students begin in the left ventricle carrying a red balloon, which represents an oxygenated blood cell.
2. They travel through the aorta.
3. After passing through the aorta, students carry oxygenated blood to the muscles and exchange the red balloon for a blue one.
4. From the muscles, students carry blood loaded with CO_2 to the right atrium.
5. From the right atrium, students travel into the right ventricle.
6. Students travel through the pulmonary artery.
7. From the pulmonary artery, students travel into the lungs, where they exchange their CO_2 for oxygen (exchange blue balloons for red ones).
8. Carrying oxygenated blood (red balloons), students enter the left atrium and are ready to begin again or hand off their balloons.

Walk one student at a time through the pathway; then have the teams send students through in a relay race.

Self-Check

How are the structures of arteries and veins related to their functions? *(See page 212 to check your answer.)*

Going with the Flow

As you read earlier, one important function of your blood is to supply the cells of your body with oxygen. Where does blood get this oxygen? It gets it from your lungs during pulmonary circulation. **Pulmonary circulation** is the circulation of blood between your heart and lungs.

When oxygen-rich blood returns to the heart from the lungs, it must be pumped to the rest of the body. The circulation of blood between the heart and the rest of the body is called **systemic circulation**. Both types of circulation are diagramed below.

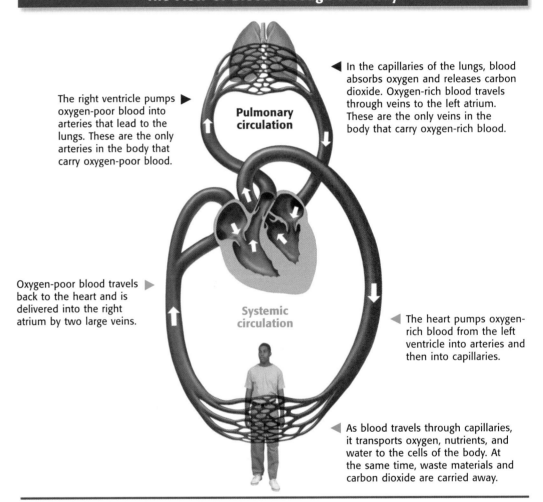

The Flow of Blood Through the Body

- The right ventricle pumps oxygen-poor blood into arteries that lead to the lungs. These are the only arteries in the body that carry oxygen-poor blood.
- In the capillaries of the lungs, blood absorbs oxygen and releases carbon dioxide. Oxygen-rich blood travels through veins to the left atrium. These are the only veins in the body that carry oxygen-rich blood.
- Oxygen-poor blood travels back to the heart and is delivered into the right atrium by two large veins.
- The heart pumps oxygen-rich blood from the left ventricle into arteries and then into capillaries.
- As blood travels through capillaries, it transports oxygen, nutrients, and water to the cells of the body. At the same time, waste materials and carbon dioxide are carried away.

Answers to Self-Check

The hollow tube shape of arteries and veins allows blood to reach all parts of the body. Valves in the veins prevent blood from flowing backward.

MISCONCEPTION ALERT

Students may think that blood from the heart enters one lung and leaves from the other. Actually, each lung is serviced by vessels carrying blood to and from the heart.

Blood Flows Under Pressure

When you run water through a hose, you can feel the hose stiffen as the water pushes against the inside of the hose. Blood has the same effect on your blood vessels. The force exerted by blood on the inside walls of a blood vessel is called **blood pressure.**

Like the man shown in **Figure 8,** many people get their blood pressure checked on a regular basis. Blood pressure is reported in millimeters (mm) of mercury, Hg. A blood pressure of 120 mm Hg means the pressure on the vessel walls is great enough to push a narrow column of mercury 120 mm high.

A normal blood pressure is about 120/80. The first number is called the systolic pressure. *Systolic pressure* is the pressure inside large arteries when the ventricles contract. As you read earlier, the surge of blood causes the arteries to bulge and produce a pulse. The second number is called the diastolic pressure. *Diastolic pressure* is the pressure in the arteries when the ventricles relax.

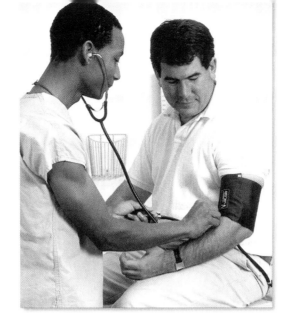

Figure 8 *This nurse is measuring a patient's blood pressure. Consistently high or low blood pressure may suggest a problem with the cardiovascular system.*

Exercise and Blood Flow

When you exercise, your muscles require much more oxygen and nutrients. To solve this problem, the heart beats faster. Physical activity causes as much as 10 times more blood to be sent to the muscles than when your body is at rest.

During exercise, some organs do not need as much blood as the skeletal muscles do. Less blood is sent to the kidneys and the digestive system so that more blood can go to the skeletal muscles, brain, heart, and lungs. This is like turning off certain water faucets in a house to allow more water to flow through other faucets.

Activity

Imagine that you are a scientist that has been chosen to explore the cardiovascular system. After being shrunk down to the size of a red blood cell, you board a miniature submarine and begin your travels. Describe where you go and what you see.

TRY at HOME

SECTION REVIEW

1. What is the function of the cardiovascular system?
2. What are the three kinds of blood vessels? Compare their functions.
3. **Identifying Relationships** How is the structure of capillaries related to their function?

Answers to Section Review

1. The cardiovascular system transports materials to and from the body's cells.
2. Arteries carry blood away from the heart to the capillaries; capillaries are small enough to allow materials in the blood to diffuse through the wall and enter the body's cells and to allow materials in the cells to diffuse into the capillaries; veins carry blood away from the capillaries back to the heart and lungs.
3. Capillary walls are only one cell thick. This simple structure allows nutrients, oxygen, and many other kinds of substances to diffuse easily through a capillary's wall to the body's other cells.

RETEACHING

If students have difficulty understanding the pulmonary circulation systems, have the class make a life-size model of the two loops. Invite two volunteers to draw a body silhouette. One student should be the model; the other should be the tracer. Then draw shapes to represent the heart and lungs in the chest. Have students take turns drawing and labeling the two circulatory loops on the silhouette. When the life-size model is complete, display it so that students can refer to it as they review this page. Sheltered English

DISCUSSION

Blood Pressure Tell students that blood pressure varies depending on the activities a person is performing. Write the following activities on the board or overhead projector, and ask students to discuss whether each activity is likely to cause an increase or a decrease in a person's blood pressure: sleeping (decrease), exercise (increase), waking up in the morning (increase), smoking (increase), tensing up (increase).

Encourage students to suggest additional activities and to speculate about the effects those have on blood pressure.

Answer to Activity

Make sure that students describe only the circulatory path, and encourage them to mention real structures and organs they will encounter. They might encounter a clot, a hardening of the arteries, an arrhythmia, or a heart murmur.

Teaching Transparency 83 "The Flow of Blood Through the Body"

3) Extend

Going Further

Nobel Prize–winning scientist Karl Landsteiner (1868–1943) discovered that some mixtures of blood are compatible and others are not. He found that he could divide the population into different groups based on how their blood reacted with blood from other people. Encourage interested students to research how Landsteiner's experiments led to the discovery of the ABO and Rh blood groups.

Using the Table

Review the material in the table to reinforce the material presented in the text. Help students compare the shapes of the antigens on each red blood cell with the shapes of the antibodies that bind to the antigens. Then ask the following questions:

- Which antibodies will a person with type AB blood produce? (none)
- Can a person with type O blood receive blood from a person with blood type AB? Why or why not? (No; the type O blood will have antibodies that attack the A and B antigens in the AB blood.)
- Can a person with type O blood donate blood to someone with type A blood? Why or why not? (Yes; the type O red blood cells lack antigens.)

Answer to APPLY

You could bring back type A, B, or O.

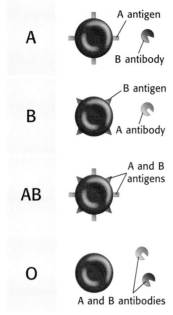

Figure 9 *This table shows which antigens and antibodies may be present in each blood type.*

Blood Delivery

A young woman is brought into the emergency room and needs a blood transfusion. Her blood type is AB. You call the blood bank to order AB blood, but you are told the bank is out of that type. What other type or types could the blood bank deliver for her transfusion?

What's Your Blood Type?

When a person loses a lot of blood, the person is given blood that has been donated from someone else. The person receiving the blood cannot be given blood from just anyone because people have different blood types. It's safe to mix some blood types, but mixing others causes a person's RBCs to clump together. The clumped cells may form blood clots, which block blood vessels, causing death.

Every person has one of the following blood types: A, B, AB, or O. Your blood type refers to the type of chemicals you have on the surface of your RBCs. These chemicals are called *antigens*. Type A blood has A antigens; type B has B antigens; and type AB has both A and B antigens. Type O blood has neither the A nor B antigen.

To Mix or Not to Mix Different blood types not only have different chemicals on their RBC surfaces but also may have different chemicals in their plasma, the liquid part of blood. These chemicals are *antibodies*. When antibodies bind to RBCs, they cause the RBCs to clump together.

As shown in **Figure 9**, the body makes antibodies against the antigens that are not on its own RBCs. For example, people with type B blood make A antibodies, which attack any blood cell with an A antigen on it. Therefore, people with type B blood can't be given type A or AB blood. Type O blood can be given to anyone because its RBCs don't have any A or B antigens on their surface. A person with type O blood is therefore said to be a *universal donor*. People with type AB blood are *universal recipients*, meaning they can receive any type of blood because they do not make any antibodies against A or B antigens.

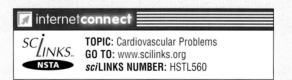

TOPIC: Cardiovascular Problems
GO TO: www.scilinks.org
sciLINKS NUMBER: HSTL560

Is That a Fact!

The first heartbeat occurs in the human embryo at four weeks of age. Although the heart looks like a simple tube at this stage, it begins to beat. This starts blood circulation through the first blood vessels, which eventually grow and develop into the entire circulatory system.

Cardiovascular Problems

When something is wrong with a person's cardiovascular system, the person's health will be affected. Some cardiovascular problems involve the heart and the blood vessels, while other problems affect the blood. Cardiovascular problems can be caused by smoking, high levels of cholesterol in blood, stress, heredity, and other factors.

Atherosclerosis The leading cause of death in the United States is a cardiovascular disease called *atherosclerosis* (ATH uhr OH skluh ROH sis). Atherosclerosis occurs when fatty materials, such as cholesterol, build up on the inside of blood vessels. The fatty buildup causes the blood vessels to become narrower and less elastic. **Figure 10** shows how the pathway through a blood vessel can become clogged. When a major artery that supplies blood to the heart becomes blocked, a person has a heart attack, and part of the heart can die.

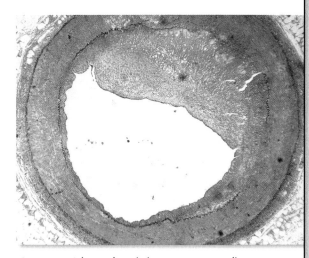

Figure 10 *Atherosclerosis is a common cardiovascular problem. Fatty deposits build up on the inside of blood vessels and block blood flow.*

A Point About Pressure Atherosclerosis also promotes *hypertension*, which is abnormally high blood pressure. Hypertension is dangerous because it overworks the heart and can weaken vessels and make them rupture. If a blood vessel in the brain becomes clogged or ruptures, certain parts of the brain will not receive oxygen and nutrients and may die. This is called a *stroke*.

The Beat Goes On

Your heart beats about 100,800 times per day. How many times does it beat per year?

SECTION REVIEW

1. Where does blood travel to and from during pulmonary circulation? during systemic circulation?
2. What happens to the oxygen level in blood as it moves through the lungs?
3. **Applying Concepts** Billy has type A blood.
   a. What kind of antigens does he have on his RBCs?
   b. What blood-type antibodies can Billy make?
   c. Which blood types could be given to Billy if he needed a transfusion?

internet connect

sciLINKS
NSTA

TOPIC: The Cardiovascular System, Cardiovascular Problems
GO TO: www.scilinks.org
*sci*LINKS NUMBER: HSTL555, HSTL560

Answers to Section Review

1. Blood travels to and from the lungs in pulmonary circulation. Blood travels to and from the body in systemic circulation.
2. The oxygen level increases in blood as the blood travels through the lungs.
3. a. A antigens
   b. B antibodies
   c. A and O

SECTION 2

Focus

The Lymphatic System

This section introduces the lymphatic system. Students will also learn about the relationship between blood and lymph.

Bellringer

Draw the following shapes on the board or on an overhead projector:

a circle, a triangle, a straight line, and a cluster of several dots

Ask students to choose the shape(s) that best represents a circulatory system and to explain in their ScienceLog. (The circle and the triangle best represent a circulatory system because they both form a continuous loop.)

Answers to Self-Check

Like blood vessels, lymph capillaries receive fluid from the spaces surrounding cells. The fluid absorbed by lymph capillaries flows into lymph vessels. These vessels drain into large neck veins instead of into an organ, such as the heart. Lymph does not deliver oxygen and nutrients.

1) Motivate

DISCUSSION

Ask students if a doctor has ever felt around their neck when they were sick. Encourage students who have had this experience to share it with the class. Then invite students to explore the purpose of this type of examination. **Sheltered English**

SECTION 2

READING WARM-UP

Terms to Learn

lymphatic system thymus
lymph   spleen
lymph nodes   tonsils

What You'll Do

- Discuss the functions of the lymphatic system.
- Identify the relationship between lymph and blood.
- Describe the organs of the lymphatic system.

Self-Check

How are the lymphatic system and the cardiovascular system similar? How are they different? *(See page 212 to check your answer.)*

The Lymphatic System

Your cardiovascular system is not the only circulatory system in your body. As blood flows through your cardiovascular system, fluid leaks out of the capillaries and mixes with the fluid that bathes your cells. Most of the fluid is reabsorbed by the capillaries, but some is not. To deal with this, your body's **lymphatic system** collects the excess fluid and returns it to your blood.

In addition to collecting the excess fluid surrounding your cells and returning it to your blood, your lymphatic system helps your body fight pathogens. Pathogens are microorganisms and viruses that make you sick.

Vessels of the Lymphatic System

The fluid collected by the lymphatic system is transported through vessels. The smallest vessels of the lymphatic system are *lymph capillaries*. From the spaces between cells, lymph capillaries absorb fluid and any particles too large to enter the blood capillaries. Some of these particles are dead cells or cells that are foreign to the body. The fluid and particles absorbed into lymph capillaries are called **lymph**.

As shown in **Figure 11**, lymph capillaries carry lymph into *lymphatic vessels*, which are larger vessels that have valves. Lymph is not pushed by a pump. Instead, the squeezing of skeletal muscles provides the force to move lymph through vessels, and valves help prevent backflow. Lymph travels through your lymphatic system and then drains into large neck veins of the cardiovascular system.

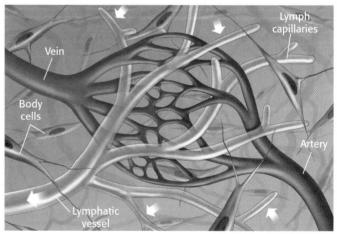

Figure 11 *The white arrows show the movement of lymph into lymph capillaries and through lymphatic vessels.*

SCIENTISTS AT ODDS

The Danish physician Thomas Bartholin (1616–1680), known as the Elder, is credited with being the first person to describe the lymphatic system. His Swedish contemporary Olof Rudbeck (1630–1702) also studied the lymphatic system. Rudbeck was furious when French anatomist Jean Pecquet and Thomas Bartholin published their findings before he did. Rudbeck spent the rest of his life studying botany.

Lymphatic Organs

In addition to vessels and capillaries, a variety of other organs are part of the lymphatic system, as shown in **Figure 12.**

Lymph Nodes As lymph travels through lymphatic vessels, it passes through lymph nodes. **Lymph nodes** are small bean-shaped organs where particles, such as pathogens or dead cells, are removed from the lymph.

Lymph nodes contain many white blood cells. Some of these cells engulf pathogens. Other WBCs produce chemicals that become attached to pathogens and mark them for destruction. When the body becomes infected with bacteria or viruses, the WBCs multiply and the nodes sometimes become swollen and painful.

Thymus Your **thymus,** which is located just above your heart, releases WBCs. The WBCs travel to other areas of the lymphatic system.

Spleen The largest lymph organ is your spleen, which is located in the upper left side of your abdomen. The **spleen** filters blood and, like the thymus, releases WBCs. When red blood cells are squeezed through the spleen's capillaries, the older and more fragile cells rupture. The RBCs are broken down, and some of their parts are reused. For this reason, the spleen can be thought of as the red-blood-cell recycling center.

Tonsils **Tonsils** are made up of groups of lymphatic tissue located at the back of your nasal cavity, on the inside of your throat, and at the back of your tongue. WBCs in the tonsils defend the body against infection. Tonsils sometimes become badly infected and must be removed.

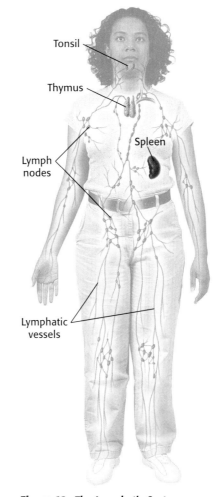

Figure 12 The Lymphatic System

SECTION REVIEW

1. What are the main functions of the lymphatic system?
2. Where does lymph go when it leaves the lymphatic system?
3. **Identifying Relationships** How are lymph nodes similar to the spleen?

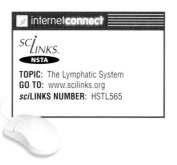

TOPIC: The Lymphatic System
GO TO: www.scilinks.org
*sci*LINKS NUMBER: HSTL565

▼ Answers to Section Review

1. The lymphatic system collects excess fluid surrounding cells and returns it to the bloodstream. It also helps the body fight pathogens.
2. Lymph drains into large veins in the neck.
3. Large white blood cells in the spleen and lymph nodes engulf dead cells and remove them from the blood.

2) Teach

RESEARCH

Writing — Have students use library and on-line resources to find out about the role of the lymphatic system in protecting the body against disease. Have them write and present reports to the class based on their findings.

3) Close

Quiz

1. What is one function of the tonsils? (The tonsils contain white blood cells, which help fight pathogens.)
2. How is the lymphatic system a circulatory system? (The lymphatic system collects fluid that is not reabsorbed by the capillaries and returns it to the bloodstream.)

ALTERNATIVE ASSESSMENT

Have students make a life-size model of the lymphatic system. Provide students with butcher paper, dried beans, glue, and markers. Then have students use their model to demonstrate the interaction of the lymphatic system with the cardiovascular system. **Sheltered English**

Directed Reading Worksheet Section 2

TOPIC: The Lymphatic System
GO TO: www.scilinks.org
*sci*LINKS NUMBER: HSTL565

SECTION 3

Focus

The Respiratory System

This section introduces the structures and functions of the respiratory system. Students will learn to trace the flow of air through the respiratory system and to recognize problems in respiratory function. Students will also find out how the respiratory system and the circulatory system are related.

Bellringer

Write the following on the board or overhead projector:

In your ScienceLog, explain whether the following statements are true or false:

- Breathing and respiration are the same thing. (False; breathing is only one part of respiration.)
- The nose is the primary doorway into and out of the respiratory system. (true)
- The vocal cords are located in the trachea. (False; they are located in the larynx.)

1 Motivate

ACTIVITY

Have students place their hands on either side of their rib cage and breathe deeply several times. Then ask students to describe what they felt while they breathed in and out. (Students should feel their rib cage moving up and expanding during inhalation and moving down and returning to its initial size during exhalation.)

Explain to students that this section focuses on what happens inside the body during breathing.

Sheltered English

SECTION 3
READING WARM-UP

Terms to Learn

respiration
respiratory system
pharynx
larynx
trachea
bronchi
alveoli

What You'll Do

- Describe the flow of air through the respiratory system.
- Discuss the relationship between the respiratory system and the cardiovascular system.
- Identify respiratory disorders.

The Respiratory System

Breathing. You do it all the time. You're doing it right now. You hardly ever think about it, though, unless your ability to breathe is suddenly taken away. Then it becomes all too obvious that you must breathe in order to survive. Why is breathing important?

Out with the Bad Air; In with the Good

Your body needs a continuous supply of oxygen in order to obtain energy from the foods you eat. That's where breathing comes in handy. The air you breathe is a mixture of several gases. One of these gases is oxygen. When you breathe, your body takes in air and absorbs the oxygen. Then carbon dioxide from your body is added to the air, and the stale air is exhaled.

The words *breathing* and *respiration* are often thought to mean the same thing. However, breathing is only one part of respiration. **Respiration** is the entire process by which a body obtains and uses oxygen and gets rid of carbon dioxide and water. Respiration is divided into two parts: breathing, which involves inhaling and exhaling, and cellular respiration, which involves the chemical reactions that release energy from food.

Breathing: Brought to You by Your Respiratory System

Breathing is made possible by the respiratory system. The **respiratory system** consists of the lungs, throat, and passageways that lead to the lungs. **Figure 13** shows the parts of the respiratory system.

Nose Your nose is the primary passageway into and out of the respiratory system. Air is inhaled through the nose, where it comes into contact with warm, moist surfaces. Air can also enter and leave through the mouth.

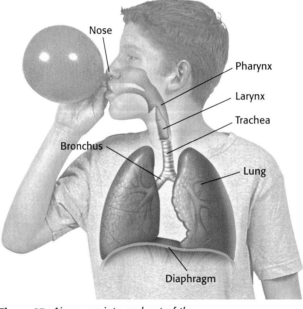

Figure 13 *Air moves into and out of the body through the respiratory system.*

TOPIC: The Respiratory System
GO TO: www.scilinks.org
sciLINKS NUMBER: HSTL570

IS THAT A FACT!

You sneeze when your breathing muscles respond to mucus or dirt that irritates the lining of your nasal cavity. A sneeze consists of a deep breath followed by a 160 km/h surge of air out of the nose!

40   Chapter 2 • Circulation and Respiration

Pharynx From the nose, air flows into the **pharynx** (FER ingks), or throat. You can use a mirror to see the walls of your pharynx behind your tongue. In addition to air, food and drink also travel through the pharynx on the way to the stomach. The pharynx branches into two tubes. One leads to the stomach and is called the esophagus. The other leads to the lungs and is called the larynx.

Larynx Tilt your head up slightly, and rub a finger up and down the front of your throat. The ridges you feel are the outside of the larynx (LER ingks). The **larynx,** or voice box, contains the vocal cords. The vocal cords are a pair of elastic bands that are stretched across the opening of the larynx. Muscles attached to the larynx control how much the vocal cords are stretched. When air flows between the vocal cords, they vibrate and make sound.

Trachea The larynx guards the entrance to a large tube called the **trachea** (TRAY kee uh), or windpipe. The trachea is the passageway for air traveling from the larynx to the lungs.

Bronchi The trachea splits into two tubes called **bronchi** (BRAHNG kie) (singular, *bronchus*). One bronchus goes to each lung and branches into thousands of tiny tubes called *bronchioles*.

Lungs Your body has two large spongelike lungs. In the lungs, each bronchiole branches to form thousands of tiny sacs called **alveoli** (singular, *alveolus*). Capillaries surround each alveolus. **Figure 14** shows the arrangement of the tubes in the respiratory system.

Chemistry CONNECTION

When people who live at low elevations travel to the mountains, they usually find that they have difficulty exerting themselves. This is because the concentration of oxygen in the air at high elevations is lower than that at low elevations. Until they become used to the change, people have to take more breaths to supply their bodies with the oxygen they need.

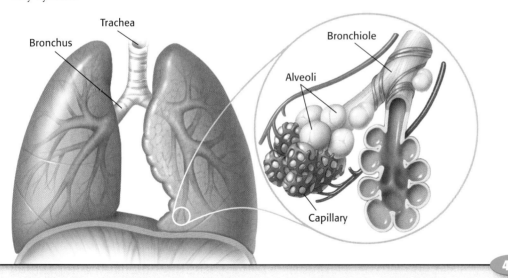

Figure 14 *Inside your lungs, the bronchi branch into bronchioles. The bronchioles lead to tiny sacs called alveoli.*

Multicultural CONNECTION

Suppose a newcomer to a Peruvian village in the Andes Mountains gets headaches, feels nauseated, and is very short of breath. What is going on? The newcomer is suffering from a lack of oxygen. The villagers don't have these problems; they are extremely well adapted to this high elevation. Because there is very little oxygen in the air, these people have developed lungs and chests that are much larger than those of the newcomer. They also carry more blood in their bloodstream.

2) Teach

ACTIVITY

Investigating Speech Ask students to place their hands lightly on their neck near the larynx and to say, "ah." Have students keep their hands in place and alternate between blowing as they would blow out candles on a birthday cake and saying, "ah." Then ask:

- What happened when you said, "ah"? (The neck vibrated.)
- What happened when you blew without saying anything? (The neck stopped vibrating, but air still rushed out of the mouth.)
- What caused the sound and the vibrations when you said "ah"? (Air rushing past the vocal cord muscles in the larynx caused the muscles to vibrate. This caused the sound.)

Point out that speech is made up of sounds produced both ways, with voicing and without voicing. If students are unconvinced, have them say *sssssss* (the snake sound) while touching their neck near the larynx. Then have them say *zzzzz* (buzz like a bee) while touching their neck. Students should find that their neck vibrates when they pronounce *z*, which is voiced, but does not vibrate when they pronounce *s*, which is not voiced.
Sheltered English

Teaching Transparency 84
"The Respiratory System"

Directed Reading Worksheet Section 3

3) Extend

ACTIVITY

Concept Mapping Have students create a concept map using the following terms:

respiration, respiratory system, pharynx, larynx, cellular respiration, trachea, bronchi, alveoli, diaphragm

RESEARCH

Writing Ask students if they have ever seen a professional football player inhale oxygen using an oxygen mask. Encourage students to find out why football players do this and to write the results of their research in a short report. As a class, discuss whether the practice is useful. (Although panting from physical exertion is a sign of an oxygen deficit, the additional oxygen does not alleviate the problem because the oxygen deficit occurs in the muscles, not in the lungs. In other words, the oxygen deficit is not a result of inadequate respiratory function but of the muscles' inability to take in more oxygen.)

CONNECT TO PHYSICAL SCIENCE

Lead a discussion of how the body creates changes in air pressure to make breathing possible. The following Teaching Transparency, "Air Pressure and Breathing," is a helpful illustration.

Teaching Transparency 225 "Air Pressure and Breathing"

LINK TO PHYSICAL SCIENCE

How Do You Breathe?

When you breathe, air is sucked in or forced out of your lungs. However, your lungs do not contain muscles that force air in and out. Instead, breathing is done by rib muscles and the *diaphragm,* a dome-shaped muscle underneath the lungs. When the diaphragm contracts and moves down, it increases the chest cavity's volume. At the same time, some of your rib muscles contract and lift your rib cage, causing it to expand. Air is sucked in.

What Happens to the Oxygen?

When oxygen has been absorbed by red blood cells, it is transported through the body by the cardiovascular system. Oxygen diffuses inside cells, where it is used in an important chemical reaction known as cellular respiration. During *cellular respiration,* oxygen is used to release energy stored in molecules of carbohydrates, fats, and proteins. When the molecules are broken apart during the reaction, energy is released along with carbon dioxide and water. The carbon dioxide and water leave the cell and return to the bloodstream. The carbon dioxide is carried to the lungs and exhaled. **Figure 15** shows how breathing and blood circulation are related.

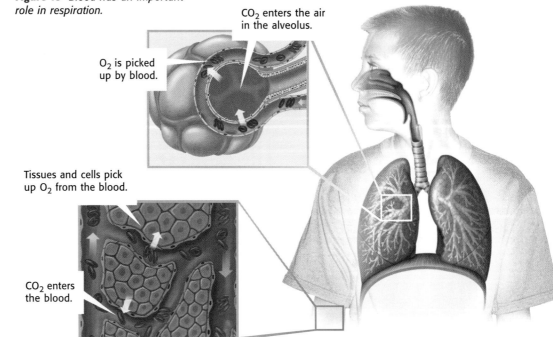

Figure 15 Blood has an important role in respiration.

CO_2 enters the air in the alveolus.

O_2 is picked up by blood.

Tissues and cells pick up O_2 from the blood.

CO_2 enters the blood.

internetconnect

TOPIC: Respiratory Disorders
GO TO: www.scilinks.org
*sci*LINKS NUMBER: HSTL575

IS THAT A FACT!

The lungs contain about 300 million alveoli. The alveoli provide a tremendous surface area for gas exchange. In fact, a person can breathe easily with only one lung.

42 Chapter 2 • Circulation and Respiration

Respiratory Disorders

Millions of people suffer from respiratory disorders. There are many types of respiratory disorders, including asthma, bronchitis, pneumonia, and emphysema.

In *asthma,* irritants cause tissue around the bronchioles to constrict and secrete large amounts of mucus. As the bronchiole tubes get narrower, the person has difficulty breathing. *Bronchitis* can develop when something irritates the lining of the bronchioles. *Pneumonia* is caused by bacteria or viruses that grow inside the bronchioles and alveoli and cause them to become inflamed and filled with fluid. If the alveoli are filled with too much fluid, the person may suffocate.

The Hazards of Smoking You probably already know that smoking cigarettes is bad for your health. In fact, smoking is the leading cause of cardiovascular diseases and lung diseases, such as *emphysema* and *lung cancer.* People with emphysema have trouble getting the oxygen they need because their lung tissue erodes away, as shown in **Figure 16**.

Why Do People Snore?

Get a **15 cm² sheet of wax paper.** Hum your favorite song. Then take the wax paper, press it against your lips, and hum the song again. Now answer the following questions:

1. How was your humming different when wax paper was pressed to your mouth?
2. Use your observations to guess what might cause snoring.

TRY at HOME

Figure 16 The photo on the left shows healthy lungs. The photo on the right shows the lungs of a person who had emphysema.

SECTION REVIEW

1. Describe the path that air travels as it moves through the respiratory system.
2. What is the difference between respiration and cellular respiration?
3. **Identifying Relationships** How is the function of the respiratory system related to that of the cardiovascular system?

TOPIC: The Respiratory System, Respiratory Disorders
GO TO: www.scilinks.org
*sci*LINKS NUMBER: HSTL570, HSTL575

4) Close

QuickLab

MATERIALS

FOR EACH STUDENT:
15 cm² sheet of wax paper

Answers to QuickLab

1. A strong vibrating sound was made when the paper was pressed against the student's lips.
2. Accept all logical answers. Most snoring is caused by soft tissues from the mouth blocking a person's airway and vibrating while the person sleeps.

Quiz

Ask students whether these statements are true or false.

1. Pneumonia is caused by air pollution. (false)
2. There are only two bronchi. (true)
3. The lungs are not made of any muscle. (true)

ALTERNATIVE ASSESSMENT

Have students make models to represent healthy lungs and lungs damaged by smoking. Photographs of damaged and healthy lungs can be found in literature from the American Lung Association, the American Cancer Society, and in various science and health textbooks.
Sheltered English

▼ **Answers to Section Review**

1. Air comes into the nose or mouth and then travels through the pharynx, larynx, trachea, and bronchi to reach the lungs.
2. Respiration involves inhaling and exhaling air, as well as cellular respiration. Cellular respiration involves the chemical reactions that release energy inside the cell.
3. The respiratory system brings in oxygen and expels carbon dioxide, and the cardiovascular system transports those materials to and from the lungs.

Making Models Lab

Build a Lung
Teacher's Notes

Time Required
One 45-minute class period

Lab Ratings

- TEACHER PREP
- STUDENT SET-UP
- CONCEPT LEVEL
- CLEAN UP

MATERIALS

You may want to build a model first to use as a reference for students. If so, you may want to substitute a bag smaller than the one that students use to model the diaphragm.

Answers

4. The balloon will inflate when the plastic bag is pulled down.

5. The balloon represents a lung, the plastic bag represents a diaphragm, and the straw represents a trachea. The bottle represents a body cavity.

6. Air enters the lungs when the diaphragm moves down and creates more space inside the chest cavity. Air is forced out of the lungs when the diaphragm moves up. This should be demonstrated by moving the plastic bag up and down.

Making Models Lab

Build a Lung

You have learned that when you breathe, you actually pull air into your lungs because your diaphragm muscle causes your chest to expand. You can see this is true by placing your hands on your ribs and inhaling slowly. Did you feel your chest expand?

In this activity, you will build a model of a lung, using some common materials. You will see how the diaphragm muscle works to inflate your lungs. Refer to the diagrams as you construct your model.

MATERIALS

- small balloon
- plastic drinking straw
- 2 rubber bands
- golf-ball-sized piece of clay
- metric ruler
- top half of a 2 L bottle
- small plastic trash bag
- transparent tape

Procedure

1. Attach the balloon to the end of the straw with a rubber band. Make a hole through the clay, and insert the other end of the straw through the hole. Be sure at least 8 cm of the straw extends beyond the clay. Squeeze the ball of clay gently to seal the clay around the straw.

2. Insert the balloon end of the straw into the neck of the bottle. Use the ball of clay to seal the straw and balloon into the bottle.

3. Place the trash bag over the cut end of the bottle, and secure it with one rubber band. Reinforce the seal with tape. Gather the excess material of the bag into your hand, and press toward the inside of the bottle slightly. This will push the excess air out of the bottle. Tape the bag to the bottle with the bag in this position.

Analysis

4. How can you make your model "lung" inflate?

5. What do the balloon, the plastic bag, and the straw represent in your model?

6. Using your model, demonstrate how air enters the lung and how air exits the lung.

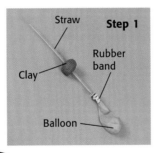

Step 1 — Straw, Clay, Rubber band, Balloon

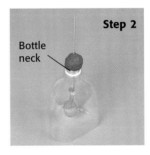

Step 2 — Bottle neck

Step 3

 Datasheets for LabBook

Yvonne Brannum
Hine Junior High School
Washington, D.C.

Skill Builder Lab

Carbon Dioxide Breath

Plants take in carbon dioxide and give off oxygen as a byproduct of photosynthesis. Animals, including you, use this oxygen and release carbon dioxide as a byproduct of respiration.

In this activity, you will explore your own carbon dioxide exhalation. Phenol red turns yellow in the presence of carbon dioxide. You will use it to detect carbon dioxide in your breath.

- 150 mL graduated cylinder
- 100 mL of water
- 150 mL Erlenmeyer flask
- eyedropper
- phenol red indicator solution
- plastic drinking straw
- paper towel
- clock with a second hand or a stopwatch
- protective gloves

Procedure

1. Place 100 mL of water into a 150 mL flask. Using an eyedropper, carefully place four drops of phenol red indicator solution into the water. The water should turn orange.

2. Place a plastic drinking straw into the solution of phenol red and water. Drape a paper towel over the beaker to prevent splashing. Carefully blow through the straw into the solution.
   Caution: Do not inhale through the straw. Do not drink the solution, and do not share a straw with anyone.

3. Have your partner time how long it takes for the solution to change color. Begin timing when you start blowing. Record the time in your Science-Log. What color does the solution become?

Analysis

4. Compare your data with those of your classmates. What was the longest length of time it took to see a color change? the shortest? How do you account for the difference?

5. Is there a relationship between the time it takes to change the solution from orange to yellow and the person's physical characteristics, such as gender or whether the tester has an athletic build?

Going Further

Do jumping jacks or sit-ups for three minutes, and then repeat the experiment. Did the timing change? Describe and explain any change.

Yvonne Brannum
Hine Junior High School
Washington, D.C.

Answers

4. Answers will depend on students' observations. It is typical for the solution to change color faster when the student is breathing fast after exercise.

5. In general, an athlete at rest will take the longest time to generate a color change in the indicator solution. There should be little difference observed between genders. There are exceptions, and all answers will depend on students' observations.

Skill Builder Lab

Carbon Dioxide Breath
Teacher's Notes

Time Required
One 45-minute class period

Lab Ratings

TEACHER PREP 🧪🧪
STUDENT SET-UP 🧪🧪
CONCEPT LEVEL 🧪🧪
CLEAN UP 🧪🧪

You may wish to substitute bromothymol blue indicator solution for the phenol red indicator. The bromothymol blue will turn green in the presence of CO_2. Clear plastic cups (6 oz or 8 oz size) may be used instead of 150 mL beakers if glassware is in short supply or if you have concerns about breakage.

Safety Caution
Remind students to review all safety cautions and icons before beginning this lab activity.

Lab Notes
Tell students that carbon dioxide is in the air of the classroom. They may need to cover their indicator solution to delay the reaction with the air. Tell them not to leave the indicator solution sitting exposed for several minutes before it is used.

 Datasheets for LabBook

Chapter Highlights

VOCABULARY DEFINITIONS

SECTION 1

cardiovascular system a collection of organs whose primary function is to transport blood to and from your body's cells; the organs in this system include the heart, the arteries, and the veins

blood a connective tissue made up of platelets, white blood cells, red blood cells, and plasma

arteries blood vessels that carry blood away from the heart

capillaries the smallest blood vessels in the body

veins blood vessels that direct blood to the heart

pulmonary circulation the circulation of blood between the heart and lungs

systemic circulation the circulation of blood between the heart and the body (excluding the lungs)

blood pressure the amount of force exerted by blood on the inside walls of a blood vessel

SECTION 2

lymphatic system a collection of organs whose primary function is to collect extracellular fluid and return it to the blood; the organs in this system include the lymph nodes and the lymphatic vessels

lymph fluid and particles absorbed into lymph capillaries

lymph nodes small, bean-shaped organs that contain small fibers that work like nets to remove particles from the lymph

thymus a lymph organ that produces lymphocytes

spleen an organ that filters blood and produces lymphocytes

tonsils small masses of soft tissue located at the back of the nasal cavity, on the inside of the throat and at the back of the tongue

Chapter Highlights

SECTION 1

Vocabulary
cardiovascular system (p. 30)
blood (p. 30)
arteries (p. 33)
capillaries (p. 33)
veins (p. 33)
pulmonary circulation (p. 34)
systemic circulation (p. 34)
blood pressure (p. 35)

Section Notes

- The cardiovascular system delivers oxygen and nutrients to the body's cells, takes away the cells' waste products, and helps the body stay healthy. The cardiovascular system is made up of blood, the heart, and blood vessels.

- Blood is a connective tissue made of plasma, red blood cells, white blood cells, and platelets. The heart is a muscular organ that pumps blood through blood vessels.

- Blood moves away from the heart through arteries and then enters capillaries. After leaving capillaries, blood is carried back to the heart through veins.

- In pulmonary circulation, blood vessels carry blood from the heart to the lungs and back to the heart. In systemic circulation, blood flows from the heart to the rest of the body and then back to the heart.

- People have different blood types. Blood type is determined by the presence of certain chemicals on red blood cells.

SECTION 2

Vocabulary
lymphatic system (p. 38)
lymph (p. 38)
lymph nodes (p. 39)
thymus (p. 39)
spleen (p. 39)
tonsils (p. 39)

Section Notes

- The lymphatic system returns excess fluid to the cardiovascular system and helps the body fight infections.

- The lymphatic system includes lymph, lymph capillaries, lymphatic vessels, lymph nodes, the spleen, tonsils, and the thymus.

☑ Skills Check

Math Concepts

A CONTINUOUS BEAT Your heart beats about 100,800 times per day. That means that your heart beats about 4,200 times every hour.

100,800 beats ÷ 24 hours = 4,200 beats

That also means that your heart beats about 70 times every minute.

4,200 beats ÷ 60 minutes = 70 beats

Visual Understanding

AIR PASSAGEWAYS Take another look at Figure 13 on page 40. With your finger, trace the path air takes to reach the lungs. As you do this, reconsider what roles the nose, pharynx, trachea, bronchi, lungs, and diaphragm play in respiration.

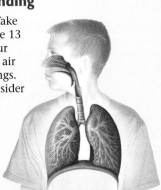

Lab and Activity Highlights

Build a Lung PG 44

Carbon Dioxide Breath PG 45

 Datasheets for LabBook (blackline masters for these labs)

SECTION 3

Vocabulary
- respiration (p. 40)
- respiratory system (p. 40)
- pharynx (p. 41)
- larynx (p. 41)
- trachea (p. 41)
- bronchi (p. 41)
- alveoli (p. 41)

Section Notes

- The respiratory system moves air into and out of the body. The respiratory system includes the nose, the mouth, the pharynx, the larynx, the trachea, and the lungs.
- Air enters the lungs through bronchi and travels to the alveoli, which are gas-filled sacs surrounded by capillaries of the cardiovascular system.
- The blood in the capillaries of the lungs absorbs oxygen and releases carbon dioxide. The carbon dioxide is exhaled. The oxygen is carried by the blood to the heart and then on to the cells of the body.
- The body's cells must have oxygen to carry out cellular respiration. Cellular respiration is a chemical process that releases the energy in carbohydrates, fats, and proteins and makes the energy available to the cells.
- Inhaling and exhaling are caused by the contraction and relaxation of the diaphragm and the muscles of the rib cage.

VOCABULARY DEFINITIONS, continued

SECTION 3

respiration the exchange of gases between living cells and their environment; includes breathing and cellular respiration

respiratory system a collection of organs whose primary function is to take in oxygen and expel carbon dioxide; the organs of this system include the lungs, the throat, and the passageways that lead to the lungs

pharynx the upper portion of the throat

larynx the area of the throat that contains the vocal cords

trachea the air passageway from the larynx to the lungs

bronchi the two tubes that connect the lungs with the trachea

alveoli tiny sacs that form the bronchiole branches of the lungs

 Vocabulary Review Worksheet

 Blackline masters of these Chapter Highlights can be found in the **Study Guide.**

---

internetconnect

 GO TO: go.hrw.com

 GO TO: www.scilinks.org

Visit the **HRW** Web site for a variety of learning tools related to this chapter. Just type in the keyword:

KEYWORD: HSTBD2

Visit the **National Science Teachers Association** on-line Web site for Internet resources related to this chapter. Just type in the *sci*LINKS number for more information about the topic:

TOPIC:	*sci*LINKS NUMBER:
The Cardiovascular System	HSTL555
Cardiovascular Problems	HSTL560
The Lymphatic System	HSTL565
The Respiratory System	HSTL570
Respiratory Disorders	HSTL575

Lab and Activity Highlights

LabBank

 Whiz-Bang Demonstrations
- Get the Beat!
- Take a Deep Breath

 Long-Term Projects & Research Ideas, Getting to the Heart

 EcoLabs & Field Activities, There's Something in the Air

Chapter Review Answers

USING VOCABULARY

1. red blood cells
2. arteries
3. Blood
4. larynx
5. alveoli

UNDERSTANDING CONCEPTS

Multiple Choice

6. b
7. d
8. a
9. c
10. b
11. a

Short Answer

12. Pulmonary circulation carries blood through the lungs and back to the heart. Systemic circulation carries blood from the heart to the rest of the body.
13. The first number is the systolic pressure, or the pressure inside the artery when the ventricles contract. The second number, the diastolic pressure, is the pressure in the arteries when the ventricles relax.
14. Carbon dioxide is a product of cellular respiration which occurs in the body's cells.

Chapter Review

USING VOCABULARY

To complete the following sentences, choose the correct term from each pair of terms listed below:

1. Oxygen is delivered to the cells of the body by __?__. (*white blood cells* or *red blood cells*)
2. Blood is carried away from the heart in __?__. (*arteries* or *veins*)
3. __?__ carries nutrients to the body's cells. (*Lymph* or *Blood*)
4. The __?__ contains the vocal cords. (*trachea* or *larynx*)
5. The pathway of air through the respiratory system ends at the tiny sacs called __?__. (*alveoli* or *bronchi*)

UNDERSTANDING CONCEPTS

Multiple Choice

6. Blood from the lungs enters the heart at the
   a. left ventricle.
   b. left atrium.
   c. right atrium.
   d. right ventricle.

7. Blood cells are made
   a. in the heart.
   b. from plasma.
   c. from lymph.
   d. in the bones.

8. Which of the following is not part of the lymphatic system?
   a. trachea
   b. lymph node
   c. thymus
   d. spleen

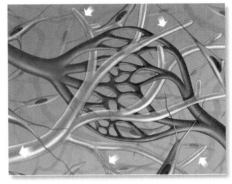

9. Alveoli are surrounded by
   a. veins.
   b. muscles.
   c. capillaries.
   d. lymph nodes.

10. What prevents blood from flowing backward in veins?
    a. platelets
    b. valves
    c. muscles
    d. cartilage

11. Air moves into the lungs when the diaphragm muscle
    a. contracts and moves down.
    b. contracts and moves up.
    c. relaxes and moves down.
    d. relaxes and moves up.

Short Answer

12. What is the difference between pulmonary circulation and systemic circulation in the cardiovascular system?
13. Walton has a blood pressure of 110/65. What do the two numbers mean?
14. What body process produces the carbon dioxide you exhale?

Concept Map

15. Use the following terms to create a concept map: blood, oxygen, alveoli, capillaries, carbon dioxide.

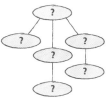

CRITICAL THINKING AND PROBLEM SOLVING

Write one or two sentences to answer the following questions:

16. Why do you think there are hairs in your nose?

17. When a person is not feeling well, sometimes a doctor will examine samples of the person's blood to see how many white blood cells are present. Why would this information be useful?

18. How is the function of the lymphatic system related to the function of the cardiovascular system?

MATH IN SCIENCE

19. After a person donates blood, the blood is stored in one-pint bags until it is needed for a transfusion. A healthy person normally has 5 million RBCs in each cubic millimeter (1 mm^3) of blood.
    a. How many RBCs are there in 1 mL of blood? One milliliter is equal to 1 cm^3 and to 1,000 mm^3.
    b. How many RBCs are there in 1 pt? One pint is equal to 473 mL.

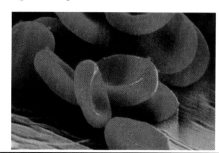

INTERPRETING GRAPHICS

The diagram below shows how the human heart would look in cross section. Examine the diagram, and then answer the questions that follow:

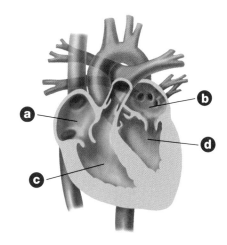

20. Which letter identifies the chamber that receives blood from systemic circulation? What is this chamber's name?

21. Which letter identifies the chamber that receives blood from the lungs? What is this chamber's name?

22. Which letter identifies the chamber that pumps blood to the lungs? What is this chamber's name?

 Take a minute to review your answers to the Pre-Reading Questions found at the bottom of page 28. Have your answers changed? If necessary, revise your answers based on what you have learned since you began this chapter.

Concept Mapping

15. An answer to this exercise can be found at the front of this book.

CRITICAL THINKING AND PROBLEM SOLVING

16. The hairs catch dust and other foreign particles. This helps keep your lungs as clean as possible.
17. Doctors will need to know if the immune system is producing large numbers of white blood cells; this would indicate an infection.
18. The lymphatic system returns leaked fluids back to your blood.

MATH IN SCIENCE

19. a. 5 billion (5,000,000,000) cells
    b. 2.365 trillion (2,365,000,000,000) cells

INTERPRETING GRAPHICS

20. *a*, the right atrium
21. *b*, the left atrium
22. *c*, the right ventricle

 Concept Mapping Transparency 23

 Blackline masters of this Chapter Review can be found in the **Study Guide.**

WEIRD SCIENCE
Catching a Light Sneeze

Activity
Divide the class into pairs. Have students use a flashlight to make their partner's pupils contract. (Be careful not to let them look directly into a bright light for too long.) They should conduct their test for the photic sneeze reflex as a scientific experiment. Have them write a hypothesis, test the hypothesis, and report their results.

CATCHING A LIGHT SNEEZE

Do you sneeze when you come out of a dark movie theater into bright sunlight? If not, look around you next time. Chances are several people will sneeze.

Reflex Gone Wrong
For some reason, about one in five people sneeze when they step from a dimly lit area into a brightly lit area. In fact, some may sneeze a dozen times or more! Fortunately, the sneezing usually stops after a few times. This reaction is called a *photic sneeze reflex*. No one knows for certain why it happens.

Normal sneezing is a reflex, which means you do it without thinking about it. Most people sneeze when something tickles the inside of their nose. They sneeze, and moving air pushes the tickling intruder out. For instance, if you get dust in your nose, sneezing pushes the dust out. In the case of people with the photic sneeze, it's a reflex gone wrong.

ACHOO!
A few years ago, some geneticists studied the photic sneeze reflex. They named it the Autosomal Dominant Compelling Helio-ophthalmic Outburst syndrome, or the ACHOO syndrome. Scientists know that the ACHOO syndrome runs in families. So the photic sneeze can be passed from parent to child. Sometimes even the number of times in a row that each person sneezes is the same throughout a family.

Possible Answers
Some scientists have offered a possible explanation for the ACHOO syndrome. First, everyone's pupils contract when they

▲ *Do you sneeze when you see bright light after exiting a dark room?*

encounter bright light. And the nerves from the eyes are right next to the nerves from the nose. Thus, people with the ACHOO syndrome may have their wires slightly crossed: bright light triggers the pupil reflex, and it also triggers the sneeze reflex!

Sneeze Fest
Sunlight is not the only strange trigger for sudden sneezes. Some people sneeze when they rub the inner corner of their eye. Others sneeze when tweezing their eyebrows or brushing their hair. In rare individuals, even eating too much has been known to cause sneezing fits!

Research the Facts
▶ Yawning is also a reflex. Do some research to find out why we yawn.

Answer to Research the Facts
Sample answer: Yawning is a response to a buildup of carbon dioxide or the sight of another person yawning.

Health Watch

Goats to the Rescue

They're called transgenic (tranz JEHN ik) goats because their cells contain a human gene. They look just like any other goats, but because of their human gene they produce a chemical that can save lives.

Lifesaving Genes

Heart attacks are the number one cause of death in the United States. Many heart attacks are triggered when large blood clots interfere with the flow of blood to the heart. Human blood cells produce a chemical called *tissue plasminogen activator* (TPA) that dissolves small blood clots. If TPA is given to a person having a heart attack, it can often dissolve the blood clot, stop the attack, and save the person's life. But TPA is difficult to produce in large quantities in the laboratory. This is where the goats come in. Researchers at Tufts University, in Grafton, Massachusetts, have genetically engineered goats to produce this lifesaving drug.

Hybrid Goats

Producing transgenic goats is a complicated process. First, fertilized eggs are surgically removed from normal female goats. The eggs are then injected with hybrid genes that consist of human TPA genes "spliced" into genes from the mammary glands of a goat. Finally, the altered eggs are surgically implanted into female goats, where they develop into young goats, or kids. Some of the kids actually carry the hybrid gene. When the hybrid kids mature, the females' milk contains TPA. Technicians then separate the TPA from the goats' milk for use in heart-attack victims.

The Research Continues

Transgenic research in farm animals such as goats, sheep, cows, and pigs may someday produce drugs faster, cheaper, and in greater quantities than are possible using current methods. The way we view the barnyard may never be the same.

Find Out for Yourself

▶ Using chemicals produced by transgenic animals is just one of many gene therapies. Do some research to find out more about gene therapy, how it is used, and how it may be used in the future.

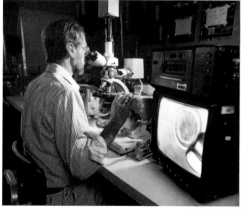

▲ *A scientist at Tufts University injects human TPA genes into fertilized goat eggs.*

Answer to Find Out for Yourself

Another form of gene therapy is to obtain healthy genes and insert them into the cells of a person with a genetic disorder. Such methods are still highly experimental.

HEALTH WATCH

Goats to the Rescue

Background

Many proteins that are produced naturally by the human body could be produced by other animals through genetic engineering. This raises the possibility of turning animals into "drug factories." Genetically altered pigs have been used to produce blood and to grow organs that can be used for human transfusions and transplants. Cows have been altered to produce milk that is more like human milk. A dozen different human proteins have been produced in the milk of various transgenic animals.

However, there are serious safety issues concerning these new animal products. Host animals may carry pathogens that are dangerous to humans. For example, some sheep and goats carry a brain disease known as scrapie. Research companies that offer transgenic drugs must be sure that pathogens are not present in their products. They must ensure that any animal proteins that might trigger allergic reactions or other side effects are removed during the purification process. In addition, these companies must be able to prove that their products will function in the same way as their natural counterparts.

Chapter Organizer

CHAPTER ORGANIZATION	TIME MINUTES	OBJECTIVES	LABS, INVESTIGATIONS, AND DEMONSTRATIONS
Chapter Opener pp. 52–53	45	National Standards: UCP 2, SAI 1, SPSP 5, HNS 2, 3, LS 1e	**Start-Up Activity,** Changing Foods, p. 53
Section 1 The Digestive System	90	▶ Describe the parts and functions of the digestive system. ▶ Compare mechanical digestion with chemical digestion. ▶ Describe some disorders of the digestive system. UCP 1, 2, 5, SAI 1, SPSP 1, 5, HNS 1, 3, LS 1a, 1d–1f, 3b; Labs UCP 2, 5, SAI 1, LS 1a	**Demonstration,** Measuring Saliva, p. 54 in ATE **QuickLab,** Break It Up! p. 55 **Demonstration,** Peristalsis, p. 57 in ATE **Skill Builder,** As the Stomach Churns, p. 66 **Datasheets for LabBook,** As the Stomach Churns **Discovery Lab,** Enzymes in Action, p. 180 **Datasheets for LabBook,** Enzymes in Action **Whiz-Bang Demonstrations,** Liver Let Live
Section 2 The Urinary System	90	▶ Describe the parts and functions of the urinary system. ▶ Explain how the kidneys filter blood. ▶ Describe some disorders of the urinary system. UCP 1–5, SAI 1, HNS 1, LS 1a, 1c–1f, 3a, 3b	**Demonstration,** Kidney Structure, p. 63 in ATE **Long-Term Projects & Research Ideas,** Copying the Kidney

See page **T23** for a complete correlation of this book with the

NATIONAL SCIENCE EDUCATION STANDARDS.

TECHNOLOGY RESOURCES

 Guided Reading Audio CD English or Spanish, Chapter 3

 One-Stop Planner CD-ROM with Test Generator

 CNN. Science, Technology, & Society, Animal Organ Transplants, Segment 32

Chapter 3 • The Digestive and Urinary Systems

CLASSROOM WORKSHEETS, TRANSPARENCIES, AND RESOURCES	SCIENCE INTEGRATION AND CONNECTIONS	REVIEW AND ASSESSMENT
Directed Reading Worksheet **Science Puzzlers, Twisters & Teasers**		
Directed Reading Worksheet, Section 1 **Transparency 86,** Enzymes Break Down Proteins **Transparency 108,** Mohs' Hardness Scale **Transparency 87,** The Stomach **Transparency 87,** The Small Intestine **Critical Thinking Worksheet,** Frankenstein's Food	**Multicultural Connection,** p. 54 in ATE **Connect to Earth Science,** p. 56 in ATE **Multicultural Connection,** p. 57 in ATE **Connect to Chemistry,** p. 58 in ATE **Math and More,** p. 59 in ATE **Environment Connection,** p. 60 **Health Watch:** A Voiceless Companion, p. 73	**Homework,** pp. 55, 60, 64 in ATE **Section Review,** p. 57 **Self-Check,** p. 59 **Section Review,** p. 61 **Quiz,** p. 61 in ATE **Alternative Assessment,** p. 61 in ATE
Directed Reading Worksheet, Section 2 **Transparency 88,** How the Kidneys Filter the Blood **Reinforcement Worksheet,** Annie Apple's Amazing Adventure	**MathBreak,** How Much Water? p. 64 **Apply,** p. 64 **Across the Sciences:** Quench Your Thirst! p. 72	**Homework,** p. 64 in ATE **Section Review,** p. 65 **Quiz,** p. 65 in ATE **Alternative Assessment,** p. 65 in ATE

END-OF-CHAPTER REVIEW AND ASSESSMENT

Chapter Review in Study Guide
Vocabulary and Notes in Study Guide
Chapter Tests with Performance-Based Assessment, Chapter 3 Test
Chapter Tests with Performance-Based Assessment, Performance-Based Assessment 3
Concept Mapping Transparency 24

 Holt, Rinehart and Winston On-line Resources
go.hrw.com

For worksheets and other teaching aids related to this chapter, visit the HRW Web site and type in the keyword: **HSTBD3**

 National Science Teachers Association
www.scilinks.org

Encourage students to use the *sci*LINKS numbers listed in the internet connect boxes to access information and resources on the **NSTA** Web site.

Chapter Resources & Worksheets

Visual Resources

TEACHING TRANSPARENCIES

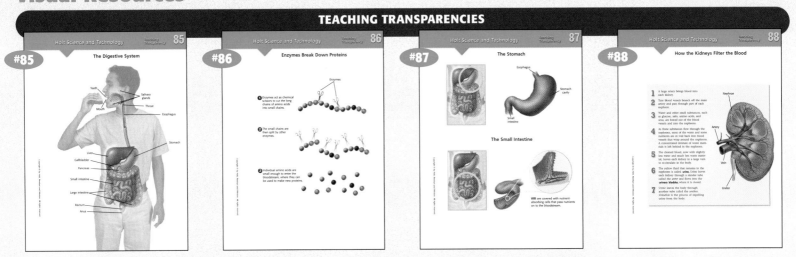

TEACHING TRANSPARENCIES

CONCEPT MAPPING TRANSPARENCY

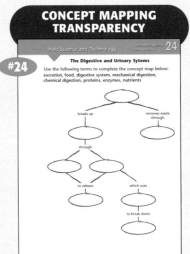

Meeting Individual Needs

DIRECTED READING

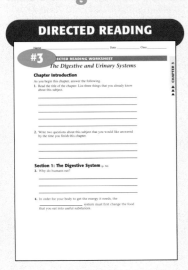

REINFORCEMENT & VOCABULARY REVIEW

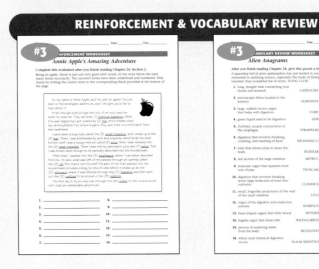

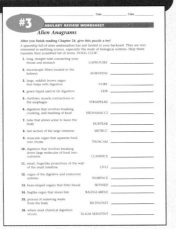

SCIENCE PUZZLERS, TWISTERS & TEASERS

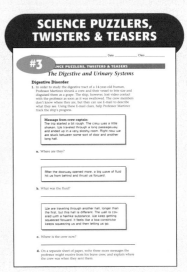

51C Chapter 3 • The Digestive and Urinary Systems

Chapter 3 • The Digestive and Urinary Systems

Review & Assessment

STUDY GUIDE

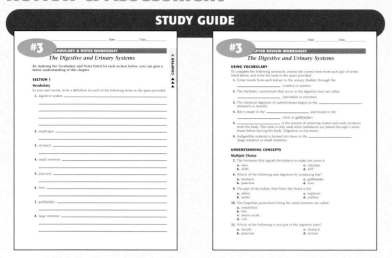

CHAPTER TESTS WITH PERFORMANCE-BASED ASSESSMENT

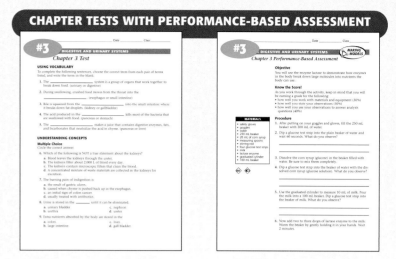

Lab Worksheets

LONG-TERM PROJECTS & RESEARCH IDEAS

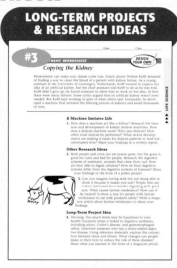

WHIZ-BANG DEMONSTRATIONS

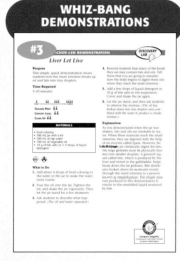

DATASHEETS FOR LABBOOK

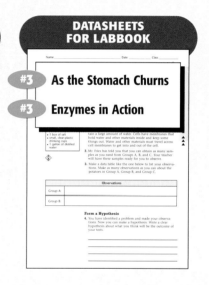

Applications & Extensions

CRITICAL THINKING & PROBLEM SOLVING

SCIENCE TECHNOLOGY

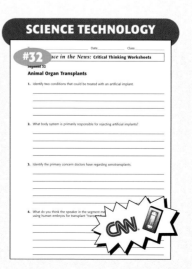

Chapter 3 • Chapter Resources & Worksheets   51D

Chapter Background

SECTION 1

▶ The Digestive System
The digestive system is composed of two sets of organs: those that make up the digestive tract and those that are called accessory organs. The digestive tract is a continuous tube consisting of the mouth, including the teeth and the tongue, the pharynx, the esophagus, the stomach, the small intestine, and the large intestine.

- The accessory digestive organs include the salivary glands, the gallbladder, the liver, and the pancreas.

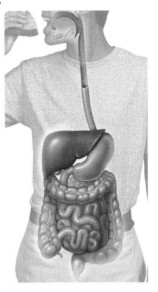

IS THAT A FACT!
- The digestive tract in a cadaver is about 9 m long. However, because of muscle tone, the digestive tract of a living person is considerably shorter.

▶ The Stomach
In short people, the stomach is found high on the abdomen and runs horizontally. In tall people, the stomach tends to run vertically, forming a J shape.

IS THAT A FACT!
- It takes about 4–8 seconds for food to pass from the top of the throat to the stomach.
- It takes less than 2 seconds for liquids to pass from the top of the throat to the stomach.

▶ The Intestines
The small intestine is the longest portion of the digestive tract. It is about 6 m long and 2.5 cm in diameter.

- Unlike the small intestine, the large intestine does not have villi, nor does it secrete digestive enzymes.

▶ The Appendix
The appendix is a 9 cm long offshoot of the large intestine. Like the tonsils, the appendix contains a large amount of lymphoid tissue.

- Doctors have begun an appendectomy on patients thought to have appendicitis. Instead of appendicitis, however, the patients had holes in the lining of their intestines made by parasitic worms. The incidence of intestinal parasitic worms is increasing as eating raw fish, such as sushi and sashimi, becomes more popular.

▶ Food Poisoning
Food poisoning is characterized by nausea, vomiting, abdominal cramps, and diarrhea. The two most common culprits are the bacteria *Salmonella* and *Shigella*.

- *Shigella* is commonly found among people who have visited developing countries. It is spread via food, feces, fingers, flies, and contaminated public bodies of water, such as swimming pools.

- *Salmonella* is typically spread by contaminated eggs and feces-contaminated hands. *Salmonella* infects the microvilli in the intestinal lining, causing blisters. If *Salmonella* poisoning is not treated promptly, the bacteria may spread throughout the body.

▶ Eating Disorders
Anorexia nervosa and bulimia are two common eating disorders that are characterized by an obsessive fear of becoming too fat. There are often underlying psychological factors that initiate these disorders.

- Anorexia nervosa is an eating disorder characterized by self-induced starvation. People with this condition refuse to eat and often exercise obsessively. Although the disorder can affect anyone, adolescent girls and young women are commonly afflicted. Boys, especially those involved in weight-conscious sports such as wrestling, can also become anorexic.

Chapter 3 • The Digestive and Urinary Systems

- Bulimia nervosa is a binge-purge syndrome in which those afflicted binge on enormous amounts of food and then regurgitate it or use laxatives to eliminate it. The binge-purge cycle may be repeated several times per day.

- Although bulimics are generally at or slightly above a normal body weight and appear to be healthy, they are not healthy. Bulimics tend to have swollen salivary glands, pancreatitis, and liver and kidney problems. In addition, they are at risk of heart failure due to electrolyte imbalance and stomach rupture, both of which can result in death. Excessive vomiting damages the esophagus and the stomach and wears away tooth enamel.

- Treatment for these eating disorders includes control of diet, often by means of hospitalization, behavior modification, nutrition education, and use of anti-depressants.

SECTION 2

▶ The Urinary System

The urinary system functions primarily to remove metabolic, nitrogen-containing wastes, such as urea, from the body in the form of urine and to maintain the correct balance of salts and water.

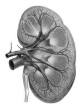

- The digestive system, the circulatory system, and the respiratory system also excrete wastes. The digestive system excretes undigested food in the form of feces. The circulatory and respiratory systems work together to rid the body of carbon dioxide.

- The yellow color of urine comes from the yellow pigment urochrome.

▶ The Kidney

The kidneys are surrounded and kept in their proper place in the body by fat. People who become too thin risk damage to the kidneys and related urinary problems. As much as 1,200 mL of blood passes through the nephrons each minute.

IS THAT A FACT!

- The volume of liquid that passes through the kidneys in a single day is equivalent to 60 times the body's entire volume of plasma.

- The kidneys require up to one-fourth of the body's oxygen supply to carry out their functions.

- The kidneys filter about 180 L of fluid from the blood each day. About 99 percent of this fluid is returned to the bloodstream. The other 1 percent leaves the body in the form of urine.

- A person who donates a kidney can maintain normal kidney functions. The remaining kidney enlarges and carries out the functions previously performed by two kidneys.

▶ The Urinary Bladder

Like the stomach, the size of the bladder varies with the amount of its contents. The bladder can hold 300 mL of urine without increasing its internal pressure significantly. At 500 mL of urine, the bladder is fairly full and may be 12.5 cm in length.

IS THAT A FACT!

- Urinary bladders of average size can contain as much as 1 L of urine!

For background information about teaching strategies and issues, refer to the *Professional Reference for Teachers*.

The Digestive and Urinary Systems

 Pre-Reading Questions

Students may not know the answers to these questions before reading the chapter, so accept any reasonable response.

Suggested Answers

1. The stomach squeezes the food to break it down mechanically. Acids and enzymes digest the food chemically.
2. Both your urinary system and your skin excrete wastes.

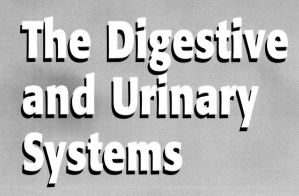

The Digestive and Urinary Systems

Sections

1. The Digestive System .. 54
   QuickLab 55
   Internet Connect 57
   Environment
   Connection 60
   Internet Connect 61
2. The Urinary System 62
   MathBreak 64
   Apply 64
   Internet Connect 65

Chapter Lab 66
Chapter Review 70
Feature Articles 72, 73
LabBook 180–183

Pre-Reading Questions

1. What happens to food when it reaches your stomach?
2. What does your urinary system have in common with your skin?

internet connect

 HRW On-line Resources
go.hrw.com
For worksheets and other teaching aids, visit the HRW Web site and type in the keyword: **HSTBD3**

 sciLINKS NSTA
www.scilinks.com
Use the *sci*LINKS numbers at the end of each chapter for additional resources on the **NSTA** Web site.

Smithsonian Institution
www.si.edu/hrw
Visit the Smithsonian Institution Web site for related on-line resources.

 CNN fyi.com
www.cnnfyi.com
Visit the CNN Web site for current events coverage and classroom resources.

AHHH!!!

You can easily see the stream of water going into this girl's mouth. What you cannot see is where that refreshing gulp of water goes after she swallows it. How do water and food travel through the body and get absorbed for use by the cells? What are the body systems for getting rid of excess water or wastes? In this chapter, you will learn about the organs of the digestive and urinary systems and what happens inside your body to the food and liquids you eat and drink.

CHANGING FOODS

During digestion, the stomach squeezes and relaxes as food passes through it. You can see the role this squeezing motion plays in the digestion of food by using a plastic bag to model your stomach.

Procedure

1. Add **200 mL of flour** and **100 mL of water** to a **resealable plastic bag.** Mix **100 mL of vegetable oil** with the flour and water.

2. Seal the plastic bag, and shake it until the flour, water, and oil are well mixed.

3. Remove as much air from the bag as possible, and reseal the bag carefully.

4. Squeeze the bag with your hands for 5 minutes. Record your observations in your ScienceLog. Be careful to keep the bag sealed.

Analysis

5. Describe the mixture before and after you kneaded the bag.

6. How might the changes you saw in the mixture relate to what happens to food you eat?

7. Do you think this is a good model of how your stomach works? Why or why not?

CHANGING FOODS

MATERIALS
FOR EACH STUDENT: • 200 mL flour • 100 mL water • clear resealable plastic bag • 100 mL vegetable oil

Safety Caution

Tell students not to ingest any of the materials used in this investigation. Strong or double bags work best. Have paper towels on hand, and clean up any spills immediately. Spilled liquids are a slipping hazard.

Answers to START-UP Activity

4. Students should observe that the oil separates from the flour mixture. Tell students that in this exercise, the oil represents fats consumed in a meal. In the human body, these fats separate from the food mixture, and bile breaks them down into smaller particles.

5. Before the students squeezed the bag, the contents were well dispersed. Once the students began squeezing the bag, the oil separated from the flour mixture.

6. The stomach is a muscular, baglike organ that is involved in the physical and chemical digestion of a meal. By squeezing the bag, the students are modeling the muscular contractions of the stomach. Note: This activity does not model the chemical digestion that occurs in the stomach.

7. Answers will vary.

SECTION 1

Focus

The Digestive System

This section introduces the structures and functions of the digestive system. Students will compare mechanical digestion with chemical digestion and will learn to trace the path of food through the digestive system. Students will also learn about some ailments of the digestive system and their causes.

Bellringer

Ask students to answer the following questions in their ScienceLog:

How does your circulatory system obtain the nutrients that it brings to your cells? (from the digestive system)

Describe as best you can the process that turns food into nutrients that cells can use. (Accept all reasonable responses.)

1) Motivate

DEMONSTRATION

Measuring Saliva Show students a 2 L beaker or other transparent container. Pour 1.2 L (2.6 pt) of water into the container, and explain that this amount of water represents the amount of saliva an average person produces in 1 day. Then ask students to discuss what roles saliva might play in the body, specifically in the process of digestion. (Answers might include lubrication of the mouth and related organs, moistening food, and breaking down molecules into nutrients that cells can use.)

Terms to Learn

digestive system
esophagus
stomach
small intestine
pancreas
liver
gallbladder
large intestine

What You'll Do

- Describe the parts and functions of the digestive system.
- Compare mechanical digestion with chemical digestion.
- Describe some disorders of the digestive system.

The Digestive System

It's your last class before lunch, and you're starving! You are so hungry you can hardly concentrate. Finally the bell rings and you get to eat your peanut butter and jelly sandwich. Yum!

You feel hungry because your brain receives signals that your cells need energy, but eating is only the beginning of the story. Your body must change a meal into substances it can use. Your **digestive system** is a group of organs that work together to digest food so that it can be used by the body.

Digestive System at a Glance

The most obvious part of your digestive system is the *digestive tract*, a series of tubelike organs that are joined end to end. The digestive tract includes your mouth, throat, esophagus, stomach, small intestine, large intestine, rectum, and anus. The human digestive tract may be more than 9 m long! The food you eat is digested as it passes through these organs. The liver, gallbladder, pancreas, and salivary glands are also part of the digestive system because they secrete substances that are used in digestion. The digestive system is shown in **Figure 1**.

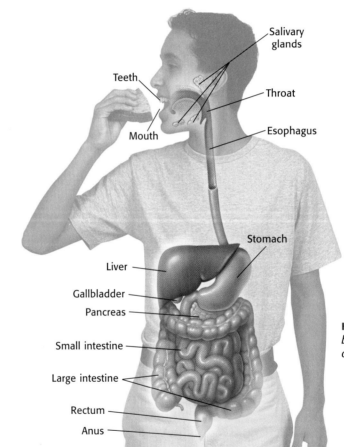

Figure 1 The digestive tract is basically a long tube with an opening at each end.

Multicultural CONNECTION

Encourage students to research the diets of people in countries such as Japan, India, Israel, Egypt, Mexico, and Russia. Encourage students to discover strategies that people in these countries use to obtain a nutritious diet, and have them compare the diets of people in each country with the diets of people in the United States. You may also want to have students compare the diets based on fat or cholesterol intake.

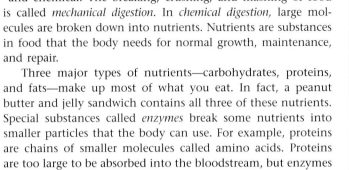

The Journey of a Sandwich

Digestion is the process of breaking down food, such as a peanut butter and jelly sandwich, into a form that can pass from the digestive tract into the bloodstream. There are two types of digestion—mechanical and chemical. The breaking, crushing, and mashing of food is called *mechanical digestion*. In *chemical digestion*, large molecules are broken down into nutrients. Nutrients are substances in food that the body needs for normal growth, maintenance, and repair.

Three major types of nutrients—carbohydrates, proteins, and fats—make up most of what you eat. In fact, a peanut butter and jelly sandwich contains all three of these nutrients. Special substances called *enzymes* break some nutrients into smaller particles that the body can use. For example, proteins are chains of smaller molecules called amino acids. Proteins are too large to be absorbed into the bloodstream, but enzymes chop up the chain into amino acids. These amino acids are small enough to pass into the bloodstream. This process is illustrated in **Figure 2.**

Break It Up!

1. Drop **one piece of hard candy** into a **clear plastic cup of water.**
2. Wrap an **identical candy** in a **towel,** and crush it with a **hammer.** Drop the crushed candy into a **second clear cup of water.**
3. The next day, examine both cups. What is different about the two candies?
4. What part of digestion is represented by breaking the hard candy?
5. How does chewing your food help the process of digestion?

Figure 2 *Enzymes in the stomach and small intestine break down proteins.*

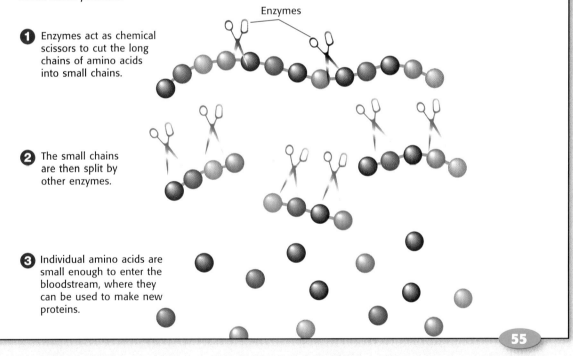

❶ Enzymes act as chemical scissors to cut the long chains of amino acids into small chains.

❷ The small chains are then split by other enzymes.

❸ Individual amino acids are small enough to enter the bloodstream, where they can be used to make new proteins.

55

Homework

Provide students with the following list of activities of the digestive system:

The teeth cut and grind food, the stomach churns food, enzymes in saliva break down carbohydrates in food, and enzymes break down proteins in food.

Ask students to indicate which activities represent mechanical digestion and which represent chemical digestion. (mechanical—the teeth cutting and grinding food, the stomach churning food; chemical—enzymes in saliva breaking down carbohydrates in food, enzymes breaking down proteins in food)

2) Teach

QuickLab

MATERIALS

FOR EACH STUDENT:
- safety goggles
- 2 identical pieces of hard candy
- 2 identical clear plastic cups of water
- towel
- hammer

Safety Caution: Tell students not to ingest or taste any of the materials, including the candy, in this QuickLab. Students should wear safety goggles while crushing the candy. Students should strike hammers away from their face and hands. Instruct students to be sure their hands and their classmates' hands are out of the way before pounding the candy with the hammer.

Answers to QuickLab

3. The crushed candy dissolved in the water. The whole candy did not dissolve as much.
4. Breaking the candy represents chewing, one form of physical digestion.
5. Chewing breaks the large chunks of food into smaller pieces. Greater surface area is then available for enzymes to digest the food.

 Teaching Transparency 86 "Enzymes Break Down Proteins"

 Directed Reading Worksheet Section 1

Section 1 • The Digestive System **55**

2) Teach, continued

READING STRATEGY

Activity Before students read the text on this page, have them read aloud the headings. Then ask students to formulate one question that they expect the text under each heading to answer. Have students write the questions in their ScienceLog. Possible questions include the following:

- "What happens in the mouth?"
- "Why is chewing important?"
- "What makes the stomach's environment harsh?"

CONNECT TO EARTH SCIENCE

Tooth enamel is an extremely hard material made primarily of carbonated calcium hydroxyapatite. Cavities (also called dental caries) form when acids produced by the bacteria on teeth demineralize the enamel, weakening it. Fluoride compounds, taken internally while the teeth are growing, strengthen the enamel by forming the enamel out of fluoroapatite, an apatite material more acid resistant than natural enamel. Applying fluoride to the outside of teeth after the teeth have formed can also help strengthen the enamel by changing the natural hydroxyapatite in the tooth to fluorapatite. Use the following Teaching Transparency to illustrate that in geology, apatite is in the middle of the hardness scale for minerals.

Teaching Transparency 108 "Mohs' Hardness Scale" LINK TO EARTH SCIENCE

Digestion Begins in the Mouth

Why is chewing so important? There are two reasons. First, chewing creates small, slippery pieces of food that are easier to swallow than big, dry pieces. Second, small pieces of food are easier to digest.

Through the Teeth Teeth are very important organs for mechanical digestion. With the help of strong muscles and your jaw bones, teeth are able to break and grind food. The outermost layer of a tooth, the *enamel*, is the hardest material in the body. Enamel protects nerves and softer material inside the tooth. **Figure 3** shows the major parts of the tooth.

Have you ever noticed that your teeth have different shapes? Look at **Figure 4** to locate the different kinds of teeth. The *molars* in the back are well suited for grinding food. The *premolars* are perfect for mashing food. The sharp teeth at the front of your mouth, the *incisors* and *canines*, are for shredding food.

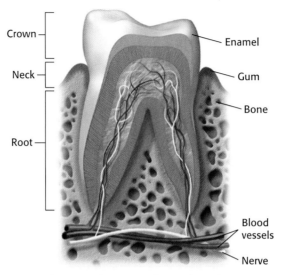

Figure 3 The crown of a tooth, such as this molar, is visible above the gum line. The root is below the gum line.

And Over the Gums As you chew, the food gets mixed with a liquid called *saliva*. Saliva is made in salivary glands located in and around the mouth. Saliva contains an enzyme that begins the chemical digestion of carbohydrates. Saliva turns complex carbohydrates into simple sugars.

Look Out Stomach, Here It Comes! Once the food has been reduced to a soft mush, the tongue pushes it into the throat, which leads to a long, straight tube called the **esophagus** (i SAWF uh guhs). The esophagus squeezes the mass of food with rhythmic muscle contractions called *peristalsis* (PER uh STAHL sis). Peristalsis forces the food into the stomach.

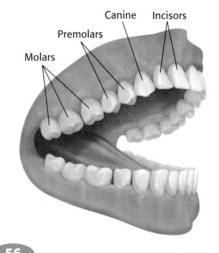

Figure 4 Most adults have 32 permanent teeth. Each type of permanent tooth has a different function in breaking up food before it is swallowed.

MISCONCEPTION ALERT

Students may mistakenly assume that teeth are made of bone. Point out that although enamel and dentin are similar to bone in appearance and calcium content, they are not bone. Bone contains blood vessels, but enamel and dentin do not.

IS THAT A FACT!

The average human eats about 455 kg (1,000 lb) of food every year.

The Stomach's Harsh Environment

The **stomach** is a muscular, baglike organ attached to the lower end of the esophagus. It is pictured in **Figure 5.** The stomach continues the physical digestion of your meal by squeezing its contents with muscular contractions. While all this squeezing is going on, tiny glands in the stomach produce enzymes and acid. These work together to break food into nutrients. Stomach acid also kills most bacteria that you might swallow with your food. After a few hours of combined physical and chemical action, your peanut butter and jelly sandwich has been reduced to a soupy mixture called *chyme* (kiem).

BRAIN FOOD

A thick substance called mucus covers the stomach's lining and offers some protection from its harsh environment. However, the acids still damage the lining, and the entire lining must be replaced every few days.

Figure 5 *The stomach grinds and mixes food for hours before it releases the mixture into the small intestine.*

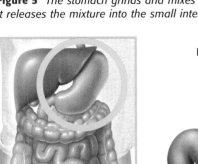
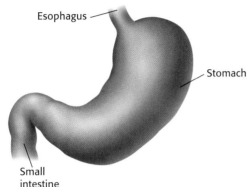

Doorway to the Small Intestine The chyme is slowly released into the small intestine through a small ring of muscle that works like a valve. This valve keeps food in the stomach until it has been thoroughly mixed with digestive fluids. Then the valve opens and closes, letting a small amount of chyme squirt into the small intestine each time. Releasing chyme slowly from the stomach gives the intestine more time to mix the chyme with fluids from the liver and pancreas.

SECTION REVIEW

1. What is the difference between mechanical digestion and chemical digestion?
2. **Inferring Conclusions** Give two reasons why the following statement is true: Digestion begins in the mouth.

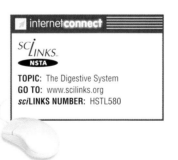

TOPIC: The Digestive System
GO TO: www.scilinks.org
*sci*LINKS NUMBER: HSTL580

2) Teach, continued

CONNECT TO CHEMISTRY

As food is digested, it undergoes both physical and chemical changes. Ask students to categorize the following activities as either physical changes or chemical changes:

- chewing food (physical change)
- enzymes in saliva breaking down carbohydrates (chemical change)
- squeezing and churning food in the stomach (physical change)
- breaking down food with pancreatic juice (chemical change)

MEETING INDIVIDUAL NEEDS

Advanced Learners Provide students with a compound light microscope and prepared slides of several different digestive organs. You might want to include cross sections of tissue from the stomach, the small intestine, the liver, and the pancreas. Review the proper use of microscopes before allowing students to operate them, and make sure that students do not work with broken or cracked slides. Have students observe the specimens, compare the different cross sections, and make sketches in their ScienceLog. Encourage students to link the structures they observe with the functions of each organ. For example, students may relate the structure of villi to the function of nutrient absorption of the small intestine.

Still hungry for news about digestion? Enzymes can help you with that steak, you know. Check it out on page 180 of the LabBook.

The Gigantic Small Intestine?

The **small intestine** is a muscular tube that is about 2.5 cm in diameter. Other than its diameter, it's really not that small at all. In fact, if you stretched it out, it would be longer than you are tall—about 6 m!

Villi If you flattened out the surface of the small intestine, it would be larger than a tennis court! How is this possible? The inside wall of the small intestine is covered with fingerlike projections called *villi*, shown in **Figure 6**. The villi are covered with tiny nutrient-absorbing cells. Because the villi extend into the chyme, these cells have greater exposure to nutrients. Once absorbed, the nutrients enter the bloodstream.

Most chemical digestion takes place in the small intestine. Chyme from the stomach moves very slowly through the small intestine by peristalsis. Proteins, carbohydrates, and fats in the chyme are digested with the help of enzymes produced in the small intestine and the pancreas.

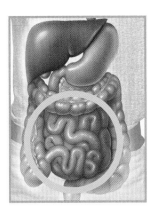
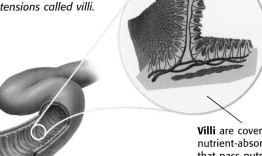

Figure 6 *The highly folded lining of the small intestine has many fingerlike extensions called villi.*

Villi are covered with nutrient-absorbing cells that pass nutrients on to the bloodstream.

The Pancreas

The **pancreas** is a fish-shaped organ located between the stomach and small intestine. It makes pancreatic juice that flows into the small intestine. This juice contains digestive enzymes and bicarbonate that neutralizes the acid in chyme. Without bicarbonate, acids would damage the lining of the intestine and prevent enzymes from doing their work. The pancreas also functions as a part of the endocrine system, making hormones that regulate blood sugar. The pancreas is shown in **Figure 7** on the next page.

SCIENCE HUMOR

A man walks into a doctor's office with a stalk of celery in one ear and a carrot in the other. The man says, "Doc, I'm just not feeling well these days." The doctor replies, "I think that's because you haven't been eating right."

The Liver and Gallbladder

The **liver** is a large reddish brown organ that helps with digestion. A human liver can be as large as a football. Your liver is located toward your right side, slightly higher than your stomach, as shown in **Figure 7**. Here are a few of the liver's important jobs:

- Your liver makes a green liquid called *bile* that is used in fat digestion
- Your liver stores nutrients
- Your liver breaks down toxic substances in the blood
- Your liver makes cholesterol for cell membranes

Bile Breaks Up Fat Although bile is made by the liver, it is temporarily stored in a small baglike organ called the **gallbladder**, shown in Figure 7. Bile is squeezed from the gallbladder into the small intestine, where it breaks up large fat droplets into very small droplets. This physical process allows more fat molecules to be exposed to digestive enzymes.

Storing Nutrients and Protecting the Body After nutrients are broken down, they are absorbed into the bloodstream and carried through the body. Nutrients that are not needed right away are stored in the liver. The liver then releases the stored nutrients into the bloodstream as needed. The liver also captures and detoxifies many substances in the body. For instance, it produces enzymes that break down alcohol and many other drugs.

BRAIN FOOD

If three-fourths of the liver were removed, the rest would go on working and would eventually grow to replace the part that was removed.

✓ Self-Check

Is bile used for chemical or mechanical digestion? Explain. *(See page 212 to check your answer.)*

Figure 7 *The liver, gallbladder, and pancreas are linked to the small intestine, but food does not move through them.*

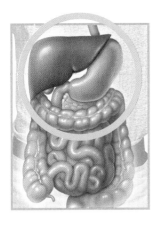

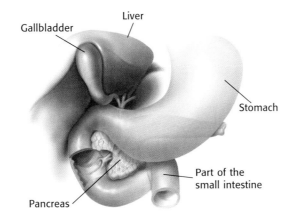

Liver
Gallbladder
Stomach
Part of the small intestine
Pancreas

Is That A Fact!

Weighing as much as 1.5 kg, the liver is the most massive internal organ in the human body.

RETEACHING

Help students remember the route food takes through the digestive tract by providing them with additional practice. Begin by showing students a model or illustration of the digestive tract. Then have students take turns naming and identifying, in order, the organs through which food travels through the digestive tract. (The proper sequence is as follows: mouth, throat, esophagus, stomach, small intestine, and large intestine.)

INDEPENDENT PRACTICE

Concept Mapping List the following terms on the chalkboard:

digestive system, mouth, stomach, liver, pancreas, gallbladder, teeth, salivary glands, saliva, digestive tract, esophagus, throat, tongue, small intestine, large intestine

Then have students copy the terms onto a piece of paper and construct a concept map using these terms and linking words between them.

Answer to Self-Check

Bile is involved in physical digestion because emulsification does not change the chemical composition of the fat molecules; it only increases the surface area of each fat droplet.

MATH and MORE

Tell students that as much as 75 percent of a person's liver can be removed or impaired before the liver stops functioning. Ask students to calculate how much of a 1 kg liver must remain for it to function.

(25 percent × 1,000 g = 250 g)

3) Extend

RESEARCH

Writing Have students research and report on the effects of one of the following eating disorders on the digestive system:

bulimia, anorexia nervosa, obesity, overeating

COOPERATIVE LEARNING

Activity Have students work in groups of three to five to make a life-size model of the digestive system. First have students in each group trace the outline of one of their classmates on a piece of butcher paper. Then provide each group with an assortment of craft materials—including plastic foam, modeling clay, construction paper, and fabric—with which to make models of the digestive organs. Encourage students to make digestive organs that fit together on their body silhouette. You may want to have students practice naming the digestive organs while tracing the path of food through their model of the digestive system. **Sheltered English**

Homework

Creating Graphs Provide students with the following average lengths of the digestive organs:

esophagus—25 cm, stomach—25 cm, small intestine—6 m, large intestine—1.5 m.

Then have students prepare a bar graph or another graphic to compare the average lengths of each digestive organ listed. You might also want to have students measure and label the lengths using adding-machine tape or a skein of yarn and compare them to the illustration of the digestive system in this chapter.

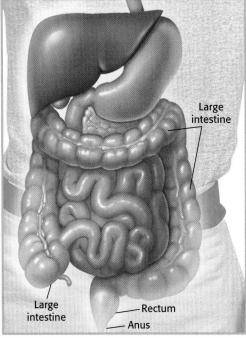

Figure 8 *The large intestine is the final organ of digestion.*

Feces and other human wastes contain microorganisms and other substances that can contaminate drinking water. Every time you flush a toilet, the water and wastes go through the sewer to a sewage treatment plant. Here the disease-causing microorganisms are removed, and the clean water is released back to rivers, lakes, and streams.

The End of the Line

Whatever can't be absorbed into the blood gets pushed into the large intestine. The **large intestine** is the organ of the digestive system that stores, compacts, and then eliminates indigestible material from the body. The large intestine, shown in **Figure 8,** is called "large" because it has a larger diameter than the small intestine. It is about 1.5 m long, and its diameter is about 7.5 cm.

In the Large Intestine Undigested material enters the large intestine as a soupy mixture. The large intestine reabsorbs most of the water in the mixture, changing the liquid into a solid mass called *feces* or *stool*.

Whole grains, fruits, and vegetables contain a carbohydrate, called cellulose, that humans cannot digest. We commonly refer to this material as fiber. Fiber keeps the stool soft and keeps things moving through the large intestine.

A Way Out The *rectum* is the last section of the large intestine. It stores feces until they can be expelled. Feces pass to the outside through an opening called the *anus*. It has taken your sandwich about 24 hours to make this journey.

Problems in the Digestive System

Disorders of the digestive system are frequently related to eating behaviors. However, digestive problems can also be caused by diseases. Some common digestive disorders are described below.

Heartburn The stomach is blocked off at either end by bands of muscle called *sphincters* (SFINGK tuhrz). Occasionally, backflow of chyme from the stomach to the esophagus causes a burning pain in the chest called heartburn. Eating too much, eating right before going to bed, and eating very acidic foods sometimes cause heartburn.

IS THAT A FACT!

In the past, medical professionals have attributed peptic ulcers to stress and heredity. However, in 1982, Australian scientists Robin Warren and Barry Marshall discovered a species of bacteria that was common to many ulcer patients. Dr. Marshall finally proved his hypothesis that the species of bacteria was responsible for ulcers by infecting himself with it, succumbing to peptic ulcers, and then curing himself with antibiotics.

Constipation and Diarrhea When the body does not get enough fiber, water, or exercise, the contents of the large intestine can become dry. Bowel movements become difficult and less frequent. This condition is called *constipation*. When bowel movements are frequent and watery, the condition is called *diarrhea*. Diarrhea occurs when too little water is removed from digested food in the large intestine. Diarrhea may cause dehydration and is especially dangerous for infants and small children, such as the girl in **Figure 9**.

Colon Cancer *Colon cancer* is a serious disease of the digestive tract that can lead to death. The colon is the long tubular portion of the large intestine. When certain colon cells divide uncontrollably, a tumor forms. Tumors interfere with the normal functioning of organs. Cells from a tumor can also break away and start tumors in other areas in the body. Colon cancer can often be treated and cured if detected early.

Gastric Ulcer An open sore in the stomach lining is called a *gastric ulcer*. **Figure 10** shows stomach tissue from a gastric ulcer. Gastric ulcers are often caused by bacteria and can be treated successfully with antibiotics. A high-fat diet, smoking, caffeine, and alcohol may make this condition worse.

Figure 9 *This child is being given fluids to replace those lost to diarrhea.*

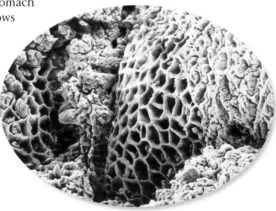

Figure 10 *This stomach lining has openings, seen in red, that indicate a gastric ulcer.*

SECTION REVIEW

1. What happens to the food that you eat when it gets to the stomach?
2. Describe the roles of the liver, the gallbladder, and the pancreas in digestion.
3. **Analyzing Relationships** How would the inability to make saliva affect digestion?

TOPIC: Problems in the Digestive System
GO TO: www.scilinks.org
*sci*LINKS NUMBER: HSTL585

Answers to Section Review

1. The contractions of the stomach break food into smaller pieces. Chemical digestion by enzymes, water, and acids breaks bonds in the food.
2. The liver stores nutrients and makes bile, which is used in the digestion of fat. Bile is stored in a small, baglike organ called the gallbladder. The pancreas makes pancreatic juice, which contains enzymes for digesting proteins, starches, and fats and contains bicarbonate for neutralizing the acid in chyme.
3. Without saliva, swallowing would be more difficult and the digestion of carbohydrates would not begin until the food reached the stomach.

4) Close

Quiz

Ask students whether these statements are true or false.

1. Digestion begins when food reaches the stomach. (false)
2. Breaking, crushing, and mashing food is an example of chemical digestion. (false)
3. Saliva contains enzymes, which begin the chemical digestion of food. (true)
4. The esophagus connects the mouth with the small intestine. (false)
5. Undigested materials are changed from a liquid to a solid mixture in the large intestine. (true)

ALTERNATIVE ASSESSMENT

Writing Have students develop an owner's guide for their digestive system. The guide should include information about the structures of the digestive system and a diagram of the location of these body structures. Students should also include information about the function of each structure and the disorders that affect the digestive system. Encourage students to share their guide with the class in the form of a presentation or a poster.

Critical Thinking Worksheet
"Frankenstein's Food"

TOPIC: Problems in the Digestive System
GO TO: www.scilinks.org
*sci*LINKS NUMBER: HSTL585

Section 1 • The Digestive System

SECTION 2

Focus

The Urinary System
This section introduces the structures and the functions of the urinary system. Students will learn how the kidneys filter blood. Students will also learn about some major problems of the urinary system and their causes.

📀 Bellringer
Tell students that the blood must be cleaned regularly. Then ask students to speculate, without looking in the textbook, how the body cleans blood. Once students have written a description of the cleaning process, allow them to check their answers against their textbook.

1 Motivate

DISCUSSION
Mongolian gerbils Tell students that Mongolian gerbils are desert animals that never drink. Ask students how they think Mongolian gerbils obtain water. (They get water from the foods they eat.)

Ask students: How have the gerbils adapted to their habitat? (These animals have adapted to use water very efficiently, losing little or none from their lungs, skin, and urine.)

Directed Reading Worksheet Section 2

internet connect
SC_LINKS **TOPIC:** The Urinary System
NSTA **GO TO:** www.scilinks.org
sciLINKS NUMBER: HSTL590

62 Chapter 3 • The Digestive and Urinary Systems

SECTION 2 READING WARM-UP

Terms to Learn
urinary system   urine
kidney   urinary bladder
nephron

What You'll Do
◆ Describe the parts and functions of the urinary system.
◆ Explain how the kidneys filter blood.
◆ Describe some disorders of the urinary system.

The Urinary System

As your body performs the chemical activities that keep you alive, waste products such as carbon dioxide and nitrogen are produced. Your body has to get rid of these waste products in order to stay healthy. *Excretion* is the process of removing wastes and excess products from the body. Three of your body systems are involved in excretion: your skin releases waste products and water when you sweat, your lungs expel carbon dioxide and water when you exhale, and the **urinary system** removes waste products from your blood. Notice that the digestive system is not involved in excretion. The term *excretion* is used only when substances must pass through a membrane in order to leave the body.

Cleaning the Blood
As blood travels through the tissues, it collects all of the waste products produced by the body's cells. Your blood is like a supply train that comes into a town to drop off supplies and take away garbage. The train has to find a way to get rid of the garbage before it can load up with more supplies. If the garbage is not removed, the townspeople will be in a very unhealthy environment. If the cells in your body cannot get rid of their waste products, they can actually be poisoned! On the next few pages, you will see how the urinary system removes waste materials from your blood so that the blood can transport nutrients again. The urinary system is shown in **Figure 11**.

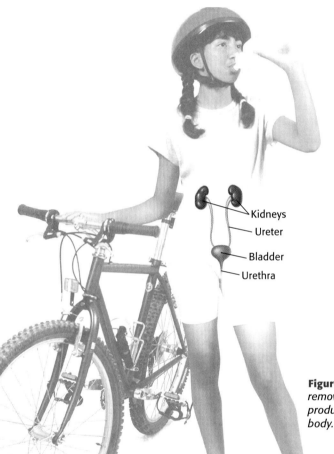

Figure 11 *The urinary system removes many of the waste products produced by the body.*

Q: What does the kidney say when it plays baseball with the other urinary organs?

A: Bladder up!

Flow-Through Filters

The **kidneys** are a pair of bean-shaped organs that constantly clean the blood. Your kidneys filter about 2,000 L of blood each day. Your body only holds 5.6 L of blood, so your blood cycles through the kidneys about 350 times a day!

Inside each kidney are more than 1 million microscopic filters called **nephrons,** shown below. Nephrons remove a variety of harmful substances from the body. Among the most important of these substances is urea, which contains nitrogen and is formed when cells use protein for energy.

Why does your mouth get so dry when the rest of you is so hot and sweaty? Turn to page 72 to find out.

How the Kidneys Filter Blood

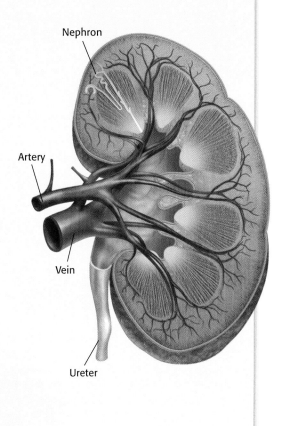

1 A large artery brings blood into each kidney.

2 Tiny blood vessels branch off the main artery and pass through part of each nephron.

3 Water and other small substances, such as glucose, salts, amino acids, and urea, are forced out of the blood vessels and into the nephrons.

4 As these substances flow through the nephrons, most of the water and some nutrients are moved back into blood vessels that wrap around the nephrons. A concentrated mixture of waste materials is left behind in the nephrons.

5 The cleaned blood, now with slightly less water and much less waste material, leaves each kidney in a large vein to recirculate in the body.

6 The yellow fluid that remains in the nephrons is called **urine.** Urine leaves each kidney through a slender tube called the *ureter* and flows into the **urinary bladder,** where it is stored.

7 Urine leaves the body through another tube called the *urethra. Urination* is the process of expelling urine from the body.

IS THAT A FACT!

In a lifetime, a person's kidneys clean more than 1 million gallons (4 million liters) of liquid. That's enough to fill a small lake.

Teaching Transparency 88
"How the Kidneys Filter the Blood"

2) Teach

MEETING INDIVIDUAL NEEDS

Learners Having Difficulty Help students with limited English proficiency with the terms on this page. Write the terms *urine* and *urinary* on the chalkboard or on a transparency. As a class, define these terms and write working definitions of the terms beside each word. You may want to allow students to use their textbook for this exercise. Then point out that the term *urine* is a noun and that the term *urinary* is an adjective. **Sheltered English**

USING THE FIGURE

Have students trace the path of blood into, through, and out of the kidney illustrated in the figure. **Sheltered English**

DEMONSTRATION

Kidney Structure Obtain a beef or sheep kidney from your local supermarket or butcher. Cut the kidney into two symmetrical halves. Allow students to observe and sketch the internal and external structures of the kidney and to compare the kidney with the kidney illustration on this page. Have students wear goggles, disposable gloves, and aprons during their examination of the kidney. Students should keep their hands away from their face and eyes during examination and should wash their hands after examining the kidney. Dispose of the kidney as you would any other biohazard. Following the examination of the kidney, ask students why this kidney is so similar to a human kidney. (Both are mammalian.) **Sheltered English**

Section 2 • The Urinary System

2) Teach, continued

Answer to MATHBREAK

You need to drink 10.6 8-oz glasses of water per day.

2500 mL ÷ 29.6 mL/oz = 84.5 oz
84.5 oz ÷ 8 oz/glass = 10.6 glasses

READING STRATEGY

Prediction Guide Before students read this page, ask them whether the following statements are true or false. Students will discover the answers as they read this page.

1. When you sweat a lot, you produce less saliva. (true)
2. When you are thirsty, your tissues need more water. (true)
3. When you are thirsty, your kidneys return more water to the blood. (true)

MISCONCEPTION ALERT

Students may think that taking large doses of vitamin C can prevent or cure a cold. In fact, vitamin C has not been shown to prevent or cure colds. However, continued doses of vitamin C of 1 g or more may eventually cause kidney stones, diarrhea, and other conditions.

Answer to APPLY

Answers will vary. Caffeine is a diuretic, a substance that causes the kidneys to make more urine, thereby decreasing the amount of water in the blood. Water is needed to produce energy for the body, so not having enough water might decrease the energy level of an athlete.

Reinforcement Worksheet
"Annie Apple's Amazing Adventure"

MATH BREAK

How Much Water?
Most adults need to consume about 2,500 mL of water per day. How many 8 oz glasses of drinking water would you need to get enough water?
Hint: 1 oz = 29.6 mL

Fortunately, your kidneys are very efficient. In fact, if you lost one to disease, you could still survive.

Beverage Ban

During football season, a football coach insists that all members of the team avoid caffeinated beverages. Many of the players are upset by the news. Pretend that you are the coach. Write a letter to the team explaining why it is better for them to drink water than drinks containing caffeine.

Homework

Poster Project Have students make a poster following the path of blood through the kidney. Have them create their own diagram with a labeled pathway using the steps and the figure on this page as a starting point.

Water In, Water Out

Our bodies take in water every day. If we did not excrete an equal amount of water, our bodies would swell up with all the excess. Losing water through sweat and urine is necessary to stay healthy. So how does the body keep the water levels in the proper balance? The balance of fluids is controlled by chemical messengers in the body called *hormones*.

Sweat and Thirst When you get hot, you lose more water in the form of sweat. The evaporation of water from your skin cools you down. As the water content of the blood drops, the salivary glands produce less saliva. This is one of the reasons you feel thirsty.

Antidiuretic Hormone When you get thirsty, other parts of your body react to the water shortage, too. A hormone called *antidiuretic* (AN tie DIE yoo RET ik) *hormone,* or *ADH,* is released. ADH signals the kidneys to take back water from the nephrons and return it to the bloodstream, thereby making less urine. When your blood is too watery, smaller amounts of ADH are released. The kidneys react by allowing more water to stay in the nephron and leave the body as urine.

Diuretics When you are thirsty, your tissues are asking for more water. Some beverages contain caffeine, which is a *diuretic* (DIE yoo RET ik). Diuretics cause the kidneys to make more urine, which decreases the amount of water in the blood. So instead of giving your body more water, caffeinated beverages cause additional water to be lost in urine.

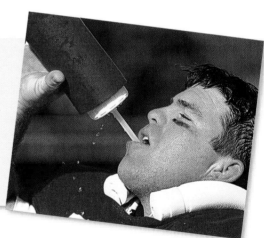

Although urea can be toxic to the body and must be eliminated, it also has a number of medical uses. It is used in some creams to treat dry-skin ailments, such as psoriasis. Urea is also used as a diuretic to reduce pressure caused by glaucoma or edema in the brain.

Urinary System Ailments

Since the urinary system regulates body fluids and removes wastes from the blood, any malfunction can become life-threatening. Some common urinary system disorders are described below.

Bacterial Infections Bacteria can get into the bladder and ureters through the urethra and cause painful infections. It is important to treat such an infection early because it could spread to the kidneys and lead to permanent damage to the nephrons.

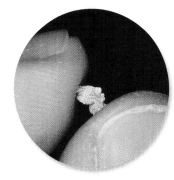

Figure 12 *This kidney stone had to be removed from a patient's urinary system.*

Kidney Stones Sometimes salts and wastes collect inside the kidneys and form kidney stones, like the one in **Figure 12.** Kidney stones interfere with urine flow and cause pain. Most kidney stones pass naturally from the body, but sometimes a medical procedure is necessary. For example, shockwaves can be used to break the stones into pieces small enough to pass through the urethra.

Kidney Disease Damage to nephrons can prevent normal kidney functioning, leading to kidney disease. If the kidneys do not function properly, a kidney machine can be used to filter waste from the blood. As shown in **Figure 13,** blood is pumped from an artery in the forearm or wrist to a kidney machine, where it is filtered. The cleaned blood is then pumped back into a vein in the arm.

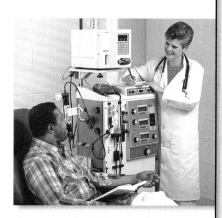

Figure 13 *The kidney machine will filter this man's blood before returning it to his body.*

SECTION REVIEW

1. Put the following statements about the urinary system in their proper order:
   a. Water is absorbed back into blood vessels.
   b. Urine leaves the kidney through the ureter.
   c. A large artery brings blood into the kidney.
   d. Water and other small substances, including glucose and urea, leave the blood vessels and enter the nephron.

2. What is the main function of the urinary system?

3. **Applying Concepts** Which contains more water, the blood going into the kidney or the blood leaving the kidney? Explain.

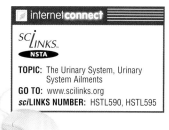

internet connect

TOPIC: The Urinary System, Urinary System Ailments
GO TO: www.scilinks.org
sciLINKS NUMBER: HSTL590, HSTL595

Answers to Section Review

1. c, d, a, b
2. The main function of the urinary system is to remove certain waste products from the blood.
3. There is more water in the blood as it enters the kidney. Though the kidney returns most of the water back to the blood vessels, some of the water leaves the bloodstream when the kidneys produce urine.

Skill Builder Lab

As the Stomach Churns
Teacher's Notes

Time Required

One 45-minute class period and another 15 minutes after 24 hours

Lab Ratings

- TEACHER PREP 🧪🧪
- STUDENT SET-UP 🧪🧪
- CONCEPT LEVEL 🧪🧪
- CLEAN UP 🧪🧪

The materials listed on the student page are enough for one student or one group of 2–4 students. The two tenderizers should be available at the grocery store. You will need to examine the different brands of powdered or liquid tenderizers. You may substitute fresh papaya or pineapple juices if they are available, if neither type of tenderizer can be found.

Safety Caution

Remind students to review all safety cautions and icons before beginning this lab activity. Caution students that HCl is a chemical they should handle very carefully. You may choose to dispense the HCl yourself. Tell students that if any HCl comes in contact with their skin, they should wash the area with plenty of water immediately and they should notify you as they wash. Tell students not to taste any materials used in this or any lab.

As the Stomach Churns

The stomach moves the food around while digestive juices—acids and enzymes—are added to begin protein digestion. Some meat tenderizers have plant enzymes that break down, or digest, proteins. You commonly can get two types of meat tenderizer at grocery stores. One type has an enzyme, called papain, from papaya. Another type has an enzyme, called bromelain, from pineapple. In this lab, you will test the effects of these two different types of meat tenderizers on beef stew meat.

- 4 test tubes
- test-tube rack
- test-tube marker
- masking tape
- 25 mL graduated cylinder
- water
- eyedropper
- $\frac{1}{4}$ tsp measuring spoon
- hydrochloric acid (0.1 M)
- meat tenderizer containing bromelain
- meat tenderizer containing papain
- 1 cm cubes of beef stew meat (3)
- protective gloves

Ask a Question

1 Which meat tenderizer will work faster? Which will make the meat more tender? Will there be a color change in the meat or in the water? What might these changes, if any, indicate? Decide what you will look for as you plan your experiment.

Form a Hypothesis

2 Look at the list of ingredients on the labels of each of the meat tenderizers. Form a hypothesis about which tenderizer will make the beef more tender.
Caution: Do not taste any of the materials in this lab.

Conduct an Experiment

3 Identify any variables and controls present in your experiment. Make a data table in your ScienceLog or on a computer to record your observations and results.

4 Label two of the test tubes with the name of the tenderizer being investigated. Label the third test tube "Control." What will this tube contain?

 Datasheets for LabBook

Yvonne Brannum
Hine Junior High School
Washington, D.C.

5 Pour 20 mL of water into each test tube.

6 With the eyedropper, add four drops of hydrochloric acid to each test tube. **Caution:** Hydrochloric acid can burn your skin. If any touches your skin, rinse the area with running water and tell your teacher immediately.

7 Using the measuring spoon, add $\frac{1}{4}$ tsp of a meat tenderizer to the tube with that label.

8 Add one piece of beef to each test tube.

9 Record your observations of each test tube immediately, after 5 minutes, after 15 minutes, after 30 minutes, and again after 24 hours.

Analyze the Results

10 Did you notice any differences in the beef in the three test tubes right away? At what time interval did you notice a significant difference in the appearance of the beef in the test tubes?

11 Did one meat tenderizer perform better than the other? Explain how you determined which tenderizer was more efficient.

Draw Conclusions

12 Was your hypothesis supported? Explain your answer.

13 Many stinging animals have venom composed of proteins. Explain how applying meat tenderizer to the wound helps relieve the pain of such a sting.

Skill Builder Lab

Answers

4. The control tube will contain everything that goes into the other tubes except the meat tenderizer.

10. There should be very little difference in the three test tubes at first. Differences will be mild until after 24 hours, when a significant difference will be noticed between the control tube and the experimental tubes.

11. Both tenderizers should work in a similar way. If one tenderizer is noted to be more efficient, students should be able to describe how they tested the stew meat to demonstrate the difference.

12. Answers will vary, but students should be able to explain how their hypothesis was supported or how their experimental results did not support their hypothesis.

13. Students may conclude that certain digestive enzymes break down proteins. Breaking down the protein in venom often makes the venom less harmful and less painful. Explain to students that using meat tenderizer on an insect bite is first aid only and is not intended to substitute for medical attention. Emphasize that snakebites require immediate medical attention.

Background

Students know that papaya and pineapple fruits are the source of the meat tenderizers they will be using in this lab. They also know about gelatin desserts made with fruit. You might want to show them a dessert made with canned pineapple and one made (or attempted) with fresh pineapple. Tell them that the ingredient that makes the gelatin gel is an animal protein. Ask them to explain why the gelatin with fresh pineapple didn't gel.

Chapter 3 • Skill Builder Lab

Chapter Highlights

VOCABULARY DEFINITIONS

SECTION 1

digestive system a collection of organs that break down food so that it can be used by the body; the organs in this system include the stomach, the pancreas, the liver, the gallbladder, the small intestine, and the large intestine

esophagus a long, straight tube that connects the mouth and throat to the stomach

stomach a muscular, baglike organ of the digestive tract; attached to the lower end of the esophagus

small intestine a muscular tube about 6 m long; the site of most chemical digestion

pancreas an organ between the stomach and small intestine that produces enzymes for chemical digestion

liver a large, reddish organ that produces bile and stores nutrients

gallbladder a small, baglike organ that stores bile

large intestine a large organ that reabsorbs water from the digestive tract and stores, compacts, and eliminates indigestible material from the body

Chapter Highlights

SECTION 1

Vocabulary
- **digestive system** (p. 54)
- **esophagus** (p. 56)
- **stomach** (p. 57)
- **small intestine** (p. 58)
- **pancreas** (p. 58)
- **liver** (p. 59)
- **gallbladder** (p. 59)
- **large intestine** (p. 60)

Section Notes
- Your digestive system is a group of organs that work together to digest food so that it can be used by the body.
- The breaking, crushing, and mashing of food is called mechanical digestion. Chemical digestion is the process in which large molecules are broken down to simpler molecules.
- Chewed food is pushed through the digestive tract by rhythmic contractions called peristalsis.
- The stomach mixes the food with enzymes and acid to break down nutrients. The mixture is called chyme.
- In the small intestine, pancreatic juice and bile are mixed with chyme.
- From the small intestine, nutrients enter the bloodstream and are circulated to the body's cells.
- The large intestine receives undigested material from the small intestine. As water is absorbed back into the body, this material becomes a solid mass called feces.

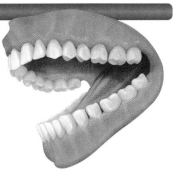

- Digestive system disorders include heartburn, constipation, diarrhea, colon cancer, and gastric ulcers.

Labs
Enzymes in Action (p. 180)

✅ Skills Check

Math Concepts

DRINK UP In the MathBreak on page 64, you determined how many glasses of water you need in order to drink 2,500 mL each day. First you must determine how many ounces are in 2,500 mL:

$$\frac{2{,}500 \text{ mL}}{1 \text{ day}} \times \frac{1 \text{ oz}}{29.6 \text{ mL}} = 84.5 \text{ oz of water}$$

You need to drink 84.5 oz of water each day.

Each glass contains 8 oz, so:

$$\frac{84.5 \text{ oz}}{1 \text{ day}} \times \frac{1 \text{ glass}}{8 \text{ oz}} = 10.6 \text{ glasses}$$

You need to drink 10.6 8-oz glasses of water each day.

Visual Understanding

KIDNEY FUNCTION Look at the illustration on page 63 to review how the kidneys filter the blood.

Lab and Activity Highlights

As the Stomach Churns PG 66

Enzymes in Action PG 180

Datasheets for LabBook
(blackline masters for these labs)

68 Chapter 3 • The Digestive and Urinary Systems

SECTION 2

Vocabulary
- **urinary system** (p. 62)
- **kidney** (p. 63)
- **nephron** (p. 63)
- **urine** (p. 63)
- **urinary bladder** (p. 63)

Section Notes
- Your skin, lungs, and urinary system are all involved in excretion.
- The urinary system cleans the blood and removes liquid waste as urine. The filtering structures in the kidneys are called nephrons.
- Most of the water and some nutrients that enter nephrons are moved back into the blood vessels.
- When urine leaves the kidneys, it passes into the urinary bladder through a tube called the ureter. The urinary bladder stores the urine until it can be eliminated.
- Urine travels from the urinary bladder to the outside through a tube called the urethra.
- Some disorders of the urinary system include bacterial infections, kidney stones, and kidney disease.

VOCABULARY DEFINITIONS, *continued*

SECTION 2

urinary system a collection of organs that remove waste from the blood; the organs in this system include the kidneys, ureters, urethra, and urinary bladder

kidney a bean-shaped organ that removes many harmful substances from the blood

nephron a microscopic filter in the kidney that removes a variety of harmful substances from the blood

urine a concentrated mixture of waste materials that forms in the nephrons of the kidney

urinary bladder a baglike organ that stores urine until it can be eliminated through the urethra

 Vocabulary Review Worksheet

 Blackline masters of these Chapter Highlights can be found in the **Study Guide.**

---

internet connect

GO TO: go.hrw.com

Visit the **HRW** Web site for a variety of learning tools related to this chapter. Just type in the keyword:

KEYWORD: HSTBD3

GO TO: www.scilinks.org

Visit the **National Science Teachers Association** on-line Web site for Internet resources related to this chapter. Just type in the *sci*LINKS number for more information about the topic:

TOPIC:	sciLINKS NUMBER:
The Digestive System	HSTL580
Problems in the Digestive System	HSTL585
The Urinary System	HSTL590
Urinary System Ailments	HSTL595
Tapeworms	HSTL600

Lab and Activity Highlights

LabBank

 Whiz-Bang Demonstrations, Liver Let Live

Long-Term Projects & Research Ideas, Copying the Kidney

Chapter Review Answers

USING VOCABULARY

1. ureters
2. peristalsis
3. mouth
4. liver, gallbladder
5. Excretion
6. large intestine

UNDERSTANDING CONCEPTS

Multiple Choice

7. b
8. d
9. c
10. d
11. b
12. a

Short Answer

13. Bicarbonate neutralizes the acidic chyme coming in from the stomach. This allows pancreatic enzymes and enzymes made in the intestine to work. By neutralizing the acid, the bicarbonate also protects the lining of the intestine.
14. The long length, folds, and villi increase the surface area that comes into contact with the food that is being digested in the small intestine.
15. A diuretic is a substance that causes the kidneys to produce more urine.

Chapter Review

USING VOCABULARY

To complete the following sentences, choose the correct term from each pair of terms listed below:

1. Urine travels from each kidney to the urinary bladder through the __?__. (*urethra* or *ureter*)
2. The rhythmic contractions that occur in the digestive tract are called __?__. (*peristalsis* or *enzymes*)
3. The chemical digestion of carbohydrates begins in the __?__. (*stomach* or *mouth*)
4. Bile is made in the __?__ and stored in the __?__. (*liver* or *gallbladder*)
5. __?__ is the process of removing wastes and waste products from the body. This term is only used when substances are passed through a membrane before leaving the body. (*Digestion* or *Excretion*)
6. Indigestible material is formed into feces in the __?__. (*large intestine* or *small intestine*)

UNDERSTANDING CONCEPTS

Multiple Choice

7. The hormone that signals the kidneys to make less urine is
   a. urea.
   b. ADH.
   c. cellulase.
   d. ATP.
8. Which of the following aids digestion by producing bile?
   a. stomach
   b. pancreas
   c. gallbladder
   d. liver
9. The part of the kidney that filters the blood is the
   a. artery.
   b. ureter.
   c. nephron.
   d. urethra.
10. The fingerlike projections lining the small intestine are called
    a. emulsifiers.
    b. fats.
    c. amino acids.
    d. villi.
11. Which of the following is not part of the digestive tract?
    a. mouth
    b. pancreas
    c. stomach
    d. rectum
12. The soupy mixture of food, enzymes, and acids in the stomach is called
    a. chyme.
    b. villi.
    c. urea.
    d. vitamins.

Short Answer

13. Give two reasons why it is important that the pancreas releases bicarbonate into the small intestine.
14. How does the structure of the small intestine improve its nutrient absorption?
15. What is a diuretic?

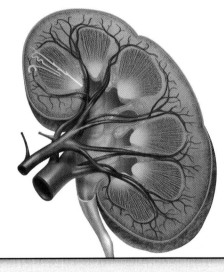

70 Chapter 3 • The Digestive and Urinary Systems

Concept Mapping

16. Use the following terms to create a concept map: teeth, stomach, digestion, bile, saliva, mechanical digestion, gallbladder, chemical digestion.

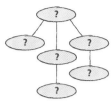

CRITICAL THINKING AND PROBLEM SOLVING

Write one or two sentences to answer the following questions:

17. How would digestion be affected if the liver were damaged?

18. Think about what happens when you put a piece of carbohydrate-dense food, such as bread, potato, or a cracker, in your mouth. If you let a small piece sit near the tip of your tongue, it might begin to taste sweet. What digestive process would explain this change?

MATH IN SCIENCE

19. Mr. Jones has lost all of his molars and two of his premolars. How many teeth does Mr. Jones have?

20. During a one-day water-balance study, a woman drank 1,500 mL of water. The food that she ate contained 750 mL of water, and 250 mL of water was produced internally during normal body processes. She lost 900 mL of water by sweating, 1,500 mL in urine, and 100 mL in feces. Overall, how much water did she gain or lose during the day?

INTERPRETING GRAPHICS

The bar graph below shows how long the average meal spends in each portion of your digestive tract. Use this graph to answer the questions below.

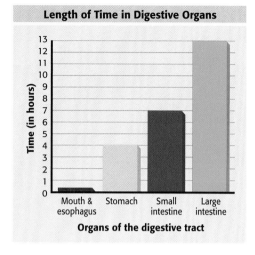

21. Where does the food spend the longest amount of time?

22. On average, how much longer does food stay in the small intestine than in the stomach?

23. Which organ mixes food with special substances to make chyme? Approximately how long does food remain in this organ?

24. Bile breaks up large fat droplets into very small droplets. How long is the food in your body before it comes into contact with bile?

 Take a minute to review your answers to the Pre-Reading Questions found at the bottom of page 52. Have your answers changed? If necessary, revise your answers based on what you have learned since you began this chapter.

Concept Mapping

16. An answer to this exercise can be found at the front of this book.

CRITICAL THINKING AND PROBLEM SOLVING

17. Sample answer: Digestion would be poor because the liver makes bile, which aids in the digestion of fats.

18. Sample answer: Saliva causes chemical digestion of the carbohydrate-dense food, breaking it down to become simple sugars, which taste sweet.

MATH IN SCIENCE

19. 18 remaining teeth
20. 0 gain, 0 loss

INTERPRETING GRAPHICS

21. in the large intestine
22. 3 hours
23. The stomach mixes food with enzymes and acid to make a soupy mixture called chyme. Food stays in the stomach for approximately 4 hours.
24. just over 4 hours

 Concept Mapping Transparency 24

 Blackline masters of this Chapter Review can be found in the **Study Guide**.

Across the Sciences
Chemistry
Quench Your Thirst!

Background

A portion of the brain called the hypothalamus contains nerve cells that monitor the concentration of sodium in the blood. When the concentration of sodium is high, these cells perform two important functions. First, they send messages to the pituitary gland, which releases antidiuretic hormone (ADH). This hormone signals the kidneys to return water to the bloodstream. Aside from signaling the pituitary gland, these nerve cells also signal the salivary glands to produce less saliva. With less saliva, the mouth becomes dry. This creates a sensation of thirst and signals the person to get a drink.

As body heat rises, the body begins to lose potassium. This is due mainly to the fact that a hormone is released that increases the amount of potassium in urine. This hormone, called aldosterone, signals the kidneys to force potassium out of the blood vessels and into the nephron tubes. The potassium is then eliminated from the body in the urine. Sports drinks help replace this essential ion in the blood. But water is the most important component of these drinks, and plain water is often just as helpful as a sports drink to a person exercising.

ACROSS THE SCIENCES

LIFE SCIENCE • CHEMISTRY

Quench Your Thirst!

Have you ever been really thirsty after a hard workout? Playing sports, riding a bike, and doing other physical activities can make you thirsty—but why? The first reason is sweat. When you are physically active, you lose a lot of water by sweating. This keeps your body from overheating. But what is going on in your body to make your mouth feel so dry? And which is better to quench your thirst, water or a sports drink?

▲ Activities that make you hot and sweaty make you want to take a drink, but why?

Thirsty Chemistry

When you lose water, your blood becomes more concentrated. Think about how you make a powdered drink, such as lemonade. If you use the same amount of powder in 1 L of water as you do in 2 L, the drinks will taste different. The lemonade made with 1 L of water will be stronger because it is more concentrated.

Losing water to sweat increases the concentration of sodium and potassium in your blood. The kidneys force the extra potassium out of the blood vessels and into nephrons. From the nephrons, the potassium is eliminated from the body in urine. Nerve cells in your brain react to the high concentration of sodium by sending out two important messages. One message tells the pituitary gland to release antidiuretic hormone. This hormone signals the kidneys to return water to the bloodstream. The second message signals the salivary glands to produce less saliva. With less saliva, your mouth becomes dry, and then you know it's time to get that drink!

With Flavor or Without?

But which is better to drink—water or a sports drink? If you have been exercising, you might think a sports drink is probably better. But studies by Kathy Grunewald, a professor at Kansas State University, indicate that sports drinks may not be necessary unless you have exercised very hard or for more than an hour and a half.

When you drink fluids, you lower the concentration of all the minerals in your blood by adding more water. This is like adding water to the 1 L of strong lemonade. The body also needs to replace the potassium that was lost from the blood. A sports drink can help replace this potassium. But if you drink water, the kidneys will eventually return the potassium concentration to normal. The most important reason for drinking fluids after physical exercise is to get water to your tissues. So whatever physical activity you choose, drink up!

Going Further

▶ If you want to investigate how much you need to drink, weigh yourself before and after your next strenuous activity. Every kilogram that you lose represents about 1 L of water. You should make sure that you drink at least as much water as you lose.

Answer to Going Further

Make sure students know that dehydration is very dangerous. Remind them to drink plenty of fluids during and after strenuous activity.

Health Watch

A Voiceless Companion

If you decided to eat the last piece of pizza in the refrigerator, someone just might ask you for a bite, right? But what if you found out that you had a constant mealtime companion who didn't want just a bite but wanted it all? And what if that companion never asked for your permission?

How to Be a Host

This constant mealtime companion might be a tapeworm. Tapeworms are invertebrate flatworms. These flatworms are parasites. A parasite is an organism that obtains its food by living in or on another living organism. The organism in which a parasite makes its home is called a host. People, cows, pigs, fish, cats, dogs, and many insects are the perfect hosts for tapeworms. Without a host, tapeworms can't survive.

Food broken down in the stomach continues to be broken down in the small intestine. Since a tapeworm doesn't have a digestive tract of its own, it borrows its host's. By attaching itself to the inside of its host's small intestine with clamps and suckers, a tapeworm can eat as much as it likes.

Although tapeworms aren't much thicker than a ribbon, they can grow to more than 6 m in length! They do this by adding one postage-stamp-sized segment at a time. Each segment has both male and female reproductive organs and can be filled with thousands of eggs.

Saying "Goodbye" and Avoiding "Hello"

When an egg-filled segment breaks off, it passes through the rest of the host's digestive tract and ends up in the feces. If another animal eats or drinks something contaminated with these feces, the eggs grow into worms in that animal's intestines. The eggs can then spread to muscle tissues (called meat in animals used for food). If humans eat this meat but don't cook it thoroughly enough to kill tapeworm larvae, the cycle begins all over again.

Getting rid of a tapeworm requires removing the head, or scolex. If the scolex is left behind, it simply produces new segments, and the tapeworm regrows itself. Sometimes humans don't realize they have a tapeworm, even though they suffer from symptoms such as weight loss and nausea. And occasionally there are no symptoms at all.

The best way to avoid these parasites is to avoid eating undercooked beef, pork, and fish. If you do this, you won't have any uninvited guests at your next meal!

Think About It

▶ Doctors prescribe certain medications to get rid of tapeworms. Research the different ways people got rid of tapeworms before modern medicines were available.

▲ *What is 10 m long, looks like it's made of postage stamps, and eats your dinner after you do?*

Chapter Organizer

CHAPTER ORGANIZATION	TIME MINUTES	OBJECTIVES	LABS, INVESTIGATIONS, AND DEMONSTRATIONS
Chapter Opener pp. 74–75	45	National Standards: SAI 1, HNS 3, LS 1e	**Start-Up Activity,** Act Fast, p. 75
Section 1 The Nervous System	90	▶ Explain how neurons in the nervous system work together. ▶ Compare the central nervous system with the peripheral nervous system. ▶ Describe the major functions of the brain and the spinal cord. UCP 1, 2, 4, 5, SAI 1, 2, SPSP 1, 5, LS 1a, 1d–1f, 3a–3c	**Demonstration,** Simulating Neuronal Impulses, p. 78 in ATE **QuickLab,** Knee Jerks, p. 82
Section 2 Responding to the Environment	90	▶ List four sensations that are detected by receptors in the skin. ▶ Describe how light relates to vision. ▶ Explain the functions of photoreceptors, taste buds, and olfactory cells. UCP 2, 4, 5, SPSP 1, 5, LS 1a, 1d 1e, 3a; Labs UCP 2, SAI 1, LS 3a	**Quick Lab,** Where's the Dot? p. 85 **Skill Builder,** You've Gotta Lotta Nerve, p. 92 **Datasheets for LabBook,** You've Gotta Lotta Nerve **Whiz-Bang Demonstrations,** Now You See It, Now You Don't **Labs You Can Eat,** A Salty Sweet Experiment
Section 3 The Endocrine System	90	▶ Explain the function of the endocrine system. ▶ List the glands of the endocrine system and describe some of their functions. ▶ Describe how feedback controls stop and start hormone release. UCP 1, 3, 4, SPSP 5, LS 1e, 1f, 3a–3c	**Long-Term Projects & Research Ideas,** Man Versus Machine

See page **T23** for a complete correlation of this book with the

NATIONAL SCIENCE EDUCATION STANDARDS.

TECHNOLOGY RESOURCES

 Guided Reading Audio CD
English or Spanish, Chapter 4

 One-Stop Planner CD-ROM with Test Generator

 CNN. Science, Technology & Society,
Correcting Colorblindness, Segment 2
Brain Cell Visuals, Segment 23
Learning from Frog Ears, Segment 26
Easy Touch Toys, Segment 33

Chapter 4 • Communication and Control

CLASSROOM WORKSHEETS, TRANSPARENCIES, AND RESOURCES	SCIENCE INTEGRATION AND CONNECTIONS	REVIEW AND ASSESSMENT
Directed Reading Worksheet **Science Puzzlers, Twisters & Teasers**		
Transparency 89, The Neuron **Directed Reading Worksheet,** Section 1 **Transparency 90,** What's in a Nerve? **Reinforcement Worksheet,** This System Is Just "Two" Nervous! **Transparency 91,** Regions of the Brain **Transparency 92,** The Spinal Cord	**Math Break,** Time to Travel, p. 77 **Real-World Connection,** p. 79 in ATE **Cross-Disciplinary Focus,** p. 81 in ATE **Eureka!** Pathway to a Cure, p. 99	**Homework,** p. 80 in ATE **Self-Check,** p. 81 **Section Review,** p. 82 **Quiz,** p. 82 in ATE **Alternative Assessment,** p. 82 in ATE
Directed Reading Worksheet, Section 2 **Transparency 281,** Wavelength **Reinforcement Worksheet,** The Eyes Have It **Critical Thinking Worksheet,** There's a Microchip in My Eye!	**Real-World Connection,** p. 84 in ATE **Connect to Physical Science,** p. 84 in ATE **Math and More,** p. 86 in ATE **Physics Connection,** p. 86 **Science, Technology, and Society:** Light on Lenses, p. 98	**Section Review,** p. 87 **Quiz,** p. 87 in ATE **Alternative Assessment,** p. 87 in ATE
Directed Reading Worksheet, Section 3 **Reinforcement Worksheet,** Every Gland Lends a Hand	**Apply,** p. 89 **Math and More,** p. 90 in ATE	**Homework,** p. 90 in ATE **Section Review,** p. 91 **Quiz,** p. 91 in ATE **Alternative Assessment,** p. 91 in ATE

END-OF-CHAPTER REVIEW AND ASSESSMENT

Chapter Review in Study Guide
Vocabulary and Notes in Study Guide
Chapter Tests with Performance-Based Assessment, Chapter 4 Test
Chapter Tests with Performance-Based Assessment, Performance-Based Assessment 4
Concept Mapping Transparency 25

---

 internet**connect**

 Holt, Rinehart and Winston On-line Resources
go.hrw.com

For worksheets and other teaching aids related to this chapter, visit the HRW Web site and type in the keyword: **HSTBD4**

 National Science Teachers Association
www.scilinks.org

Encourage students to use the *sci*LINKS numbers listed in the internet connect boxes to access information and resources on the **NSTA** Web site.

Chapter 4 • Chapter Organizer   **73B**

Chapter Resources & Worksheets

Visual Resources

TEACHING TRANSPARENCIES

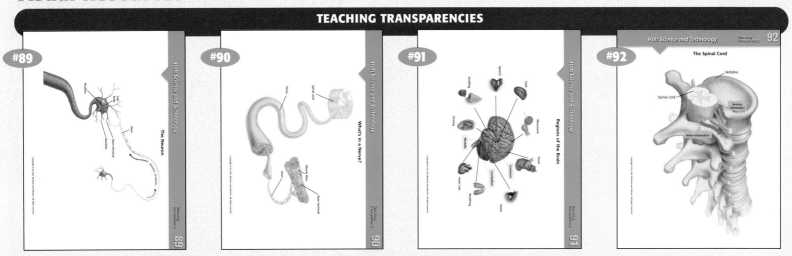

TEACHING TRANSPARENCIES

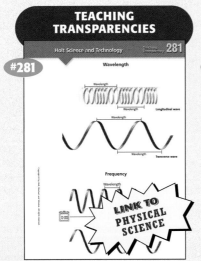

CONCEPT MAPPING TRANSPARENCY

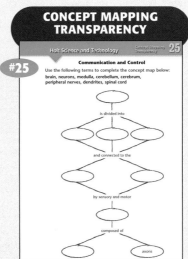

Meeting Individual Needs

DIRECTED READING

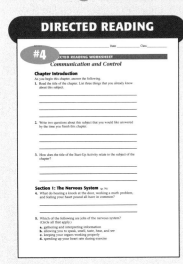

REINFORCEMENT & VOCABULARY REVIEW

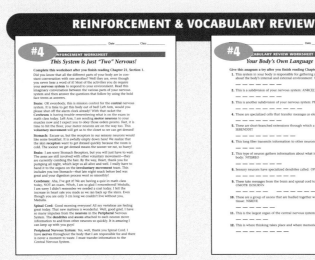

SCIENCE PUZZLERS, TWISTERS & TEASERS

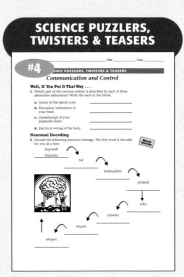

73C Chapter 4 • Communication and Control

Chapter 4 • Communication and Control

Review & Assessment

STUDY GUIDE

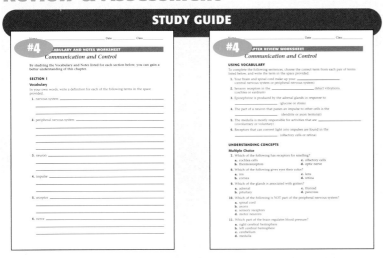

CHAPTER TESTS WITH PERFORMANCE-BASED ASSESSMENT

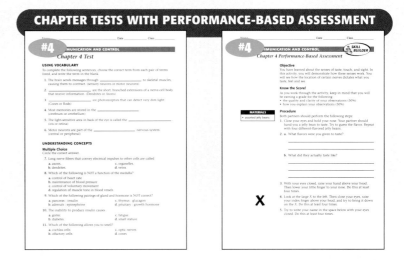

Lab Worksheets

WHIZ-BANG DEMONSTRATIONS

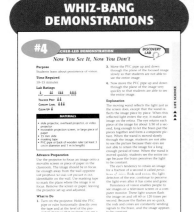

LABS YOU CAN EAT

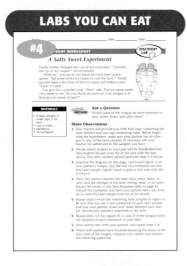

LONG-TERM PROJECTS & RESEARCH IDEAS

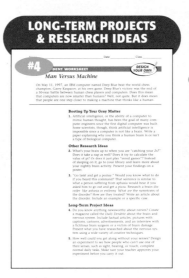

DATASHEETS FOR LABBOOK

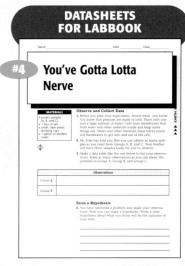

Applications & Extensions

CRITICAL THINKING & PROBLEM SOLVING

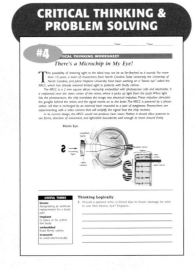

SCIENCE TECHNOLOGY

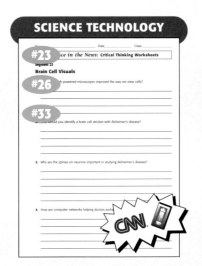

Chapter 4 • Chapter Resources & Worksheets

Chapter Background

SECTION 1

The Nervous System

▶ **The Brain**
The cerebrum's surface area is a more significant indicator of an organism's ability to think and perform than is the brain's volume or weight. The surface area of a porpoise's cerebrum relative to its body size is second to that of the human cerebrum.

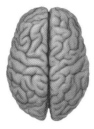

- Computerized scanning techniques allow physicians to take pictures of the brain to detect abnormalities. Scanning techniques include CT scanning, MRI scanning, radionucleotide scanning, ultrasound scanning, and PET scanning.

IS THAT A FACT!

- The cerebral cortex makes up more than 80 percent of the total human brain mass.

▶ **The Spinal Cord**
Like the brain, the spinal cord contains both gray matter and white matter. The center of the spinal cord is made up of neuron cell bodies and is called gray matter. The outer layer of the spinal cord is made up of axons that traverse the spinal cord. This part of the spinal cord is called white matter.

- The spinal cord is protected by 25 bones, the vertebrae and the sacrum. These bones are connected by joints and separated by cartilaginous disks.

SECTION 2

Responding to the Environment

▶ **Hearing Loss**
There are two principal kinds of deafness: conductive deafness and sensorineural deafness. Conductive deafness results when transmission of sound from the outer ear to the inner ear fails. It may occur as a result of earwax buildup or damage to the middle ear.

- Sensorineural deafness results when sounds reach the inner ear but are not transmitted to the brain due to either damaged inner ear structures or damaged nerves that carry information from the ear to the brain.

IS THAT A FACT!

- Sensorineural deafness occurs in 1 out of every 1,000 babies.

▶ **The Eye Doctor**
Ophthalmologists are physicians who specialize in the eyes. An ophthalmologist can examine eyes, prescribe corrective lenses, treat eye disorders, and perform eye surgery.

- An optician may only fit and adjust glasses and contact lenses.

- An optometrist can examine and test eyes and prescribe corrective lenses in the form of glasses or contact lenses.

IS THAT A FACT!

- The idea of using contact lenses to correct poor vision was first recorded by Leonardo da Vinci (1452–1519) in 1508.

- The first contact lens was made of glass. It covered the entire frontal surface of the eyeball. This first lens was made by Adolf Fick in 1887.

▶ **The Sense of Taste**
Saliva dissolves chemicals in the food and drink we consume. After passing through pores in the taste buds, these chemicals stimulate small nerve endings, which send messages to the brain. These messages form our sense of taste.

Chapter 4 • Communication and Control

- People often lose their sense of taste when they lose their sense of smell. This occurs when olfactory bulbs are damaged or when the person has a stuffy nose. It is rare for a person to maintain the sense of smell and to lose the sense of taste.

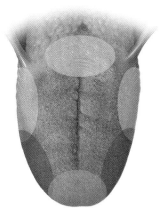

Section 3

The Endocrine System

▶ Endocrine Glands
There are two main types of glands in the body: exocrine glands and endocrine glands. Exocrine glands, such as sweat glands and salivary glands, secrete substances through ducts to a local area. Unlike exocrine glands, endocrine glands secrete substances directly into the bloodstream (no ducts are involved). The substances secreted by endocrine glands are carried, often to distant parts of the body, by the bloodstream and have effects throughout the body.

▶ The Pituitary Gland
The pituitary gland is often called the master gland because its secretions regulate several other endocrine glands.

- About 10 percent of brain tumors affect the pituitary gland. Although usually benign, these tumors can have a great effect on the body because they can affect the production of the pituitary hormones.

- Because of the pituitary gland's location in the brain, enlargement of the gland can cause vision disorders by creating pressure on the nearby optic nerve.

▶ Diabetes
The term *diabetes* refers to more than one disorder. There is diabetes insipidus (a rare condition) and diabetes mellitus, of which there are two types. Type I requires regular injections of insulin. Type II can often be controlled by changes in diet and exercise, although insulin injections might also be necessary.

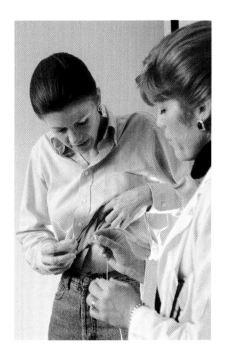

IS THAT A FACT!

- As many as 200 of every 100,000 people in the United States suffer from Type I diabetes.

- About 2,000 of every 100,000 people in the United States have Type II diabetes.

For background information about teaching strategies and issues, refer to the *Professional Reference for Teachers*.

Communication and Control

Pre-Reading Questions

Students may not know the answers to these questions before reading the chapter, so accept any reasonable response.

Suggested Answers

1. Answers will vary, but some students may suggest that the senses provide the brain with information that can help the individual respond appropriately to environmental changes.

2. Answers will vary, but some students may suggest that during a fight-or-flight response, your body needs extra blood supply to either run or fight for survival.

3. Lenses help bring images into focus. A concave lens corrects nearsightedness, and a convex lens corrects farsightedness.

CHAPTER 4

Communication and Control

Sections

1. The Nervous System ... 76
   - MathBreak 77
   - QuickLab 82
2. Responding to the Environment 83
   - QuickLab 85
   - Physics Connection ... 86
   - Internet Connect 87
3. The Endocrine System .. 88
   - Apply 89
   - Internet Connect 91
- Chapter Lab 92
- Chapter Review 96
- Feature Articles 98, 99

Pre-Reading Questions

1. What are your senses? How do senses help us survive?
2. Why does your heart beat faster when something frightens you?
3. How do eyeglasses and contact lenses help some people see better?

Outta Sight!

This may look like a flower garden or an oceanic reef. But it's really something much closer to home. It's the human tongue (magnified thousands of times, of course). You know these round bumps as *taste buds*. You use taste and other senses to gather information about your surroundings. This information helps your body respond to its environment. In this chapter, you will find out how the human body senses the world and controls its own functions.

internet connect

HRW On-line Resources

go.hrw.com

For worksheets and other teaching aids, visit the HRW Web site and type in the keyword: **HSTBD4**

www.scilinks.com

Use the *sci*LINKS numbers at the end of each chapter for additional resources on the **NSTA** Web site.

Smithsonian Institution

www.si.edu/hrw

Visit the Smithsonian Institution Web site for related on-line resources.

www.cnnfyi.com

Visit the CNN Web site for current events coverage and classroom resources.

ACT FAST!

If you want to catch an object, your brain sends a message to your arm's muscles. In this exercise, you will see how long that takes.

Procedure

1. Sit in a **chair** with one arm in a "handshake" position. Your partner should stand facing you, holding a **meterstick** vertically. The stick should be positioned to fall between your thumb and fingers.

2. Tell your partner to let go of the meterstick without warning. Catch the stick between your thumb and fingers. Your partner should catch the meterstick if it tips over.

3. Record the number of centimeters the stick dropped before you caught it. That distance represents your reaction time.

4. Repeat steps 1–3 three times. Calculate the average distance.

5. Repeat steps 1–4 with your other hand.

6. Trade places with your partner, and repeat steps 1–5.

Analysis

7. Compare the reaction times of your own hands. Why might one hand react faster?

8. Compare your results with your partner's. Why might one person react faster than another?

START-UP Activity

ACT FAST!

MATERIALS

For Each Group:
- meterstick

Safety Caution

Remind students to handle the meterstick carefully and to keep it away from their face and far away from the faces and eyes of their classmates. Allow students to have a practice trial. Instruct students to look only at the ruler and not at their partner.

Teacher's Notes

Point out to students that looking at their partner could distort the results of the investigation because the partner might give a clue about when he or she will drop the meterstick. Remind students to use the correct units of measurement when calculating average distances.

Answers to START-UP Activity

7. Answers will vary, but some students might notice a difference between the reaction time of the left hand and that of the right hand. We use each hand differently, so each has different abilities.

8. Student answers will vary, but possible inferences include how rested an individual feels or if distractions occur in the room.

SECTION 1

Focus

The Nervous System

This section introduces the structures and functions of the nervous system. Students will learn the differences between the central nervous system, which includes the brain and spinal cord, and the peripheral nervous system, which consists of nerves that connect every area of the body to the central nervous system.

Bellringer

Ask students to list as many different functions of the brain as they can in their ScienceLog. After students complete their list, you may want to make a master list on the board or on a transparency. Explain that in this chapter, students will learn how the brain coordinates these many different activities.

1) Motivate

DISCUSSION

Reacting to Stimuli Invite students to describe a time when they reacted quickly. Encourage students to describe not only what happened but also how quickly they were able to react and what they were thinking about as they reacted. Sample experiences include jerking a hand away from a hot object, quickly catching a falling object, and extending one's hand out to brace for a fall. Based on students' experiences, lead a discussion about how quickly the nervous system is able to respond to a stimulus.

Sheltered English

SECTION 1
READING WARM-UP

Terms to Learn

central nervous system
peripheral nervous system
neuron
impulse
receptor
nerve
brain
reflex

What You'll Do

- Explain how neurons in the nervous system work together.
- Compare the central nervous system with the peripheral nervous system.
- Describe the major functions of the brain and the spinal cord.

The Nervous System

What do the following events have in common? You hear a knock at the door, you write a book report, you feel your heart pounding after a run, you work a math problem, you are startled by a loud noise, and you enjoy eating a sweet mango. These events are all activities of your nervous system. The nervous system gathers and interprets information about the body's internal and external environments and responds to that information. The nervous system keeps your organs working properly and allows you to speak, smell, taste, hear, see, move, think, and experience emotions.

Two Systems Within a System

Your nervous system controls and coordinates many things that happen in your body. It acts as a central command post, collecting and processing information and making sure appropriate information gets sent to all parts of the body. These tasks are accomplished by two subdivisions of the nervous system, the *central nervous system* and the *peripheral nervous system*. The **central nervous system** (CNS) includes your brain and spinal cord. It processes all incoming and outgoing messages. The **peripheral nervous system** (PNS) consists of communication pathways, or *nerves*, that connect all areas of your body to your CNS. **Figure 1** shows the major divisions of the nervous system.

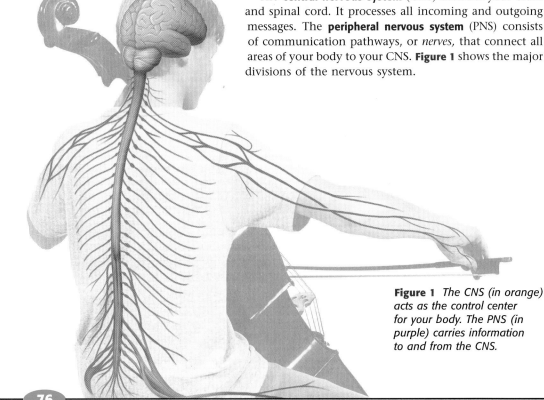

Figure 1 The CNS (in orange) acts as the control center for your body. The PNS (in purple) carries information to and from the CNS.

Human skulls from 20,000 years ago provide evidence that ancient humans cut and drilled holes into each other's heads. This ancient practice is called trephining, and it may have been intended to release evil spirits believed to cause mental problems or illnesses such as migraines or epilepsy. Evidence of this surgery has been found in skulls from Europe, North Africa, parts of Asia, New Zealand, some Pacific Islands, and South America.

76   Chapter 4 • Communication and Control

The Peripheral Nervous System

How long does it take for a light to come on when you flip a light switch? The light seems to come on immediately. In a similar way, specialized cells called **neurons** transfer messages throughout your body in the form of fast-moving electrical energy. A typical neuron is shown in **Figure 2**. The electrical messages that pass along the neurons are called **impulses.** Impulses may travel as fast as 150 m/s or as slow as 1 m/s.

Neuron Structure A neuron consists of a cell body, dendrites, and axons. The enlarged region called the cell body contains a nucleus and cell organelles. Look again at Figure 2. The neuron generally receives information from other cells through short, branched extensions called *dendrites*. A neuron may have many dendrites, allowing it to receive impulses from thousands of other cells.

From the cell body, information is transmitted to other cells by a fiber called an *axon*. Axons can be very short or quite long. You have some really long axons that extend almost 1 m from your lower back to your toes. The end of an axon often has branches that allow information to pass to yet more cells. The tip of each branch is called an *axon terminal*.

MATH BREAK

Time to Travel

To calculate how long it takes for an impulse to travel a certain distance, you can use the following equation:

$$\text{Time} = \frac{\text{distance}}{\text{speed}}$$

If an impulse travels 100 m/s, about how long would it take for an impulse to travel 10 m?

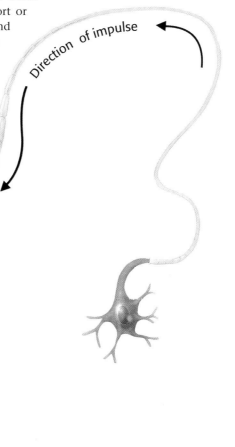

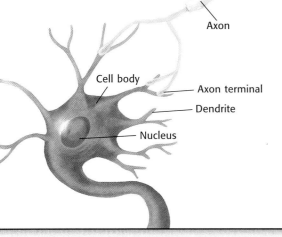

Figure 2 *Neurons are special cells that transfer electrical messages throughout the body.*

IS THAT A FACT!

Unlike humans, canaries replace old brain cells with new neurons each year. Male canaries even sing a new song every spring. Their brain-cell clusters associated with vocalization grow larger during the spring, when males compose their new melodies and females learn to recognize them. In the fall, the brain clusters shrink, neurons die, and the males forget what they sang. Scientists theorize that this happens so that these birds can acquire new information without having to carry around a large and heavy brain.

2) Teach, continued

MEETING INDIVIDUAL NEEDS

Learners Having Difficulty
To help students visualize the central and peripheral nervous systems, have them work in pairs to trace the outline of one student's body on butcher paper. Next have each pair fill in the outline, using different colors for each of the nervous systems. Models should include the brain and the spinal cord (the central nervous system), and sensory and motor neurons throughout the body (the peripheral nervous system). **Sheltered English**

DEMONSTRATION

Simulating Neuronal Impulses
Ask students to form a circle and hold hands. Explain that each person in the circle represents a neuron. Every left hand represents a dendrite, every body represents a cell body, and every right hand represents an axon. Join the circle, and initiate a nerve impulse by gently squeezing the hand of the student to your right. Instruct students to pass the nerve impulse to the person to their right by gently squeezing his or her hand. Once students understand the mechanics of the activity, have them call out *dendrite, cell body,* and *axon* as the impulse is passed along the circle. **Sheltered English**

Teaching Transparency 90
"What's in a Nerve?"

Reinforcement Worksheet
"This System Is Just 'Two' Nervous!"

BRAIN FOOD
The number of neurons in your brain is about 100 billion, which is about the same as the number of stars in the Milky Way galaxy!

Information Collection Special neurons called *sensory neurons* gather information about what is happening in and around your body and send this information on to the central nervous system for processing. Sensory neurons have specialized dendrites called **receptors** that detect changes inside and outside the body. For example, receptors in your eyes detect the light around you. Receptors in your stomach let your brain know when your stomach is full or empty.

Delivering Orders Neurons that send impulses from the brain and spinal cord to other systems are called *motor neurons*. Motor means "to move"; when muscles get impulses from motor neurons, they respond by contracting. For example, motor neurons cause the muscles around your eyes to move when the sensory neurons in your eyes detect bright light. This movement makes you squint, which reduces the amount of light entering the eye. Motor neurons also send messages to your glands, such as sweat glands. These messages tell the sweat glands to release sweat.

Just a Bundle of Axons

The central nervous system is connected to the rest of your body by nerves. **Nerves** are axons bundled together with blood vessels and connective tissue. Nerves extend throughout your body. Most nerves contain the axons of both sensory and motor neurons. **Figure 3** shows the structure of a nerve. The axon in this nerve transmits information from the spinal cord to muscle fibers.

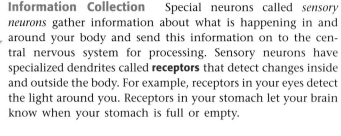

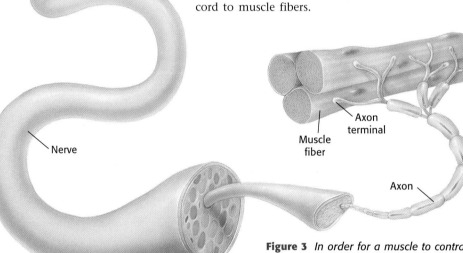

Figure 3 *In order for a muscle to contract, a message must travel from the spinal cord to the muscle. The message travels along the axon of a motor neuron inside the nerve.*

internet connect
SC LINKS **TOPIC:** The Nervous System
GO TO: www.scilinks.org
*sci*LINKS NUMBER: HSTL605

SCIENCE HUMOR
Q: How do nerves shop?

A: They buy only on impulse.

The Central Nervous System

The central nervous system works closely with the peripheral nervous system. It receives information from the sensory neurons and responds by sending messages to various parts of the body via motor neurons.

Mission Control The **brain,** part of your central nervous system, is the nervous system's largest organ. It has hundreds of different jobs. Many of the processes that the brain controls happen automatically and are referred to as *involuntary.* For example, you couldn't stop digesting the food you have eaten even if you tried. Other activities controlled by your brain are *voluntary.* When you want to move your arm, your brain sends signals along motor neurons to muscles in your arm. This causes the muscles to contract and your arm to move. The brain has three connected parts—the cerebrum, the cerebellum, and the medulla. Each part has its own functions.

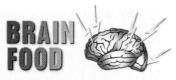

The organism with the largest brain is the sperm whale. Its brain is six times the size of a human brain!

Your Thinking Cap The largest part of your brain is called the *cerebrum.* Its shape resembles a mushroom cap over a stalk. This dome-shaped area is where you think and where most memories are stored. It controls voluntary movements and allows you to detect touch, light, sound, odors, taste, pain, heat, and cold.

The cerebrum has two halves called *hemispheres.* The left hemisphere directs the right side of the body, and the right hemisphere directs the left side of the body. This is because axons cross over to the opposite side of the body in the spinal cord. **Figure 4** gives a general model of the activities that each hemisphere controls. However, most brain activity involves both hemispheres.

Figure 4
The Cerebral Hemispheres

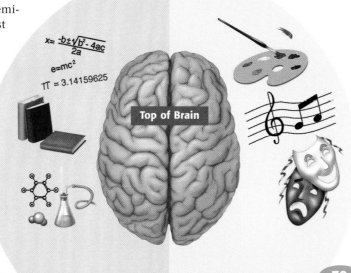

▼ The **right hemisphere** primarily controls activities that involve imagination, appreciation, and creativity.

The **left hemisphere** ▶ primarily controls activities such as speaking, reading, writing, and solving problems.

IS THAT A FACT!

What is most amazing about the human brain is not its size—a sperm whale's brain is about six times larger. It is the high proportion of brain size to body size as well as the huge surface area. The cerebral hemispheres are folded and wrinkled; but laid out flat, the brain would cover the surface of an office desk. This large surface provides room for highly complex and sophisticated connections.

READING STRATEGY

Prediction Guide Before students read this section, ask them whether the following statements are true or false.

1. The brain is the body's largest organ. (false)
2. The largest part of the brain is the cerebrum. (true)
3. The medulla is responsible for speech and balance. (false)
4. The spinal cord is about as big around as your thumb. (true)

Have students evaluate their answers after they have read these pages.

REAL-WORLD CONNECTION

Students have probably watched people faint in movies or television shows. Point out that fainting and other forms of unconsciousness, except sleeping, are usually the result of some problem in the brain. Fainting is often caused by suddenly low blood pressure and insufficient blood flow to the cerebrum.

2 Teach, continued

RETEACHING

If students have difficulty distinguishing the structures of the brain, make a life-size model of the brain as a class. Invite two volunteers to make a silhouette of the head; one student should be the model; the other, the tracer. Then draw in a brain. Have students take turns drawing and labeling the parts of the brain on the silhouette. Include the cerebrum, cerebellum, medulla, and the top of the spinal cord. Also have students label the hemisphere that is shown in the diagram. When the life-size model is complete, post it so students can refer to it as they review this section.
Sheltered English

Early anatomists Herophilus (C.335–C.280 B.C.) and Eristratus (C.276–c.194 B.C.) were experts at dissection, and they produced extensive work on human anatomy and physiology that is now on display at the Museum at Alexandria, in Egypt. Their dissections were discontinued, however, because of the Egyptian belief that the body must be kept intact for the afterlife. Another 15 centuries passed before dissection again was used to study human anatomy. You may want to have students compare the Egyptian views about dissection with modern-day views about organ donation.

Teaching Transparency 91
"Regions of the Brain"

The Balancing Act The second largest part of your brain is the *cerebellum* (SER uh BEL uhm). It lies underneath the back of your cerebrum and receives sensory impulses from skeletal muscles and joints. This allows the brain to keep track of your body's position. For example, if you begin to lose your balance, like the girl in **Figure 5,** the cerebellum sends impulses to different skeletal muscles to make them contract, keeping you upright.

The Mighty Medulla The part of your brain that connects to your spinal cord is called the *medulla* (mi DOOL uh). The medulla is only about 3 cm long, but you couldn't live without it. The medulla controls your blood pressure, heart rate, involuntary breathing, and some other involuntary activities.

Your medulla constantly receives sensory impulses from receptors in your blood vessels. It uses this information to regulate your blood pressure. If your blood pressure gets too low, the medulla sends out impulses that tell blood vessels to tighten up to increase the blood pressure. The medulla also sends impulses to the heart to make it beat faster or slower as necessary. **Figure 6** shows the location of each part of the brain and some of the functions associated with each part.

Figure 5 Your cerebellum causes skeletal muscles to make adjustments in order to keep you upright.

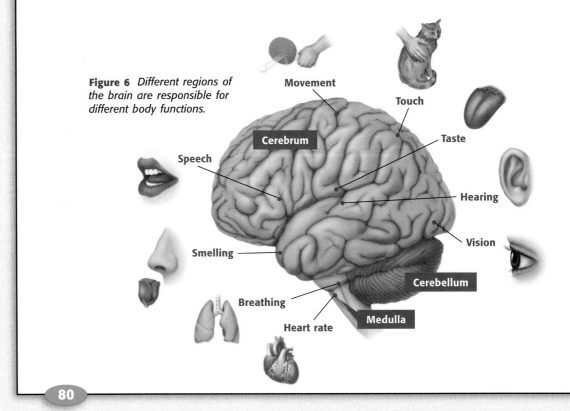

Figure 6 Different regions of the brain are responsible for different body functions.

Homework

Dream Research Dreaming, sleepwalking, and daydreaming are all phenomena of the brain. Have students research one of these topics and give group presentations before the class. They may use posters, signs, songs, skits, oral reports, and other techniques for their presentations.

IS THAT A FACT!

Synapses form in a human baby's brain at the rate of 3 billion a second. At 8 months, a baby's brain has about 1,000 trillion connections. After that, the number begins to decline. Half the connections die off by the time the child reaches age 10, leaving about 500 trillion.

The Spinal Cord

Your spinal cord, part of the central nervous system, is about as big around as your thumb. It contains neurons and bundles of axons that pass impulses to and from the brain. As shown in **Figure 7,** the spinal cord is surrounded by protective bones called *vertebrae* (VUHR tuh BRAY).

The nerve fibers in your spinal cord enable your brain to communicate with your peripheral nervous system. Sensory neurons in your skin and muscles send impulses along their axons to your spinal cord. The spinal cord then conducts impulses to your brain, where they can be interpreted as pain, heat, cold, or other sensations. Impulses moving from the brain down the cord are relayed to motor neurons, which carry the impulses along their axons to muscles and glands all over your body.

Spinal Cord Injury If the spinal cord is injured, any sensory information coming into it below where the damage occurred may be unable to travel to the brain. Likewise, any motor commands the brain sends to an area below the injury may not get through to the peripheral nerves. Thousands of people each year are paralyzed by spinal cord injuries. Many of these injuries occur in automobile accidents. Among young people, spinal cord injuries are often sports related.

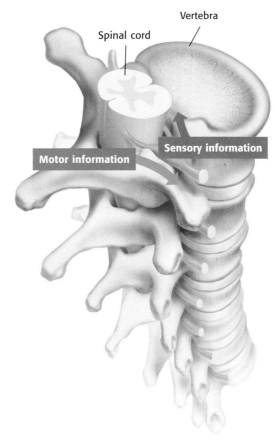

Figure 7 *The spinal cord carries information to and from the brain. It is protected by vertebrae.*

Self-Check

1. What part of the brain do you use to do your math homework?
2. What part of the brain helps a gymnast maintain balance on the balance beam?
3. What is the function of the vertebrae?

(See page 212 to check your answers.)

3) Extend

GUIDED PRACTICE

Concept Mapping
List the following terms on the board:

brain, hemispheres, cerebrum, cerebellum, medulla, spinal cord, central nervous system, peripheral nervous system, sensory neurons, motor neurons

Have students construct a concept map that uses these terms.

CROSS-DISCIPLINARY FOCUS

Language Arts Long-term memory enables us to recall events that happened to us long ago. These memories can be triggered by a chance stimulus or can be deliberately recalled. Recalling memories helps us to refresh them and makes them last a lifetime. Ask students to write down their earliest memory, their happiest recollection, or their most embarrassing moment in their ScienceLog.

Answer to Self-Check

1. cerebrum
2. cerebellum
3. to protect the spinal cord

Teaching Transparency 92
"The Spinal Cord"

IS THAT A FACT!

Sometimes it is not possible to inject anesthetic into a part of the body that needs to be anesthetized. In these cases, a *nerve block* is performed. In this procedure, anesthetic is injected into or around a nerve that feeds into the part of the body that needs to be anesthetized. Nerve blocks are often performed on nerves that carry messages away from the spine. This blocks the pain impulses before they reach the brain.

4 Close

Quiz

Ask students whether these statements are true or false.

1. Typically, a dendrite will transmit an electrical impulse to a neighboring axon. (false)
2. Most neurons are made up of either axons or dendrites; few neurons have both. (false)
3. The central nervous system is made up of the brain and the spinal cord. (true)

ALTERNATIVE ASSESSMENT

 Writing Have students develop an owner's guide for their central nervous system. The guide should include information about the various components of the central nervous system and a diagram showing their location in the body. Encourage students to share their guide with the class in the form of a presentation or a poster.

QuickLab

Safety Caution: Students should tap their partner's knee gently. There is no pain involved in a true reflex. Instruct students to tap while standing to the side of their partner to avoid being accidentally kicked.

Answers to QuickLab

2. Students should observe movement in the lower leg; no, an individual has no control over a reflex.
3. Students should indicate that the impulse traveled from the knee to the spinal cord and back to the thigh muscle that moves the lower leg.

Ouch! That Hurt!

Have you ever stepped on something sharp? You probably pulled your foot up without thinking. This quick, involuntary action is called a **reflex.** Reflexes help protect your body from damage.

When you step on a sharp object, the message "pain" travels to your spinal cord, and a message to move your foot travels back to the muscles in your leg. The muscles in your leg respond before the information ever reaches the brain. By the time your brain finds out what happened, your foot has already moved. If you had to wait for your brain to get the message, your foot might be seriously injured! The man in **Figure 8** lifted his foot before he realized he had stepped on a toy.

Figure 8 When pain impulses from your foot reach your spine, a message is sent immediately to your leg muscles to lift your foot.

QuickLab

Knee Jerks

1. Sit on the edge of a desk or table so your feet don't touch the floor.
2. While your leg is completely relaxed, have a classmate *gently* tap on your knee slightly below the kneecap with the edge of his or her hand. How did your leg respond? Did you have any control over what happened? Explain.
3. Describe the pathway taken by the impulse that started with the tap on the knee.

SECTION REVIEW

1. Make a labeled diagram that shows the path of an electrical message from one neuron to another neuron.
2. Explain how the peripheral nervous system connects with the central nervous system.
3. If a spider is crawling up your left arm, which cerebral hemisphere controls the movement that you will use to knock it off?
4. List the three major parts of the brain, and describe their functions.
5. **Applying Concepts** Describe a time when you experienced a reflex.

▼ **Answers to Section Review**

1. Diagrams will vary but should show the following path: dendrite, cell body, axon, axon terminal, dendrite.
2. The peripheral nervous system consists of communication pathways, or nerves, that connect all areas of your body to your central nervous system.
3. You would use the left hemisphere, since you would have to use your right hand to knock it off.
4. The three major parts of the brain are the cerebrum, cerebellum, and medulla. Functions are described on pages 79 and 80.
5. Answers will vary but should demonstrate an understanding that reflexes involve involuntary muscle actions.

SECTION 2

Reading Warm-Up

Terms to Learn
- retina
- rods
- cones
- iris
- lens
- cochlea

What You'll Do
- List four sensations that are detected by receptors in the skin.
- Describe how light relates to vision.
- Explain the functions of photoreceptors, taste buds, and olfactory cells.

Responding to the Environment

How do you know when someone taps you on the shoulder or calls your name? How do you feel the touch or hear the sound? Impulses from sensory receptors in your shoulder and in your ears travel to your brain, sending information about your external environment. Your brain depends on this information to make decisions that affect your survival.

Come to Your Senses

Information about your surroundings and the conditions in your body is detected by sensory receptors. This information is converted to electrical signals and sent to your brain for interpretation. Once the signals reach your brain, you become aware of them. This awareness is called a *sensation*. It is in your brain that you have thoughts, feelings, and memories about sensations.

There are many different kinds of sensory receptors in your body. For example, receptors in your eyes detect light. Receptors in your ears detect vibrations called sound waves. The taste buds on your tongue have receptors that detect chemicals in the foods you eat. You have special receptors in your nose that detect tiny particles in the air. Your skin has a variety of receptors as well. Look at **Figure 9** to see some of the different kinds of receptors in the skin.

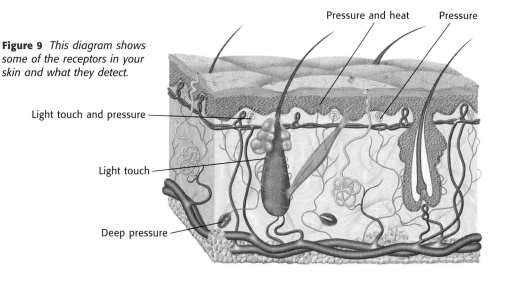

Figure 9 This diagram shows some of the receptors in your skin and what they detect.

WEIRD SCIENCE

One out of every 25,000 people has *synesthesia*. These people experience a blend of senses. They can hear colors, taste shapes, or other combinations of senses. When one of their senses is stimulated, another sense is also stimulated.

internet**connect**

SC*L*INKS
NSTA

TOPIC: The Senses
GO TO: www.scilinks.org
*sci*LINKS NUMBER: HSTL610

2) Teach

REAL-WORLD CONNECTION

Vision is commonly tested by using a Snellen's chart. This familiar chart consists of rows of letters set in decreasing size. A person who can correctly read a line of letters near the bottom of the chart from a distance of 6 m (20 ft) is said to have normal, or 20/20, vision. **Sheltered English**

MISCONCEPTION ALERT

Students may think that people who are colorblind see in black and white. People who are colorblind can usually perceive colors, but certain colors may appear very similar to one another. This is caused by a lack of at least one of the three cones in the eye. Many people do not know that they are colorblind because they have learned to distinguish other differences in their perception of colors.

CONNECT TO PHYSICAL SCIENCE

In the retina there are three types of cones that only respond to red, blue, or green light. When red cones and green cones are stimulated at the same time, the color yellow is perceived. White is perceived when all three kinds of cones are equally stimulated. Different colors of visible light have different wavelengths. Red has the longest wavelength, and violet has the shortest. Use the Teaching Transparency below to illustrate how wavelengths are measured.

Teaching Transparency 281 "Wavelength" LINK TO PHYSICAL SCIENCE

BRAIN FOOD

Carrots and other foods rich in vitamin A can improve your night vision. Vitamin A is important in maintaining proper functioning of the rods in your retina.

Something in My Eye

As you read this sentence, you are using one of your most important senses—*vision*. Vision is your awareness of light energy. Your eyes have special receptors that detect visible light, a portion of the sun's energy that reaches the Earth.

An Eyeful The eye is a complex sensory organ. Examine the eye in **Figure 10**. The outer surface of the eye is covered by the cornea, a transparent membrane that protects the eye but allows light to enter. Visible light that is reflected by objects around you enters through an opening at the front of your eye called the *pupil*. Light is detected by cells at the back of your eye in a light-sensitive layer called the **retina.**

The retina is packed with special neurons called *photoreceptors* (*photo* means "light") that convert light into electrical impulses. There are two types of photoreceptors in the retina—rods and cones. **Rods** can detect very dim light. They are important for night vision. Impulses from rods are perceived in tones of gray. In bright light, the **cones** give you a very colorful view of the world.

Light energy produces changes in photoreceptors that trigger nerve impulses. These impulses travel along axons, leaving the back of each eye through an *optic nerve*.

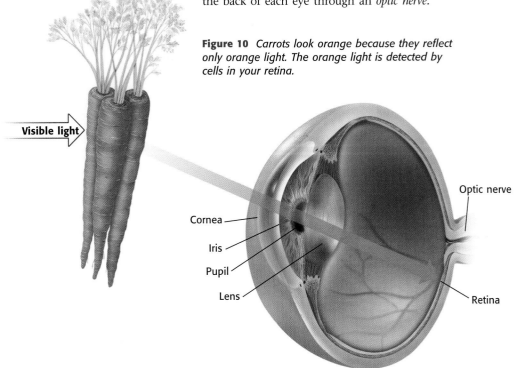

Figure 10 Carrots look orange because they reflect only orange light. The orange light is detected by cells in your retina.

 **SCIENCE**

Have students research and report on the tuatara, a reptile that has three eyes.

84 Chapter 4 • Communication and Control

Seeing the Light Light rays enter the eye through the *pupil*. Your pupil looks like a black dot in the center of your eye, but it is actually an opening. It is surrounded by the **iris,** the part of the eye that gives the eye color. A ring of muscle fibers causes the iris to open and close, making the pupil change size. This regulates the amount of light that passes to the retina. In bright light, your pupil is small, and in dim light, your pupil is large.

Hocus Focus Light travels in straight lines until it passes through the cornea and the *lens*. A **lens** is a piece of curved material behind the pupil that allows light to pass through but changes its direction. The lens focuses the light entering the eye on the retina. The lens of an eye changes shape to adjust focus. When you look at objects close to the eye, the lens becomes more curved. When you look at objects far away, the lens gets flatter.

In some eyes, the lens focuses the light just in front of the retina (resulting in nearsightedness) or just behind the retina (resulting in farsightedness). Glasses or contact lenses can usually correct these vision problems. Focus on **Figure 11** to see how corrective lenses work.

What are some other uses of lenses? Turn to Light on Lenses on page 98 to find out.

Figure 11 *A concave lens bends light rays outward to correct nearsightedness. A convex lens bends light rays inward to correct farsightedness.*

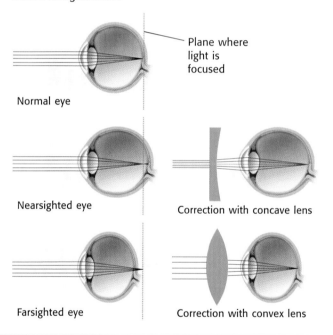

QuickLab

Where's the Dot?

1. Hold this book at arm's length, and close your right eye. Focus your left eye on the solid dot below.

2. Slowly move the book toward your face. Stay focused on the solid dot.
3. What happens to the white dot?
4. Do some research on the optic nerve to find out why this happens.

TRY at HOME

ACTIVITY

Pupil Action Ask students to write a paragraph that explains what happens to their eyes and vision when they first leave a dark movie theater on a sunny day. Students should discuss the change from dim to bright light and its effects on the pupils as well as on their ability to see.

DISCUSSION

Eye Strain For us to see objects within a distance of 6 m, the muscles in the eye must work constantly to focus. Long periods of focusing on very near objects, such as a book, can tire the eye muscles and cause eyestrain. Ask students to suggest ways to avoid straining the eye muscles during these activities. (Answers may include taking breaks from the activity once an hour and looking up and allowing the eyes to relax occasionally.)

Answers to QuickLab

3. Students should observe that the white dot "disappears" when the image is held about 10 cm from their face. The exact point where this happens will vary from person to person. Encourage the students to try this several times until they get this result.

4. Students should discover that the area where the optic nerve leaves the back of the eyeball does not contain any photoreceptors and is called the blind spot. So when the white dot was focused on the eye's blind spot, it disappeared.

internetconnect

TOPIC: The Eye
GO TO: www.scilinks.org
*sci*LINKS NUMBER: HSTL615

IS THAT A FACT!

Students may have noticed that young people are likely to hold a book very close to their face when reading, while adults tend to hold a book farther away from their face. Explain that the closest point on which a person can focus, called the near point of vision, changes with age. A typical child can focus on an object about 10 cm from his or her eyes (objects closer than the near point of vision can be seen but appear fuzzy). The change in the near point of vision is caused by the lens's decreasing elasticity over time.

Section 2 • Responding to the Environment

3) Extend

RESEARCH

Writing Prolonged exposure to loud sounds can result in a partial or complete loss of hearing. Have students research at what threshold noises are detrimental to hearing. Encourage students to arrange the information they gather into a report or a visual aid. (Students should find that sounds greater than 90 dB can cause hearing loss over long periods of time.)

MATH and MORE

Tell students that taste buds are shed approximately every 10 days. Ask students how many times a single taste bud is replaced in a year. (365 $\frac{days}{year}$ ÷ 10 $\frac{days}{cycle}$ = 36.5 replacements in 1 year)

BRAIN FOOD

Owls have great hearing. In fact, some owls actually hunt by listening for mice tunneling under the snow. They have many adaptations that help them to hear so well. Feathers on the side of an owl's head are arranged to direct sound waves to very large ear openings. Furthermore, the feathers on an owl's face are rigid, and tightly packed into two parabolic formations which also direct sound waves toward the ears. This system works so well that owls can find mice in complete darkness and, according to experiments, are great hunters even when blindfolded.

Physics CONNECTION

Sound is produced by vibrating objects. Objects that vibrate more than 20,000 times a second produce sounds too high for humans to hear. Dolphins can detect sounds from objects that vibrate up to 150,000 times per second.

Did You "Ear" That?

When a guitar string is plucked, what enables you to hear the sound? A sound begins when an object, such as the guitar string, begins to vibrate. The vibrations push on surrounding air particles. These air particles push on other air particles, transferring energy in waves away from the source. Hearing is the sensation experienced in response to these sound waves.

Journey of a Sound Wave Your ears are organs specialized for hearing. Each ear has an outer, middle, and inner part. The parts of the ear are shown in **Figure 12**. When sound waves reach the outer ear, they are funneled into the middle ear, where they cause the eardrum to vibrate. The vibrating eardrum makes tiny ear bones vibrate. One of the tiny bones vibrates against the **cochlea** (KAHK lee uh), a tiny snail-shaped organ of the inner ear. Inside the cochlea, the vibrations create waves that are similar to the waves you can make by tapping on a glass of water. Neurons in the cochlea convert these waves to electrical impulses and send them to the area of the brain that interprets sound.

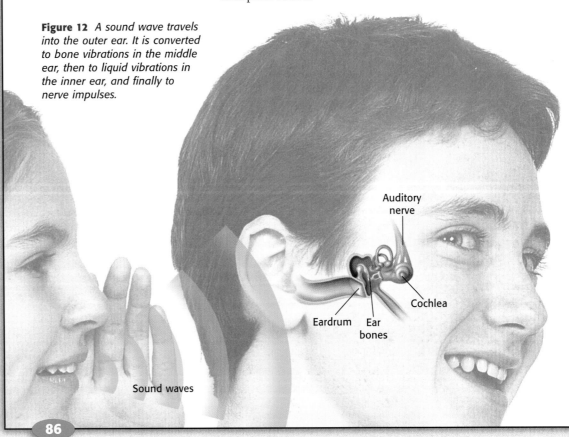

Figure 12 A sound wave travels into the outer ear. It is converted to bone vibrations in the middle ear, then to liquid vibrations in the inner ear, and finally to nerve impulses.

IS THAT A FACT!

Although many people think that the perception of taste occurs primarily in the mouth, 80 percent of the perception of taste is actually the sense of smell.

WEIRD SCIENCE

Some spicy foods, such as chile peppers, actually stimulate pain receptors in the mouth. This is why spicy foods feel like they are "burning" your mouth.

Does This Suit Your Taste?

When you put food in your mouth, your sense of what the food tastes like comes mostly from your tongue. Taste is the sensation you feel when the brain is made aware of certain dissolved chemicals in your mouth. The receptors for taste are clustered in the *taste buds*. The tongue is covered with tiny bumps called *papillae* (puh PIL ee), and the taste buds are embedded in the sides of these bumps. As shown in **Figure 13,** there are four types of taste buds. Each type responds to one of the four basic tastes: sweet, sour, salty, and bitter.

Your Nose Knows

Have you ever noticed that when you have a congested nose you can't taste food very well? Try eating a piece of peppermint while holding your nose. The mint taste is not very intense until you inhale through your nose. That's because smell and taste are closely related. The brain combines information from your taste buds and nose to give you a sense of flavor. The receptors for smell are located on *olfactory cells* in the upper part of your nasal cavity. They react to chemicals that are inhaled and dissolved in the moist lining of the nasal cavity. The woman in **Figure 14** is using her sense of smell to test the effectiveness of underarm deodorants.

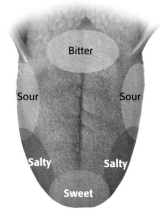

Figure 13 *Taste buds in different parts of the tongue respond to different types of chemicals.*

Figure 14 *This woman's nose is detecting chemicals in the sweat and in the deodorants used by this man. Her brain will generate opinions about the smells that she will then record in her report.*

SECTION REVIEW

1. List three sensations that receptors in the skin can detect.
2. Explain why you would have trouble seeing bright colors at a candlelit dinner.
3. How is your sense of taste similar to your sense of smell?
4. **Applying Concepts** If you can focus on objects close to you but things become blurry when they are far away, would a concave or convex lens correct your vision?

internet connect

SC*LINKS*
NSTA

TOPIC: The Senses, The Eye
GO TO: www.scilinks.org
*sci*LINKS NUMBER: HSTL610, HSTL615

SECTION 3

Focus

The Endocrine System

This section introduces students to the endocrine system. Students will learn how endocrine glands control the body's slower, long-term processes via hormones. They will also learn the location and function of specific endocrine glands. Hormonal imbalances that cause diabetes and goiter are also discussed.

Bellringer

Write the following on the board or overhead projector:

Unscramble the following words, and write them in your ScienceLog:

nalgd	(gland)
meornoh	(hormone)
noclotr	(control)

1 Motivate

DISCUSSION

Endocrine System Ask students how their pulse and breathing rate differed before and after being scared. (Both should be elevated.)

Tell students that the endocrine system is responsible for the changes that occurred in their pulse and breathing rate. Ask the students in each pair to tell their partner about a time when they were frightened and to describe the physical responses they had. Make a list on the board of the types of physical responses named. Return to this list when students read about the adrenal glands and the fight-or-flight response in this section.

SECTION 3
READING WARM-UP

Terms to Learn

endocrine system
gland
hormone
feedback control

What You'll Do

♦ Explain the function of the endocrine system.
♦ List the glands of the endocrine system and describe some of their functions.
♦ Describe how feedback controls stop and start hormone release.

The Endocrine System

You already know that the job of the nervous system is to communicate with all the other body systems. Its main role is to respond to stimuli. But it is not the only system that has this role. Your **endocrine system** is involved with the control of slower, long-term processes, such as fluid balance, growth, and sexual development. Instead of electrical messages, the endocrine system sends messages via chemicals.

Chemical Messengers

The endocrine system controls body functions with the use of chemicals that are released from endocrine glands. A **gland** is a group of cells that makes special chemicals for your body. Chemicals that are produced by the endocrine glands are called **hormones**. The chemicals made by endocrine glands are released into the bloodstream and carried to other places in the body. Because hormones act as chemical messengers, an endocrine gland near your brain can control the actions of an organ located somewhere else in your body.

Glands at Work Endocrine glands often affect many organs at one time. For example, your adrenal glands prepare your organs to deal with stress. They make the hormone *epinephrine* (ep ih NEF rihn), also known as *adrenalin*. Epinephrine speeds up your heartbeat and breathing rate to prepare your body either to run from danger or to fight for survival. This hormone effect is often referred to as the fight-or-flight response. You may have noticed these effects when you were frightened or angry.

Figure 15 When you have to move quickly to avoid danger, your adrenal glands help by making more blood glucose available for energy.

IS THAT A FACT!

Epinephrine occurs naturally in the human body, but it is also administered as a drug by doctors. It can be injected into the heart to help revive a person who has suffered a heart attack. It also dilates the bronchioles of people with asthma.

88   Chapter 4 • Communication and Control

Fight or Flight?

Maria was working late at the library. She was worried about walking home alone. As she started home, she noticed a shadowy figure walking quickly behind her. The figure was gaining on her! She could feel her heart pounding in her chest. She began to run, and then a familiar voice called out her name. It was her father. He had walked to the library to check on her. What a relief!

Maria had a fight-or-flight response. Write a paragraph describing a time when you had a fight-or-flight experience. Include in your story the following terms: *hormones*, *fight-or-flight*, and *epinephrine*.

Your body has several other endocrine glands, some with many different functions. For example, your pituitary gland stimulates skeletal growth, helps the thyroid function properly, regulates the amount of water in the blood, and stimulates the birth process in pregnant women. The names and some of the functions of this and other endocrine glands are summarized in **Figure 16.**

Figure 16 *Your endocrine glands produce chemicals called hormones that control many of your body functions.*

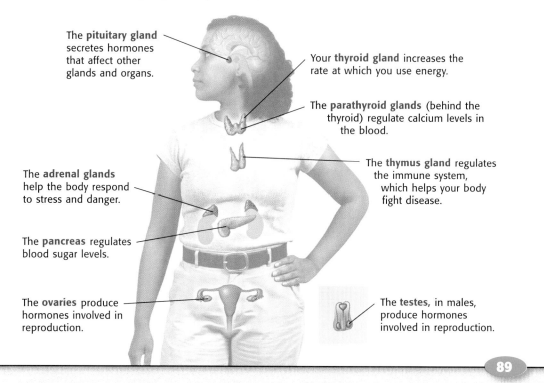

The **pituitary gland** secretes hormones that affect other glands and organs.

Your **thyroid gland** increases the rate at which you use energy.

The **parathyroid glands** (behind the thyroid) regulate calcium levels in the blood.

The **adrenal glands** help the body respond to stress and danger.

The **thymus gland** regulates the immune system, which helps your body fight disease.

The **pancreas** regulates blood sugar levels.

The **ovaries** produce hormones involved in reproduction.

The **testes**, in males, produce hormones involved in reproduction.

IS THAT A FACT!

Since the mid 1980s research at Rutgers University shows that some potent hormones in the last third of pregnancy prepare and motivate mothers to care for their young. The most important of these hormones is oxytocin. It is thought to reach the brain at the same time the mother meets her newborn, helping them to bond.

2) Teach

Answer to APPLY
Paragraphs will vary. Students should write about a time when they experienced a fight-or-flight response to stress and include the terms *hormone*, *fight-or-flight*, and *epinephrine*.

USING THE FIGURE

Write the following terms on a board or a transparency:

endocrine, pituitary, thyroid, parathyroid, thymus, adrenals, pancreas, ovaries, testes

As a class, pronounce each term. Then locate each gland shown in **Figure 16.**

BRAIN FOOD

Some cells found in tumors secrete hormones that are identical to those secreted by endocrine glands. Typically, the tumor cells secrete too much of the hormone and do not respond to feedback control. Ask students to speculate on the effects on the body of these tumor cells and the hormones they produce. (Students should recognize that when hormones are produced in improper amounts or in an uncontrolled way, the body's homeostasis will be disrupted. The specific effects will depend on the body's internal environment and on which hormones the tumor cells secrete.)

 Directed Reading Worksheet Section 3

Section 3 • The Endocrine System

3) Extend

USING THE FIGURE

Have students relate the events in **Figure 17** to the events that affect a thermostat. Ask: What happens when the temperature becomes too warm? (The air conditioner turns on and cools the room.)

What happens if the temperature becomes too cool? (The air conditioner turns off, and the room warms up.)

What is the overall effect of feedback control? (The level of the variable remains within a narrow range. It never becomes extremely high or extremely low and thus maintains homeostasis.)

MATH and MORE

After students have analyzed **Figure 17,** ask them to draw a graph that shows the changing levels of glucose in the blood over the course of the figure. (Graphs should approximate a sine curve.)

Have students label their graph "Negative feedback." Then draw a graph that shows a line at a nearly 45° angle on the board or a transparency, and label it "Positive feedback." Ask students to compare the graph with the "Negative feedback" curve they drew. Point out that the "Negative feedback" curve shows that the glucose level stays within a narrow range. The "Positive feedback" line shows a variable that constantly increases.

Controlling the Controls

How do endocrine glands know when to start and stop hormone release? They know because your body has special systems called **feedback controls** that turn endocrine glands on and off. Feedback controls work something like a thermostat on an air conditioner. Once a room reaches the required temperature, the thermostat sends a message to the air conditioner to stop sending in cold air. Much in the same way, a feedback control sends a message to an endocrine gland to stop sending in a particular hormone. **Figure 17** traces the steps of a feedback control that regulates blood sugar.

Figure 17 *In this feedback-control system, the pancreas produces hormones that help your body maintain the correct blood sugar level.*

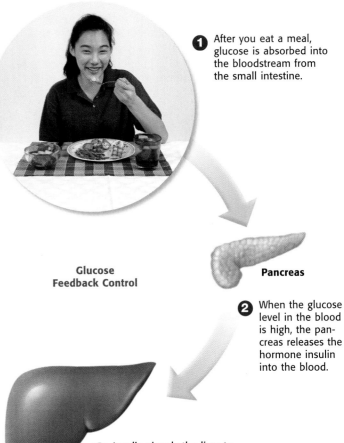

Glucose Feedback Control

1. After you eat a meal, glucose is absorbed into the bloodstream from the small intestine.
2. When the glucose level in the blood is high, the pancreas releases the hormone insulin into the blood.
3. Insulin signals the liver to take in glucose from the body, convert it to glycogen, and store it for future energy needs.
4. When the blood sugar level returns to normal, the pancreas stops releasing insulin.
5. To keep your blood sugar level from falling below normal, you must eat again.

Homework

Researching Steroids Tell students that hormones are used as medicines to treat endocrine disorders. However, other hormones, such as anabolic steroids, are abused to increase athletic prowess. Have students research the effects and dangers of abusing anabolic steroids. Have them write a brief report or prepare an oral presentation to share their findings.

Hormone Imbalances

Insulin is a hormone made by the pancreas. When the blood sugar level rises after a person has eaten something, insulin triggers the cells to take in glucose and sends a message to the liver to store glucose. A person whose pancreas cannot make enough insulin has a condition called *diabetes mellitus*. A person with diabetes mellitus may need daily injections of insulin to keep his or her blood glucose levels within safe limits. Some patients, like the woman in **Figure 18,** receive their insulin automatically from a small machine they wear next to their body.

Growth Hormone Sometimes a child may have a pituitary gland that doesn't make enough growth hormone. This causes the child's growth to be stunted. Fortunately, if this problem is detected soon enough, a doctor can prescribe hormone replacement medication and monitor the child's growth. If the pituitary makes too much growth hormone at an early age, the person becomes much taller than expected.

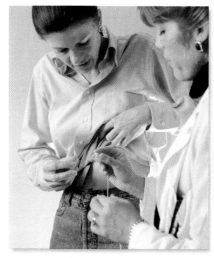

Figure 18 *This young woman has diabetes and must have daily injections of the hormone insulin.*

Thyroxine When a person doesn't get enough iodine in the diet, the thyroid gland cannot make enough of the hormone *thyroxine*. This causes the thyroid to swell up and form a mass called a *goiter*. Because thyroxine increases metabolism, this person's cells are less active than normal, causing fatigue, weight gain, and other problems.

SECTION REVIEW

1. What is the function of the endocrine system?
2. Why are feedback controls important?
3. Name four endocrine glands, and tell what each one does in the body.
4. **Applying Concepts** Epinephrine, the fight-or-flight hormone, increases the level of glucose in the blood. Why would this be important in times of stress?
5. **Illustrating Concepts** Look around your house for an example of a feedback control. Draw a diagram explaining how this feedback control works to start and stop an action.

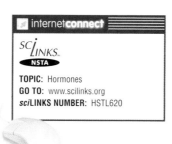

internet connect

SC_INKS_
NSTA

TOPIC: Hormones
GO TO: www.scilinks.org
*sci*LINKS NUMBER: HSTL620

4) Close

Quiz

Ask students whether these statements are true or false.

1. The endocrine system is involved with high-speed processes. (false)
2. Hormones are chemicals secreted into the bloodstream. (true)
3. All endocrine glands come in pairs. (false)
4. The body uses feedback control to regulate the secretion of hormones. (true)

ALTERNATIVE ASSESSMENT

Writing Have students write a fictional story about a new hormone that controls a function not discussed in this lesson. Possibilities include a hormone that controls a person's ability to sleep, tell jokes, roll the tongue, or perform some task. Students should describe the feedback control of the hormone and describe the conditions caused by overproduction and underproduction of the fictional hormone. Encourage students to be informative as well as creative.

Reinforcement Worksheet
"Every Gland Lends a Hand"

Answers to Section Review

1. Your endocrine system controls body processes such as fluid balance, growth, and sexual development with the use of chemicals called hormones.
2. Feedback controls stop and start hormone release.
3. Answers will vary but should include four of the glands described in **Figure 16.**
4. During stressful situations, your body functions might need to speed up, requiring more energy.
5. Answers will vary, but students might draw a diagram demonstrating how a thermostat keeps water or air within a certain temperature range.

Section 3 • The Endocrine System

Skill Builder Lab

You've Gotta Lotta Nerve
Teacher's Notes

Time Required
One 45-minute class period

Lab Ratings

TEACHER PREP
STUDENT SET-UP
CONCEPT LEVEL
CLEAN UP

The materials listed on the student page are enough for 1–2 students. Tell students that they will not be testing for pain. The protective cover on the sharp end of the dissecting pin must remain in place at all times.

Safety Caution

Remind students to review all safety cautions and icons before beginning this lab activity.

Remind students to be safe and gentle with each other in this exercise, respecting the sensitivity and comfort of their peers.

Lab Notes

This activity works best if the student whose hand is being tested looks away or is loosely blindfolded while his or her hand is being tested. Often students will say they feel something when they think they should feel something. Students should be given the choice of being blindfolded or looking away.

Skill Builder Lab

You've Gotta Lotta Nerve

Your skin has thousands of nerve receptors that detect sensations such as heat, cold, and pressure. Your brain is designed to filter out or ignore most of the input it receives from these skin receptors. If this were not the case, simply wearing clothes would trigger so many responses that you couldn't function.

Some areas of the skin, such as the back of your hand, are more sensitive than others. In this activity, you will map the receptors for heat, cold, and pressure on the back of your hand.

- fine-point washable pens or markers
- metric ruler
- graph paper
- eyedropper
- very cold water
- facial tissue
- hot tap water
- dissecting pin with a small piece of cork or a small rubber stopper covering the sharp end

Procedure

1. Form a group of three. One group member will volunteer the back of his or her hand to be tested, one will do the testing, and the third will record the results. Check with your teacher to see if you may switch roles so that each group member may play each part.

2. Use a fine-point washable marker or pen and a metric ruler to mark off a 3 cm × 3 cm square on the back of the volunteer's hand. Draw a grid within the area, spacing the lines approximately 0.5 cm apart. You will have 36 squares in the grid when you are finished. Examine the photograph below to make sure you have drawn the grid correctly.

3. Mark off three 3 cm × 3 cm areas on the graph paper. Make a grid in each area exactly as you did on the back of the volunteer's hand. Label one grid "Cold," another grid "Hot," and the third grid "Pressure."

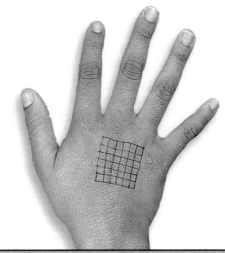

 Datasheets for LabBook

 Science Skills Worksheet
"Understanding Variables"

Christopher Wood
Western Rockingham
Middle School
Madison, North Carolina

4. Begin locating receptors on the volunteer's hand. The volunteer should not look while his or her hand is being tested! Use the eyedropper to apply one small droplet of cold water on each square in the grid. The volunteer should tell you when he or she feels a cold droplet. On your graph paper, mark an X on the "Cold" grid in the square that corresponds to where the sensation of cold was felt on the hand. You will need to carefully blot the water off your partner's hand after several drops.

5. Repeat the test using hot-water droplets. The water will cool enough as it drops from the eyedropper that it will not hurt your partner. Mark an X on the "Hot" grid to indicate where the sensation of heat was felt on the hand.

6. Repeat the test using the head—not the point!—of the pin. Touch the skin with the pinhead to detect pressure receptors. Use a very light touch. Mark an X on the "Pressure" grid to indicate where pressure was felt on the hand.

Analysis

7. Count the number of Xs in each grid. How many heat receptors are there per 3 cm^2? cold receptors? pressure receptors?

8. Do you have areas on the back of your hand where the receptors overlap? Why or why not?

9. How do you think the results of this experiment would be similar or different if you mapped an area of your forearm? the back of your neck? the palm of your hand?

10. Prepare a written report that includes a description of your investigation and a discussion of items 7–9.

Going Further

In the library or on the Internet, research what happens if a receptor is continuously stimulated. Does the kind of receptor make a difference? Does it make a difference how intense the stimulation is? Explain.

Skill Builder Lab

Answers

7. Answers will vary. Some students will have more receptors in each category than others.

8. Students may notice that the same square on the grid can feel heat, cold, and pressure. Students may explain this by noting that they have different kinds of receptors in the same spot. Some perceptive students may recognize hot and cold as variations of a single sensation—temperature. They also may perceive that the blunt tip of the dissecting pin might have been cold and that they felt the temperature instead of the pressure.

9. Different areas of the body are more sensitive than others. Results may vary because some areas are not visible to the student, thus the variable of expected sensation is eliminated.

Going Further

Students will find that different responses are made to a constant stimulus according to type and intensity. A person may become insensitive to an odor or even mild pain over time. Intense, sudden, or continued stimuli, such as heat, pain, or noise, may do great damage to a person over time.

Chapter 4 • Skill Builder Lab   **93**

Chapter Highlights

VOCABULARY DEFINITIONS

SECTION 1

central nervous system a collection of organs whose primary function is to process all incoming and outgoing messages from the nerves; includes the brain and the spinal cord

peripheral nervous system the nerves whose primary function is to exchange information from all areas of the body and the outside environment to the central nervous system and from the central nervous system to the rest of the body

neuron a specialized cell that transfers messages throughout the body in the form of fast-moving electrical signals

impulse an electrical message that passes along a neuron

receptor a specialized dendrite that detects changes inside or outside the body

nerve an axon bundled together with blood vessels and connective tissue

brain mass of nerve tissue that is the main organ of the nervous system

reflex a quick, involuntary response to a stimulus

Chapter Highlights

SECTION 1

Vocabulary
- central nervous system (p. 76)
- peripheral nervous system (p. 76)
- neuron (p. 77)
- impulse (p. 77)
- receptor (p. 78)
- nerve (p. 78)
- brain (p. 79)
- reflex (p. 82)

Section Notes
- The central nervous system includes the brain and spinal cord. The peripheral nervous system includes nerves and sensory receptors.
- A neuron receives information at branched endings called dendrites and passes information to other cells along a fiber called an axon.
- Sensory neurons detect information about the body and its environment. Motor neurons carry messages from the brain and spinal cord to other parts of the body.
- The cerebrum is the largest part of the brain and is involved with thinking, sensations, and voluntary muscle control.
- The cerebellum is the second largest part of the brain. It keeps track of the body's position and helps maintain balance.
- The medulla controls involuntary activities such as heart rate, blood pressure, and breathing.
- Pain signals can trigger a quick, involuntary action, called a reflex, in which a motor neuron sends a message to a muscle without first receiving a signal from the brain.

✓ Skills Check

Math Concepts

THE SPEED OF AN IMPULSE An impulse travels very fast. As shown in the MathBreak on page 77, to calculate the amount of time that it takes for an impulse to travel a certain distance, you must first know the speed it is traveling. Then you can divide the distance by the speed to get the time. For example, if an impulse travels 150 m/s, it would take it 0.02 seconds to travel 3 m.

$$\text{time} = \frac{3 \text{ m (distance)}}{150 \text{ m/s (speed)}} = 0.02 \text{ s}$$

Visual Understanding

PATH OF LIGHT Look back at Figure 10 on page 84 to review the path of light entering the eye. The light first passes through the transparent cornea, then through the opening called the pupil, and then through the lens. At the back of the eye, the light is detected by receptors in the retina.

Lab and Activity Highlights

You've Gotta Lotta Nerve **PG 92**

Datasheets for LabBook
(blackline masters for this lab)

94   Chapter 4 • Communication and Control

SECTION 2

Vocabulary
- retina (p. 84)
- rods (p. 84)
- cones (p. 84)
- iris (p. 85)
- lens (p. 85)
- cochlea (p. 86)

Section Notes
- Different kinds of receptors in the skin are responsible for detecting touch, pressure, temperature, and pain.
- The retina of the eye contains photoreceptors that react to light and cause impulses to be sent to the brain.
- The lens of the eye can change shape to adjust the point of focus so that the image is focused on the retina. Improper focus can usually be corrected with glasses or contact lenses.
- Special receptors inside the cochlea of the ear react to sound waves and send impulses to the brain.
- Receptors for taste are located in taste buds on the bumps of the tongue.
- Receptors for smell are on olfactory cells located in the upper part of the nasal cavity.

SECTION 3

Vocabulary
- endocrine system (p. 88)
- gland (p. 88)
- hormone (p. 88)
- feedback control (p. 90)

Section Notes
- The endocrine system communicates with other systems using chemicals called hormones.
- Hormones are made in endocrine glands.
- The adrenal glands secrete hormones that help the body cope with stress. Epinephrine is the hormone most associated with fight-or-flight situations.
- Feedback control is the body's way of turning glands on and off so that they release hormones only when necessary.

VOCABULARY DEFINITIONS, continued

SECTION 2

retina layer of light-sensitive cells in the back of the eye

rods photoreceptors that can detect very dim light

cones photoreceptors that can detect bright light and that help you see colors

iris the colored part of the eye

lens a curved, transparent object that forms an image by refracting light

cochlea an ear organ that converts sound waves into electrical impulses

SECTION 3

endocrine system a collection of organs, called glands, whose primary function is to control body fluid balance, growth, and sexual development

gland a group of cells that make special chemicals for the body

hormone a chemical messenger that carries information from one part of an organism to the other; made by the endocrine glands

feedback control the system that turns endocrine glands on or off

internet connect

GO TO: go.hrw.com

Visit the **HRW** Web site for a variety of learning tools related to this chapter. Just type in the keyword:

KEYWORD: HSTBD4

GO TO: www.scilinks.org

Visit the **National Science Teachers Association** on-line Web site for Internet resources related to this chapter. Just type in the sciLINKS number for more information about the topic:

TOPIC:	sciLINKS NUMBER:
The Nervous System	HSTL605
The Senses	HSTL610
The Eye	HSTL615
Hormones	HSTL620

Vocabulary Review Worksheet

Blackline masters of these Chapter Highlights can be found in the **Study Guide.**

Lab and Activity Highlights

LabBank

Whiz-Bang Demonstrations, Now You See It, Now You Don't

Labs You Can Eat, A Salty Sweet Experiment

Long-Term Projects & Research Ideas, Man Versus Machine

Chapter Review
Answers

Using Vocabulary
1. central nervous system
2. cochlea
3. stress
4. axon terminal
5. involuntary
6. retina

Understanding Concepts
Multiple Choice
7. c
8. a
9. c
10. a
11. d
12. c

Short Answer
13. Answers will vary but should describe dangerous or stressful situations that the student has encountered.
14. Light stimulates a ring of muscle fibers, which enable the iris to open and close.
15. A reflex is a quick, involuntary action. Because the impulse does not need to travel to the brain before a muscle can respond, this reaction can be very quick.
16. The vibrations of the eardrum make the ear bones vibrate, one of which vibrates against the cochlea, creating waves that are detected by neurons.

Concept Mapping Transparency 25

Chapter Review

USING VOCABULARY

To complete the following sentences, choose the correct term from each pair of terms listed below:

1. Your brain and spinal cord make up your __?__. (*central nervous system* or *peripheral nervous system*)

2. Sensory receptors in the __?__ detect vibrations. (*cochlea* or *eardrum*)

3. Epinephrine is produced by the adrenal glands in response to __?__. (*glucose* or *stress*)

4. The part of a neuron that passes an impulse to other cells is the __?__. (*dendrite* or *axon terminal*)

5. The medulla is mostly responsible for activities that are __?__. (*involuntary* or *voluntary*)

6. Receptors that can convert light into impulses are found in the __?__. (*olfactory cells* or *retina*)

UNDERSTANDING CONCEPTS

Multiple Choice

7. Which of the following has receptors for smelling?
   a. cochlea cells
   b. thermoreceptors
   c. olfactory cells
   d. optic nerve

8. Which of the following gives eyes their color?
   a. iris
   b. cornea
   c. lens
   d. retina

9. Which of the glands is associated with goiters?
   a. adrenal
   b. pituitary
   c. thyroid
   d. pancreas

10. Which of the following is not part of the peripheral nervous system?
    a. spinal cord
    b. axons
    c. sensory receptors
    d. motor neurons

11. Which part of the brain regulates blood pressure?
    a. right cerebral hemisphere
    b. left cerebral hemisphere
    c. cerebellum
    d. medulla

12. Which of the following is associated with the endocrine system?
    a. reflex
    b. salivary gland
    c. fight-or-flight response
    d. voluntary response

Short Answer

13. Describe several situations in which your adrenal glands might release epinephrine, causing you to have a fight-or-flight reaction.

14. What causes the size of your pupils to change?

15. What is a reflex? How does a reflex enable you to act quickly?

16. What is the function of the middle-ear bones?

Chapter 4 • Communication and Control

17. Using the terms you learned in this chapter, write down a step-by-step sequence for the path taken by an impulse, beginning at a pain receptor in your left big toe. Be sure to mention each kind of neuron and its parts as well as specific organs in the nervous system.

Concept Map

18. Use the following terms to create a concept map: the nervous system, spinal cord, medulla, peripheral nervous system, brain, cerebrum, central nervous system, cerebellum.

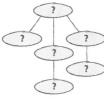

CRITICAL THINKING AND PROBLEM SOLVING

Write one or two sentences to answer the following questions:

19. Why is it important to have a lens that can change shape inside the eye?

20. Why can the nervous system have a faster effect on the body than the endocrine system?

21. Why is it important that reflexes occur without thought?

MATH IN SCIENCE

22. Sound travels about 335 m/s (1 km is equal to 1,000 m). How many kilometers would a sound travel in 1 minute?

23. Some axons can send one impulse every 0.4 milliseconds. One second is equal to 1,000 milliseconds. How many impulses could one of these axons send every second?

INTERPRETING GRAPHICS

Look at the drawing below, and answer the following questions:

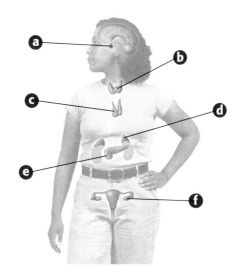

24. Which letter identifies the gland that regulates blood sugar?

25. Which letter identifies the gland that releases a hormone that stimulates the birth process in pregnant women?

26. Which letter identifies the gland that helps the body fight disease?

 Take a minute to review your answers to the Pre-Reading Questions found at the bottom of page 74. Have your answers changed? If necessary, revise your answers based on what you have learned since you began this chapter.

17. The impulse travels from sensory receptors in the toe along axons leading to the spinal cord. Motor neurons carry a new message to the muscles in the leg telling them to move the foot. The pain message continues along the spinal cord to the cerebrum.

Concept Mapping

18. An answer to this exercise can be found at the front of this book.

CRITICAL THINKING AND PROBLEM SOLVING

19. The lens must change shape to focus objects that are varying distances from the eye.
20. Electrical messages travel along nerves faster than chemicals can travel in the bloodstream.
21. The speed of a reflex might prevent serious harm to the body.

MATH IN SCIENCE

22. 20.1 km/min
23. 2,500 impulses/second

INTERPRETING GRAPHICS

24. *e* (pancreas)
25. *a* (pituitary)
26. *c* (thymus)

Blackline masters of this Chapter Review can be found in the **Study Guide.**

Science, Technology, and Society

Light on Lenses

Background

Many students may be familiar with lenses because they wear eyeglasses. Lenses mounted in frames and used to improve vision were first developed in both Europe and China, probably during the thirteenth century.

Benjamin Franklin invented bifocals in 1784. These were a new type of eyeglasses in which each lens was divided in half. One lens corrected for nearsightedness and the other for farsightedness. Today, single lenses can be ground to include both features. In Franklin's version, each bifocal lens was made of two separate pieces of glass held together by the frame.

Science, Technology, and Society

Light on Lenses

Can you see in pitch darkness? No, of course not! You need light to see. But there is something else you need in order to see. You need a lens. A **lens** is a curved transparent object that *refracts,* or bends, light.

Lenses are necessary to focus light in all kinds of applications, including in telescopes, microscopes, binoculars, cameras, contact lenses, eyeglasses, and magnifying lenses.

Light Bounces

To learn how lenses work, you must first know something about how light travels. A ray of light travels in a straight path from its source until it strikes an object. When light strikes an object, much of the light bounces off, or is reflected. The light reflects from the object at the same angle that it struck the object in the first place.

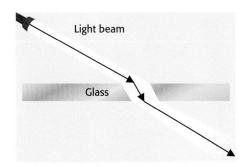

▲ *Light changes speed and direction when it passes from one material into another.*

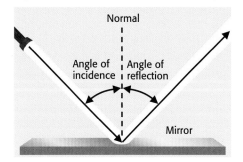

▲ *The angle formed by the incoming light (angle of incidence) always equals the angle of the reflected light (angle of reflection).*

Lenses Bend Light

A lens allows light to travel through it. However, as the light passes through the lens, it is refracted. **Refraction** is the bending of a light ray as it passes from one transparent material into another, such as when light traveling through air passes through a glass lens.

The type of lens determines how much and in which direction the light is bent. A lens that is thicker in the middle than at its edges is called a **convex lens.** This type of lens bends light toward its center. Convex lenses are used in magnifying glasses, microscopes, and telescopes. The lenses in your eyes are convex lenses.

A lens that is thinner in the middle than at its edges is called a **concave lens.** This type of lens bends light away from its center. Both convex lenses and concave lenses are often used to help correct vision. Convex lenses are also used in combination with concave lenses in cameras to focus light on the film.

Light Your Way

▶ Do some additional research to find out what a photorefractive keratectomy (PRK) is and how it works to correct a person's vision.

Answer to Light Your Way

PRK modifies the cornea to correct a person's myopic vision. A surgeon uses an excimer laser to remove a 5–9 mm diameter portion of the cornea. For mild or moderate cases of myopia, the surgeon removes about 5–10 percent of the cornea's thickness, while for extreme cases, up to 30 percent is removed. This procedure has an advantage over RK (radial keratotomy) in that the integrity of the cornea's dome is not compromised, as it is when cuts are made in the cornea during an RK procedure. In the RK procedure, deep incisions are made in a spokelike pattern. The drawbacks of this method are that it is viable only in mild cases of myopia and is ineffective at correcting hyperopia (farsightedness). The incisions in the cornea flatten it, and over time, it continues to flatten, which increases the person's farsightedness.

Eureka!
Pathway to a Cure

Do you know what would happen if your brain sent out too many impulses to the muscles in your body? First the overload would increase the number of contractions in your muscles, and it would be difficult to carry out simple movements, like scratching your arm or picking up a glass. Even when you wanted to rest, your muscles would continue to tremble. This is what happens to people with Parkinson's disease, and unfortunately, there is still no cure.

The Disease

Parkinson's disease affects the cells in the brain that regulate muscles. These cells require the chemical dopamine, which slows down the activity of nerves so they can function properly. But if the cells that supply the muscle-regulating cells with dopamine are damaged, the brain will send continuous impulses to the muscles. This results in Parkinson's disease.

Parkinson's disease is often diagnosed only after a person has already lost about 80 percent of his or her dopamine-supplying cells. Although there is no known cure for Parkinson's disease, some patients can be treated with chemicals that act like dopamine. Unfortunately, these substitutes are not as good as the real thing. Dopamine itself cannot be given because it cannot pass from the blood into the brain tissue.

Breakthrough

Dr. Bertha Madras studies the effects of drug addiction on the brain. While studying the effects of cocaine addiction, she discovered that a chemical called tropane attaches itself to the same nerves that release dopamine in the brain. This discovery may be used to detect and diagnose Parkinson's disease earlier and at a lower cost to the patient.

A Glow in the Darkness

Madras and her colleagues thought they could use tropane to study the cells that release dopamine. They added a radioactive component to the tropane to make a chemical called altropane. Altropane also attaches to the dopamine-releasing cells. But unlike tropane, altropane glows, so it shows up in a brain scan. Healthy people have large areas where the altropane attaches. Among patients with Parkinson's disease, because of the nerve loss, the altropane attaches to fewer nerves. Therefore, brain scans from these patients do not have as many glowing collections of altropane.

Using this new procedure to diagnose Parkinson's disease could allow doctors to find the disease in people before the neurons are severely damaged or completely lost.

 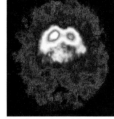

Healthy subject   Parkinson's subject

▲ *Brain scans, such as the ones above, can be used to diagnose Parkinson's disease.*

Activity

▶ Find out what a Single Photon Emission Computed Tomography (SPECT) image is and how it is used to study Parkinson's disease.

Eureka!
Pathway to a Cure

Background

Parkinson's disease was first described by James Parkinson more than 180 years ago. No one knows the cause of Parkinson's disease, although it does tend to run in families. Most people with Parkinson's disease are elderly, but young people can also have the disease. Major signs of Parkinson's disease include stiffness, trembling, and difficulty moving muscles. Dr. Madras was working on the effects of cocaine use on the brain when she discovered tropane. Tropane has the same chemical backbone as cocaine, and it binds specifically to the neurons that transmit dopamine in the brain. By radioactively labeling the tropane molecule and turning it into altropane, scientists could use the tracer molecule to visualize the status of dopamine-producing neurons in the brain.

Several techniques have been used to visualize nerve damage in the brain due to Parkinson's disease. Many of these techniques are very costly, and none of them are as good as the altropane method. The other procedures use chemicals that bind to other neurons besides the ones that produce dopamine. Altropane is very specific to the neurons that transport dopamine.

Answer to Activity

The type of brain scan done with altropane is called single photon emission computed tomography (SPECT) imaging. The other imaging method used to diagnose Parkinson's is positron emission tomography (PET). Only seven locations in the United States use the PET method, and it is a complicated procedure. A PET imaging can cost $2,500. SPECT imaging is cheaper than PET, costing about $1,000, and is more widely available at hospitals. SPECT imaging relies on the use of radioactively labeled molecules to create an image.

Chapter Organizer

CHAPTER ORGANIZATION	TIME MINUTES	OBJECTIVES	LABS, INVESTIGATIONS, AND DEMONSTRATIONS
Chapter Opener pp. 100–101	45	National Standards: UCP 3, SAI 1, LS 2a, 3a	**Start-Up Activity,** How Grows It?, p. 101
Section 1 Animal Reproduction	90	▶ Distinguish between asexual and sexual reproduction. ▶ Explain the difference between external and internal fertilization. ▶ Describe the three different types of mammalian development. UCP 5, SAI 1, LS 2a, 2b, 2d	**Demonstration,** Eggs, p. 104 in ATE
Section 2 Human Reproduction	90	▶ Describe the functions of the male and female reproductive systems. ▶ Discuss disorders and diseases that are associated with human reproduction. UCP 1, 5, SAI 1, SPSP 1, 4, LS 1a, 1d–1f, 2b	**Demonstration,** Multiple Births, p. 107 in ATE
Section 3 Growth and Development	90	▶ Summarize the processes of fertilization and implantation. ▶ Describe the course of human development. UCP 2, 3, 5, SAI 1, ST 2, SPSP 1, 5, LS 1a, 1b, 1d, 2b; Labs UCP 2, 3, SAI 1, ST 1	**QuickLab,** Life Grows On, p. 115 **Skill Builder,** It's a Comfy, Safe World! p. 116 **Datasheets for LabBook,** It's a Comfy, Safe World! **Skill Builder,** My, How You've Grown! p. 182 **Datasheets for LabBook,** My, How You've Grown! **Long-Term Projects & Research Ideas,** Get a Whiff of This!

*See page **T23** for a complete correlation of this book with the*

NATIONAL SCIENCE EDUCATION STANDARDS.

TECHNOLOGY RESOURCES

 Guided Reading Audio CD English or Spanish, Chapter 5

 One-Stop Planner CD-ROM with Test Generator

 CNN. Eye on the Environment, Eagles and DDT, Segment 5

Chapter 5 • Reproduction and Development

Chapter 5 • Reproduction and Development

CLASSROOM WORKSHEETS, TRANSPARENCIES, AND RESOURCES	SCIENCE INTEGRATION AND CONNECTIONS	REVIEW AND ASSESSMENT
Directed Reading Worksheet **Science Puzzlers, Twisters & Teasers**		
Directed Reading Worksheet, Section 1 **Reinforcement Worksheet,** Reproduction Review	**MathBreak,** Chromo-Combos, p. 103	**Homework,** pp. 102, 105 in ATE **Self-Check,** p. 103 **Section Review,** p. 105 **Quiz,** p. 105 in ATE **Alternative Assessment,** p. 105 in ATE
Transparency 93, The Male Reproductive System **Directed Reading Worksheet,** Section 2 **Transparency 94,** The Female Reproductive System **Math Skills for Science Worksheet,** Multiplying Whole Numbers **Math Skills for Science Worksheet,** Dividing Whole Numbers with Long Division	**MathBreak,** Counting Eggs, p. 107 **Cross-Disciplinary Focus,** p. 107 in ATE **Apply,** p. 108 **Math and More,** p. 108 in ATE **Chemistry Connection,** p. 109	**Section Review,** p. 109 **Quiz,** p. 109 in ATE **Alternative Assessment,** p. 109 in ATE
Directed Reading Worksheet, Section 3 **Transparency 154,** How Sonar Works **Transparency 95,** Growth Chart **Reinforcement Worksheet,** The Beginning of a Life **Critical Thinking Worksheet,** One to Grow On!	**Math and More,** p. 112 in ATE **Connect to Earth Science,** p. 113 in ATE **Cross-Disciplinary Focus,** p. 113 in ATE **Across the Sciences:** Acne, p. 122 **Science, Technology, and Society:** Technology in Its Infant Stages, p. 123	**Self-Check,** p. 111 **Section Review,** p. 115 **Quiz,** p. 115 in ATE **Alternative Assessment,** p. 115 in ATE

END-OF-CHAPTER REVIEW AND ASSESSMENT

Chapter Review in Study Guide
Vocabulary and Notes in Study Guide
Chapter Tests with Performance-Based Assessment, Chapter 5 Test
Chapter Tests with Performance-Based Assessment, Performance-Based Assessment 5
Concept Mapping Transparency 26

 Holt, Rinehart and Winston On-line Resources
go.hrw.com

For worksheets and other teaching aids related to this chapter, visit the HRW Web site and type in the keyword: **HSTBD5**

 National Science Teachers Association
www.scilinks.org

Encourage students to use the *sci*LINKS numbers listed in the internet connect boxes to access information and resources on the **NSTA** Web site.

Chapter Resources & Worksheets

Visual Resources

TEACHING TRANSPARENCIES

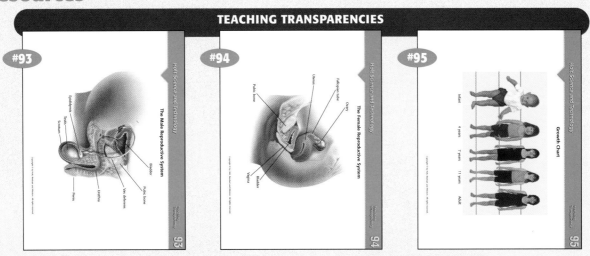

TEACHING TRANSPARENCIES

CONCEPT MAPPING TRANSPARENCY

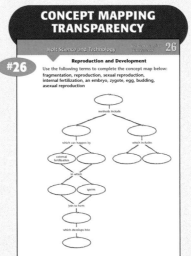

Meeting Individual Needs

DIRECTED READING

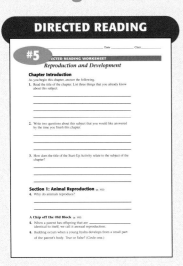

REINFORCEMENT & VOCABULARY REVIEW

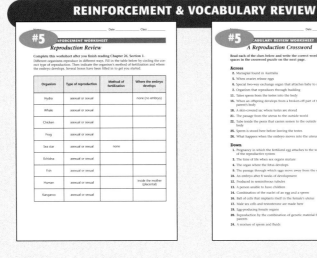

SCIENCE PUZZLERS, TWISTERS & TEASERS

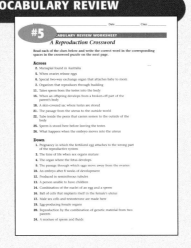

Chapter 5 • Reproduction and Development

Chapter 5 • Reproduction and Development

Review & Assessment

STUDY GUIDE

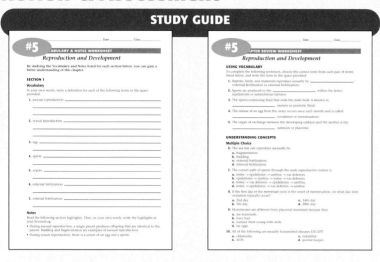

CHAPTER TESTS WITH PERFORMANCE-BASED ASSESSMENT

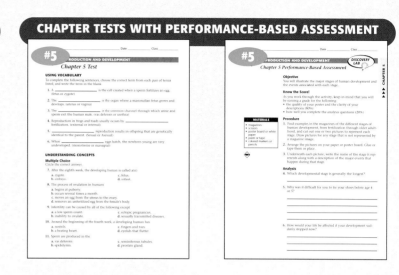

Lab Worksheets

LONG-TERM PROJECTS & RESEARCH IDEAS

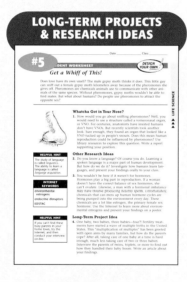

DATASHEETS FOR LABBOOK

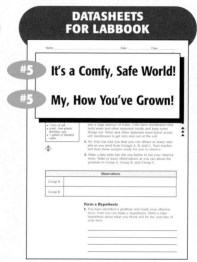

Applications & Extensions

CRITICAL THINKING & PROBLEM SOLVING

EYE ON THE ENVIRONMENT

Chapter 5 • Chapter Resources & Worksheets **99D**

Chapter Background

SECTION 1

Animal Reproduction

▶ Asexual Reproduction
The most widespread forms of asexual reproduction are binary fission, budding, and fragmentation. Binary fission is used mainly by bacteria.

- A major advantage of asexual reproduction is that it does not require a mate. Asexual reproduction also allows animals to produce many offspring in a short period of time. Many animals that do not move around, like sea sponges, reproduce asexually.

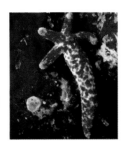

▶ Sexual Reproduction
Because sexual reproduction brings together genetic material from two parents, there is greater variation among animals that reproduce sexually versus those that reproduce asexually.

▶ Fertilization
In external fertilization, eggs can be fertilized without physical contact between the parents. Instead, chemical signals coordinate the fertilization process, ensuring that the parents release their sex cells at the appropriate time.

- Internal fertilization requires a more sophisticated reproductive system, including organs for delivering and storing sperm. Fertilized eggs can develop externally, as with birds, or internally, as with placental mammals. Internally protected embryos are more likely to survive, but placental females do not usually produce as many offspring as do egg-laying females.

IS THAT A FACT!

- Some animal species that reproduce sexually don't have separate sexes. Instead, every individual contains male and female sexual characteristics. This situation is known as *hermaphrodism*.

- Hermaphrodites can exchange sex cells with one another and some can reproduce by themselves as well.

SECTION 2

Human Reproduction

▶ The Male Reproductive System
Male reproductive functions mainly concern sperm production. The head of a sperm contains DNA, and the tail region contains mitochondria. The mitochondria are "engines" for the sperm, providing the sperm's tail with the energy to whip back and forth.

- The process of maturation of sperm from germ cells to spermatozoa takes about 74 days. Even then, they cannot yet penetrate an ovum. First they must "ripen" in the epididymis, a process that takes about 10 days. Though the maturation process is lengthy, once sperm are fully mature, they can remain viable for about 6 weeks.

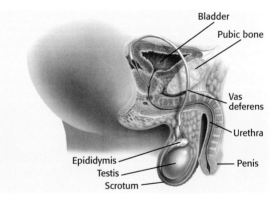

IS THAT A FACT!

- If the seminiferous tubules—the bundle of tubes that makes up each testicle—were joined together and extended, they would be more than 200 m long!

▶ The Female Reproductive System
The ovaries are the primary female reproductive organs. About the size of large almonds, the ovaries are located on either side of the uterus, each anchored by an ovarian ligament. These tiny organs secrete the hormones largely responsible for development during puberty. They are also responsible for releasing eggs.

Chapter 5 • Reproduction and Development

- Every menstrual cycle, several ova begin to ripen. In most cases, however, only one egg reaches maturity at a time. This mature, ripened ovum, encased in a Graafian follicle, travels to the surface of the ovary, where it remains until midcycle, when ovulation occurs. Then the Graafian follicle, distended with fluid, ruptures, sending the egg into the abdominal cavity. The fallopian tube then captures the ovum, and it begins the descent to the uterus.

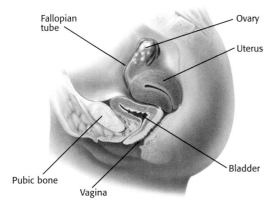

▶ Sexually Transmitted Diseases

Chlamydia is the most prevalent sexually transmitted disease (STD) in North America. Caused by an organism called *Chlamydia trachomatis*, its symptoms include a frequent desire to urinate, pain with urination, and penile or vaginal discharge.

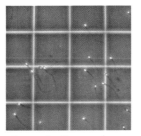

In women, there are often few symptoms in the early stages. Troublesome as the symptoms are, the consequences are worse: chlamydia is a major cause of infertility. If left untreated, it can cause pelvic inflammatory disease and, in women, the subsequent inability to conceive. In men, infection that reaches the testes results in infertility.

- Also known as salpingitis, pelvic inflammatory disease (PID) is an infection of the fallopian tubes, uterus, and cervix. While a number of bacteria can cause PID, the usual culprits are chlamydia and gonorrhea. Though these infections can be completely cured with antibiotics, PID often leaves the fallopian tubes—the conduits between the ovaries and the uterus—scarred, making conception difficult or impossible. Other potential consequences of PID include ectopic pregnancy, peritonitis, and death.

SECTION 3

Growth and Development

▶ Sex Determination

One pair of human chromosomes determines the sex of a baby. There are two types of these sex chromosomes: X and Y. Because the egg contains only the X chromosome, the gender of the baby is determined by the father's sperm, which may contain either an X or a Y chromosome. If a sperm containing an X chromosome joins with an egg, the baby will be a girl. If a sperm containing a Y chromosome fertilizes the egg, the baby will be a boy.

▶ Fetal Development

The development of a baby from a single cell progresses at an astounding rate. Early in development, the embryo resembles a tiny tadpole, with a rounded body and tail. By about 7.5 weeks of gestation, however, limb buds—with knee and elbow joints evident—form, and facial features are recognizable. By the ninth week, nerves and muscles have developed enough that the fetus can move independently. By 12 weeks, the fetus is about 7.6 cm long and has a mass of about 28 g.

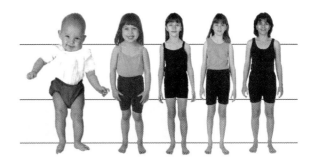

For background information about teaching strategies and issues, refer to the *Professional Reference for Teachers*.

CHAPTER 5

Reproduction and Development

Pre-Reading Questions

Students may not know the answers to these questions before reading the chapter, so accept any reasonable response.

Suggested Answers

1. No; animals that can reproduce asexually—such as sea stars—can have only one parent.
2. Adults are larger and more sexually developed than children.
3. You inherit an equal number of chromosomes from both your parents.

CHAPTER 5

Reproduction and Development

Sections

1. Animal Reproduction .. 102
   MathBreak 103
   Internet Connect 105
2. Human Reproduction .. 106
   MathBreak 107
   Apply 108
   Chemistry Connection. 109
   Internet Connect 109
3. Growth and
   Development 110
   QuickLab 115
   Internet Connect 115

Chapter Lab 116
Chapter Review 120
Feature Articles 122, 123
LabBook 182–183

Pre-Reading Questions

1. Do all animals have two parents?
2. What makes you physically different from an adult?
3. What percentage of genes do you inherit from your mother? your father?

Sneak Preview

If someone had taken your picture when your mother was about seven months pregnant with you, it would have looked very much like this photograph. By the eighth month, your eyes opened and you could see light. Can you believe how much you changed in such a short time? In this chapter, you will learn about how a single cell grows and develops into a complete person. You will also learn how you continue to change from infancy through adulthood.

internet connect

 HRW On-line Resources
go.hrw.com
For worksheets and other teaching aids, visit the HRW Web site and type in the keyword: **HSTBD5**

 sciLINKS NSTA
www.scilinks.com
Use the sciLINKS numbers at the end of each chapter for additional resources on the **NSTA** Web site.

 Smithsonian Institution
www.si.edu/hrw
Visit the Smithsonian Institution Web site for related on-line resources.

 CNNfyi.com
www.cnnfyi.com
Visit the CNN Web site for current events coverage and classroom resources.

HOW GROWS IT?

As you read this, you are aging. Your body is growing into the body of an adult. But does your body have the same proportions that an adult's body has? Do this exercise to find out.

Procedure

1. Have a classmate help you measure your total height, head height, and leg length with a **tape measure** and **meterstick**. Your teacher will tell you how to take these measurements.

2. Calculate your head-to-body proportion and leg-to-body proportion. Use the following equations:

$$\text{head proportion} = \left(\frac{\text{head height}}{\text{body height}}\right) \times 100$$

$$\text{leg proportion} = \left(\frac{\text{leg length}}{\text{body height}}\right) \times 100$$

3. Your teacher will give you the head, body, and leg measurements of three adults. Calculate their proportions. Record all the measurements and calculations.

Analysis

4. Using the direct evidence you collected, evaluate how your proportions compare with the proportions of adults.

START-UP Activity

HOW GROWS IT?

MATERIALS

FOR EACH PAIR:
- tape measure
- meterstick

Teacher's Notes

The head height can be measured by having the student stand next to a sheet of paper with one ear against it. The top of the head and bottom of the chin can be marked with a pencil. That length can then be measured. Total body height can be measured in a similar way. Length should be measured from where the leg bends when the student sits down. The student can hold a meterstick perpendicular to the floor at that level and can mark that height on paper taped to the wall or drop a tape measure down to the floor from the tip of the meterstick. Demonstrate measuring techniques for your students.

Before students come to class, write the measurements of at least three adults on the board. The measurements can be of you and two other teachers.

Answer to START-UP Activity

4. The student's head height should take up a greater proportion of his or her overall height than the adults' head height. The student's leg length should be about 50 percent of his or her overall height. This should match the leg-length proportion of the adults.

SECTION 1

Focus

Animal Reproduction

In this section, students learn about two types of asexual reproduction. Students will review meiosis and learn that sexual reproduction unites an egg and sperm using an internal or external fertilization process. The final section focuses on differences in mammalian reproduction.

Bellringer

Write the following list on the blackboard or on a transparency:

a. bird   **c.** ants
b. human **d.** sea stars

Ask students to write a paragraph explaining how they think reproduction differs among these four animals.

1) Motivate

DISCUSSION

Reproduction Lead a discussion based on students' answers to the Bellringer. Ask students to think about the similarities and differences among the ways animals reproduce. Birds and ants lay eggs, but humans and sea stars don't. Females and males mate to reproduce in humans and birds, but not in sea stars. Help them understand that the end result of reproduction is the same for all animal species, but the means differ widely.

SECTION 1
READING WARM-UP

Terms to Learn

asexual reproduction
sexual reproduction
egg
sperm
zygote
external fertilization
internal fertilization

What You'll Do

- Distinguish between asexual and sexual reproduction.
- Explain the difference between external and internal fertilization.
- Describe the three different types of mammalian development.

Animal Reproduction

The life span of some living things is very short compared with ours. For instance, a fruit fly lives only about 80 days. Other organisms live for a long time. A bristlecone pine can live for 2,000 to 6,000 years. But all living things eventually die. If a species is to survive, its members must reproduce.

A Chip off the Old Block

Some animals, particularly simpler ones, reproduce asexually. In **asexual reproduction,** a single parent has offspring that are genetically identical to itself.

One kind of asexual reproduction is called *budding*. This occurs when a small part of the parent's body develops into an independent organism. The hydra shown in **Figure 1** is reproducing asexually by budding. The young hydra is genetically identical to its parent.

Fragmentation is another type of asexual reproduction. In fragmentation, an organism breaks into two or more parts, each of which may grow into a separate individual. Sea stars can reproduce by fragmentation. Because sea stars eat oysters, people used to try to kill sea stars by chopping them into pieces and throwing the pieces back into the water. They didn't know that each arm of a sea star can grow into an entire organism! This can be seen in **Figure 2.**

Figure 1 *The hydra bud will separate from its parent. Buds from other organisms, such as coral, remain attached to the parent.*

Figure 2 *The largest arm on this sea star was a fragment, from which the rest of the sea star has grown. In time, all of the sea star's arms will grow to the same size.*

Homework

Making Tables Ask students to make a table in their ScienceLog. In one column they will list 10 mammals. In the next column they will indicate how each mammal produces young. From the information in this section, students should be able to indicate whether the animal is a monotreme, a marsupial, or a placental. Students should fill in the table to the best of their knowledge, research the correct answers, and put them in an additional column.

It Takes Two

Sexual Reproduction produces offspring by combining the genetic material of more than one parent. Most animals, including humans, reproduce sexually. Sexual reproduction most commonly involves two parents, a male and a female. The female parent produces sex cells called **eggs**. The male parent produces sex cells called **sperm**. When an egg's nucleus joins with a sperm's nucleus, a new kind of cell, called a **zygote**, is created. This joining of an egg and sperm is known as *fertilization*.

Review of Meiosis Genes are located in *chromosomes*. All human cells except egg and sperm cells contain 46 chromosomes. Eggs and sperm each contain only 23 chromosomes. Eggs and sperm are formed by a process known as *meiosis*.

In humans, meiosis involves the division of one cell with 46 chromosomes into four sex cells with 23 chromosomes each. When an egg and sperm join to form a zygote, the original number of 46 chromosomes is restored. This combination of genes from the father and mother results in a zygote that will grow into a unique individual. **Figure 3** shows how genes are intermixed through three generations.

MATH BREAK

Chromo-Combos

A cell undergoes meiosis. This cell has 6 chromosomes in 3 pairs. How many chromosomal combinations are possible in the formed sex cells? To find out, use the following formula:

2^x = possible variations

where x = the number of pairs.

2^3 (or $2 \times 2 \times 2$) = 8

Therefore, 8 variations are possible.

A typical human cell has 46 chromosomes in 23 pairs. If the cell undergoes meiosis, how many chromosomal combinations are possible in the resulting sex cells?

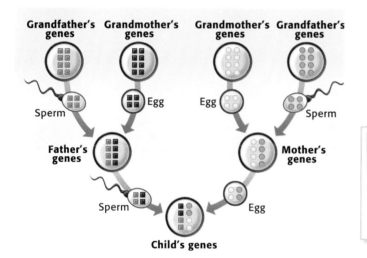

Figure 3 *Eggs and sperm contain genes. You inherit genes from both of your parents. Your parents each inherited genes from both of their parents.*

Self-Check

What is the difference between sexual and asexual reproduction? *(See page 212 to check your answer.)*

MISCONCEPTION ALERT

In certain species, a single animal can sexually reproduce alone. For example, in some species of nematodes, sperm are produced and then stored until eggs are produced. This is followed by self-fertilization.

3) Extend

GOING FURTHER

Encourage interested students to investigate the role that the endocrine system plays in reproduction. Have them draw diagrams illustrating the interactions among the hypothalamus, pituitary, and ovaries (in females) or testes (in males). Allow time for students to present their findings to the class.

DEMONSTRATION

Eggs Bird eggs are fertilized internally, but they develop externally. This requires good packaging. The shell of the egg provides the developing embryo with a stable, safe environment. Show students an ordinary chicken egg. Demonstrate for them that the shape of the egg keeps it from rolling away from the mother. Next, wrap your hand around the egg and demonstrate that the shape of the egg is quite protective against squeezing pressure from the outside. Tell students that the egg is shaped to withstand the pressure of the passage through the chicken and the pressure of incubation. It is also designed to be easy to break from within so that even a weak young chick can peck its way out.

Internal and External Fertilization

Depending on the animal, fertilization may occur either outside or inside the female's body. Some fishes and amphibians reproduce by **external fertilization,** in which the sperm fertilize the eggs outside the female's body. External fertilization must take place in a moist environment so the delicate zygotes won't dry out.

Many frogs, such as those pictured in **Figure 4,** mate every spring. The female frog releases her eggs first. The male frog then releases sperm over the eggs to fertilize them. The frogs leave the fertilized eggs to develop on their own. In about two weeks, the eggs hatch into tadpoles.

Figure 4 Frogs fertilize their eggs externally. Some species can produce more than 300 offspring in one season.

The Inside Story With **internal fertilization,** eggs and sperm join inside the female's body. Reptiles, birds, mammals, and some fishes reproduce by internal fertilization. Many animals that use internal fertilization lay fertilized eggs. The female penguin in **Figure 5,** for example, usually lays one or two eggs after internal fertilization has occurred.

In most mammals, internal fertilization is followed by the development of a fertilized egg inside the mother's body. Many mammals give birth to young that are well developed. Young zebras, like the one in **Figure 6,** can stand up and nurse almost immediately after birth.

Figure 5 Instead of leaving the eggs to develop on their own, penguin parents take turns crouching over them to keep them warm.

Figure 6 This zebra has just been born, but he is already able to stand. Within an hour, he will be able to run.

 Reinforcement Worksheet "Reproduction Review"

IS THAT A FACT!

There are only 2 or 3 days each month when fertilization can occur in a human female.

Making Mammals

All mammals reproduce sexually and nurture their young with milk. There are some differences in how mammals produce offspring, but every mammal follows one of three types of development.

Monotremes Mammals that lay eggs are *monotremes*. Two families of monotremes live today—the echidna and the platypus. After these animals lay their eggs, there is an incubation period that lasts up to 2 weeks. When the eggs hatch, the babies are very undeveloped. They crawl into a fold of their mother's skin and are nourished by the milk that oozes from her pores.

Marsupials Mammals that give birth to live young that are only partially developed are *marsupials*. There are about 260 species of marsupials. Most of them have pouches where their young develop, but some South American species do not have this feature. Marsupials with pouches have extra bones to help support the weight of their young, as can be seen in **Figure 7**. When a baby marsupial attaches itself to its mother's nipple, the nipple expands in the baby's mouth to prevent the baby from separating from its mother.

Placental Mammals There are almost 4,000 different species of placental mammals. These include whales, elephants, armadillos, bats, horses, and humans. *Placental mammals* nourish their young internally before birth. Newborn placental mammals are highly developed compared with newborn marsupials or monotremes.

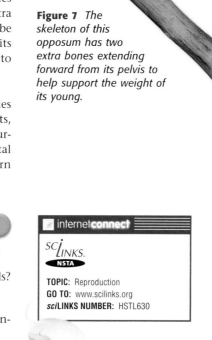

Figure 7 The skeleton of this opposum has two extra bones extending forward from its pelvis to help support the weight of its young.

SECTION REVIEW

1. How many parents are needed to reproduce asexually?
2. What is the difference between monotremes and marsupials?
3. How is a zygote formed?
4. **Applying Concepts** Birds lay eggs, but they are not considered monotremes. Explain why.

internet connect

SCILINKS
NSTA

TOPIC: Reproduction
GO TO: www.scilinks.org
*sci*LINKS NUMBER: HSTL630

---

4) Close

Quiz

1. How do the offspring created by asexual reproduction differ from those created by sexual reproduction? (Offspring from asexual reproduction are genetically identical to the parent, while offspring from sexual reproduction have genetic material from two parents.)
2. How do internal and external fertilization differ? How are they alike? (Internal fertilization occurs inside the female's body, while external fertilization occurs outside the female's body. Both result in the creation of single cells with combined genetic information from the parents.)

ALTERNATIVE ASSESSMENT

Concept Mapping Have students create a concept map using the new terms in this section and the section title.

Homework

Writing Encourage students to research an egg-laying mammal. Then have them write a short story about how the egg develops into a new individual.

▼ Answers to Section Review

1. one
2. Monotremes lay eggs and marsupials give birth to live young.
3. A zygote is formed when a sperm enters an egg and the nuclei of the egg and sperm join.
4. Birds are not mammals. They do not nurture their young with milk, and they are not covered with hair.

SECTION 2

Focus

Human Reproduction

In this section, students are introduced to the male and female reproductive systems. Students then learn about some of the irregularities and problems that affect the human reproductive system, including multiple births, ectopic births, and sexually transmitted diseases.

Bellringer

Ask students if they think that cloning human beings could be considered reproduction. Why or why not? What kind of reproduction is it? Have students write answers to these questions in their ScienceLog.

1) Motivate

DISCUSSION

Ask students to compare reproduction in birds with reproduction in humans. (Both birds and humans fertilize their eggs internally. Birds lay eggs and must protect and keep the eggs warm while obtaining food for themselves. Human mothers carry their baby inside their body, so the baby is always protected.)

Teaching Transparency 93 "The Male Reproductive System"

Directed Reading Worksheet Section 2

Terms to Learn

testes
puberty
vas deferens
semen
penis
ovaries
fallopian tube
uterus
vagina
infertile

What You'll Do

◆ Describe the functions of the male and female reproductive systems.
◆ Discuss disorders and diseases that are associated with human reproduction.

Human Reproduction

When a human sperm and egg combine, a new human begins to grow. About 9 months later, a mother gives birth to her baby. But what happens before that? Where do eggs and sperm come from?

The Male Reproductive System

The male reproductive system, shown in **Figure 8**, produces sperm and delivers it to the female reproductive system. The **testes** (singular, *testis*) make sperm and testosterone. Testosterone is the principal male sex hormone. It regulates the production of sex cells and the development of male characteristics.

Sperm Production The human body is usually around 37°C, but sperm cannot develop properly at such high temperatures. That is why the two testes rest in the *scrotum,* a skin-covered sac that hangs from the body. The scrotum is about 2 degrees cooler than the body. Inside each testis are masses of tightly coiled tubes called *seminiferous* (SEM uh NIF uhr uhs) *tubules* (TOO byoolz), where sperm are produced. A healthy adult male produces several hundred million sperm each day! This massive, continuous sperm production begins at puberty. **Puberty** is the time of life when the sex organs of both males and females become mature.

Before sperm leave a testis, they are stored in a tube called an *epididymis* (EP uh DID i mis). Another tube called a **vas deferens** (vas DEF uh RENZ) passes from each epididymis into the body. As sperm swim through the vas deferens, they mix with fluids from several glands. The mixture of sperm and fluids is called **semen.**

To leave the body, semen passes through the vas deferens into the *urethra,* the tube that runs through the penis. The **penis** transfers semen into the female's body during sexual intercourse.

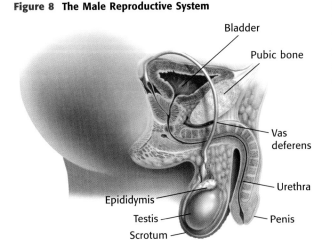

Figure 8 The Male Reproductive System

Labels: Bladder, Pubic bone, Vas deferens, Urethra, Epididymis, Testis, Scrotum, Penis

IS THAT A FACT!

Mumps, a common childhood disease, poses a risk to males who contract it during puberty or adulthood. When mumps occurs after childhood, it can cause inflammation of the testes, known as acute orchitis, which in rare cases can result in sterility.

106 Chapter 5 • Reproduction and Development

The Female Reproductive System

The female reproductive system, shown in **Figure 9**, produces eggs, nurtures fertilized eggs, and gives birth. The **ovaries** produce the eggs. The two ovaries also produce sex hormones, such as estrogen and progesterone, that regulate the release of eggs and direct the development of female characteristics.

The Egg's Journey An ovary contains eggs in various stages of development. As an egg matures, it becomes a huge cell, growing to almost 200,000 times the size of a sperm. During *ovulation,* an egg is ejected through the ovary wall. Then the egg passes into a fallopian (fuh LOH pee uhn) tube. A **fallopian tube** leads from each ovary to the uterus. The **uterus** is the organ where a baby grows and develops.

Every month starting at puberty, the lining of the uterus thickens in preparation for pregnancy. If fertilization occurs, the zygote moves down a fallopian tube and embeds in the lining of the uterus. When a baby is born, it passes from the uterus through the vagina. The **vagina** is the same passageway that received the sperm during sexual intercourse.

Menstrual Cycle To prepare for pregnancy, a female's reproductive system goes through several changes. These changes, called the menstrual cycle, usually occur every 28 days. The first day of the cycle is the beginning of *menstruation,* the monthly discharge of blood and tissue from the uterus. Menstruation lasts about 5 days. As soon as menstruation is over, the uterus's lining begins to build up again in preparation for ovulation. Ovulation typically occurs around the 14th day of the cycle. If the egg isn't fertilized by the time it reaches the uterus, it will deteriorate. Menstruation will flush the egg away, starting the cycle over again. A female's menstrual cycle begins at puberty and continues until late middle age.

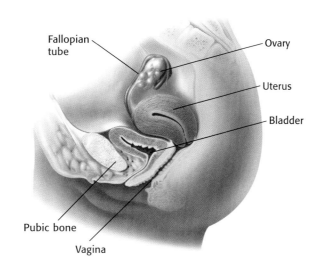

Figure 9 The Female Reproductive System

MATH BREAK

Counting Eggs

1. The average human female ovulates every month from about age 12 to about age 50. How many mature eggs can she produce during that time period?

2. A female's ovaries typically house 2 million immature egg cells. If she ovulates regularly from age 12 to age 50, what percentage of her eggs will mature?

CROSS-DISCIPLINARY FOCUS

Language Arts The words *male* and *female* actually originate from two unrelated Latin words. *Female* comes from *femella,* meaning "girl," while *male* comes from *masculus,* meaning "male."

Irregularities and Disorders

In most cases, the human reproductive system completes its functions flawlessly. However, as with any body system, there can sometimes be irregularities or disorders.

Multiple Births Have you ever seen a pair of identical twins? Sometimes they are so similar that even their parents can't tell them apart. About one pair of identical twins is born for every 250 births. Another type of twins, called fraternal twins, is also born frequently. Fraternal twins can look very different from each other.

Twins, such as those shown in **Figure 10**, are the most common type of multiple births, but humans can also have triplets (3 babies), quadruplets (4 babies), quintuplets (5 babies), and so on. These types of multiple births are extremely rare. For instance, quadruplets occur only about once in every 705,000 births. Do you know what circumstances result in a multiple birth? To find out, do the Apply exercise at the bottom of this page.

Figure 10 *Identical twins have the exact same genes. Many identical twins who are reared apart have similar personalities and interests.*

Ectopic Pregnancy In a normal pregnancy, the fertilized egg travels to the uterus and attaches itself to the uterus's wall. In an *ectopic* (ek TAHP ik) *pregnancy,* the fertilized egg attaches itself to a fallopian tube or another area of the reproductive system. Because the zygote cannot develop correctly outside of the uterus, an ectopic pregnancy can be very dangerous for both the mother and child.

Two Types of Twins

Zach and Drew are fraternal twins. They don't look much alike. Emily and Carol are identical twins. They are hard to tell apart. Why are some twins identical and others fraternal? Consider the two possibilities illustrated at right: In *A*, the mass of cells from a single fertilized egg separates into two halves early in development, and in *B* two eggs are released by an ovary and fertilized by two different sperm cells. Record the answers to the following questions in your ScienceLog:

1. Which instance, *A* or *B*, would produce identical twins? Explain your answer.
2. Could fraternal twins be (a) both boys, (b) both girls, (c) one girl and one boy, or (d) all of the above?

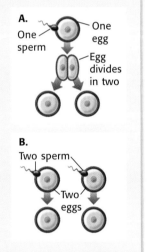

Infertility In the United States, about 15 percent of married couples have difficulty producing offspring. Many of these couples are **infertile**, which means they are unable to have children. Men may be infertile because they cannot produce enough healthy sperm. This is called a low sperm count. Women may be infertile because they do not ovulate normally. Sexually transmitted diseases can also cause infertility.

STDs *Sexually transmitted diseases* (STDs) are diseases that can pass from an infected person to an uninfected person during sexual contact. An STD you may have heard about is the acquired immune deficiency syndrome (AIDS). Other common STDs are shown in the table below.

One in four American youths catches an STD before age 21.

The Spread of STDs in the U.S.	
STD	Approximate new cases each year
Chlamydia	3–10 million
Gonorrhea	1–3 million
Genital warts	1 million
AIDS	750,000
Genital herpes	500,000
Syphilis	120,000

Cancer Cancer, the uncontrolled division of cells, sometimes occurs in the reproductive organs. The testes and the prostate gland, a gland that produces the fluid in semen, are common sites of cancer in men over age 50. In women, the ovaries and breasts are common sites of cancer.

Many chemicals in pollutants are similar to female hormones. Studies are beginning to link these chemicals with early menstruation and low sperm counts.

SECTION REVIEW

1. What is the difference between sperm and semen?
2. Can a woman become pregnant at any time of the month? Explain.
3. Define *sexually transmitted diseases*, and give three examples.
4. **Applying Concepts** How are the ovaries similar to the testes? How are they different?

internetconnect

SCI*LINKS*
NSTA

TOPIC: Reproductive System Irregularities or Disorders
GO TO: www.scilinks.org
*sci*LINKS NUMBER: HSTL640

3) Extend

GROUP ACTIVITY

Writing Have students work in small groups to research breast or prostate cancer. Students should focus on the incidence of the disease, risk factors, and early detection. Have them use the information they gather to write a public-service brochure, complete with artwork, designed to educate the public about the disease. It should emphasize the importance of early detection and treatment. Allow time for students to view the brochures of other groups.

4) Close

Quiz

1. What is puberty? (the time of life when sex organs become mature)
2. What purpose does the epididymis serve? (as a temporary storage site for sperm before they leave the testes)
3. What is menstruation? (the monthly discharge of blood and tissue from the uterus)

ALTERNATIVE ASSESSMENT

Ask students to make diagrams that illustrate the path an egg or sperm must travel before fertilization. Have them label anatomical structures and indicate, with arrows, the direction that the reproductive cell travels.
Sheltered English

▼ **Answers to Section Review**

1. A sperm is a male sex cell, while semen is the fluid that contains sperm.
2. No; usually only one egg per month is released, and it is viable for only a few days after ovulation.
3. Sexually transmitted diseases are diseases passed through sexual contact. Some examples are chlamydia, gonorrhea, genital herpes, and AIDS. (Students must list three.)
4. The ovaries and the testes release sex cells and hormones. The testes continuously produce masses of sex cells, but the ovaries release only one mature sex cell per month.

Section 2 • Human Reproduction

SECTION 3

Focus

Growth and Development

In this section, students learn about egg fertilization and implantation. Students are also introduced to the different stages of growth of a baby *in utero*, culminating in its birth. Finally, they learn about the stages of human development, from birth through adulthood.

Bellringer

Write this statement on the board or an overhead projector: Name the stages of physical development you have passed through thus far in your life.

Have students list the stages in their ScienceLog. Remind students that their growth and development began while they were still in the uterus. (Students will likely respond with the following: crawling, walking, talking, growing taller, perhaps developing lower voices for some of the boys.)

1) Motivate

DISCUSSION

Life Stages Ask students to list as many characteristics of each of the following as they can:

    infancy, childhood, adolescence, adulthood

Tell students that while there are individual differences, all people go through these stages.

SECTION 3 READING WARM-UP

Terms to Learn

embryo
implantation
placenta
umbilical cord
fetus

What You'll Do

◆ Summarize the processes of fertilization and implantation.
◆ Describe the course of human development.

Growth and Development

Every one of us starts out as a single cell that will become a complete person. We are made of millions of cells, each with its own job to do. You, of course, are no exception. You have become a very complex individual, capable of thousands of different thoughts and actions. It is hard to believe that a person as remarkable as you began your life as a single cell, but that is just what happened.

A New Life

The natural process of creating a human baby starts when a man deposits millions of sperm into a woman's vagina during sexual intercourse. Most of the sperm will die because of the vagina's acidic environment, but a few hundred are able to make it through the uterus and into the fallopian tube, as can be seen in **Figure 11**. The surviving sperm cover the egg, releasing enzymes that help dissolve the egg's outer covering. As soon as one sperm gets through, a membrane closes around the fertilized egg. This membrane keeps other sperm cells from entering.

Figure 11 Fertilization and Implantation

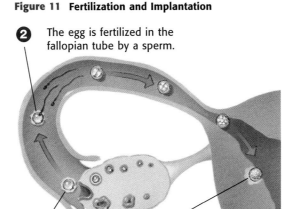

❶ The egg is released from the ovary.
❷ The egg is fertilized in the fallopian tube by a sperm.
❸ The embryo implants itself in the uterus's wall.

Implantation The fertilized egg travels down the woman's fallopian tube toward her uterus. The journey takes about 5 days. The zygote undergoes cell division many times during the trip. By the time it reaches the uterus, it is a tiny ball of cells called an **embryo**. During the next few days, the embryo must embed itself in the thick, nutrient-rich lining of its mother's uterus. This process is called **implantation**, and only about 30 percent of all embryos successfully do it. **Figure 12** shows an implanted embryo.

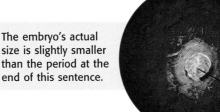

The embryo's actual size is slightly smaller than the period at the end of this sentence.

Figure 12 *This embryo has implanted in the wall of its mother's uterus.*

Embryo

 Directed Reading Worksheet Section 3

IS THAT A FACT!

The longest gestation period, 22 months, belongs to the Indian elephant. The shortest, 12 days, belongs to the Virginia opossum.

110 Chapter 5 • Reproduction and Development

Before Birth

When the embryo implants itself in a woman's uterus, the woman is officially pregnant. For the embryo to survive, a special two-way exchange organ called a **placenta** begins to grow. The placenta contains a network of blood vessels that provide the embryo with oxygen and nutrients from the mother's blood. Wastes that the embryo produces are removed by the placenta and transported to the mother's blood for her to excrete. Although the embryo's blood and the mother's blood flow very near each other inside the placenta, they never actually mix.

Self-Check

Why is it important that the embryo be implanted in the uterus and not elsewhere? *(See page 212 to check your answer.)*

First Month About 1 week after implantation, the embryo's blood cells and a heart tube form. Then the heart tube begins to twitch, starting the rhythmic beating that will continue for the individual's entire life. By the fourth week, the embryo is almost 2 mm long. Surrounding the embryo is a thin, fluid-filled membrane called the *amnion*, which is formed to protect the growing embryo from shocks. The **umbilical cord** is another new development. It connects the embryo to the placenta. The umbilical cord, amnion, and placenta can be seen in **Figure 13**.

Second Month By the time the embryo is 4 weeks old, it has the beginnings of a brain and spinal cord. It also has tiny limb buds that will eventually develop into arms and legs. Its nostrils, eyelids, hands, and feet then begin to form. Its muscles begin to develop, and for the first time in its life, its brain begins to send signals to other parts of its body. Despite all these transformations, the embryo is still only about the size of a peanut. **Figure 14** shows a 5-week-old embryo.

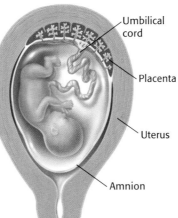

Figure 13 *The placenta, amnion, and umbilical cord are the life support system for the fetus.*

Figure 14 A 5-Week-Old Embryo

Actual size

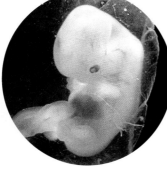

IS THAT A FACT!

The European badger and the American marten each have a 250-day gestation period. Due to delayed implantation, the embryo grows during only 50 of those days. The fertilized egg develops for a few days immediately after conception in July or August, remains dormant in the uterus until January, and then completes its growth between January and March, at which time the baby is born.

② Teach, continued

RESEARCH

Writing Tell students that the first few months of pregnancy are crucial for the healthy mental and physical development of the fetus. Drinking alcohol during this period can lead to birth defects and miscarriages. Have students research the causes and consequences of fetal alcohol syndrome (FAS) and present their findings in a short report.

USING THE FIGURE

Ask students to identify in **Figure 16** the trimester during which the major organs and body structures form. (the first trimester)

Ask them to describe the events that occur during the remainder of the pregnancy. (The structures formed earlier in pregnancy grow and mature during the second and third trimesters.)

MATH and MORE

Have students calculate the factor of increase in a fetus's body length from the eighth week, when it is about 2.5 cm long, to the fifth month, when it is about 25 cm, to birth, when it is about 50 cm long. (2.5 cm to 25 cm is an increase by a factor of 10; 25 cm to 50 cm is an increase by a factor of 2; 2.5 cm to 50 cm is an increase by a factor of 20)

My, How You've Grown!

Figure 15 A 12-Week-Old Fetus

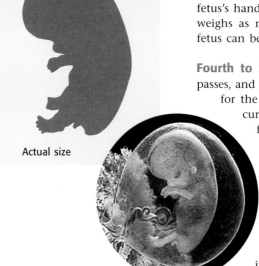

Actual size

Third Month The next stage comes as tiny movements begin to flutter through the embryo's body. The embryo stretches its legs and twitches its arms. It is now 8 weeks old and is developed enough to be called a **fetus**. Three more weeks pass, and it continues to grow at a fast rate, doubling and then tripling its size within a month. The fetus's hands are now the size of teardrops, and its body weighs as much as two pieces of paper. A 12-week-old fetus can be seen in **Figure 15**.

Fourth to Sixth Month The fetus's 13th week of life passes, and suddenly new movement! It can blink its eyes for the first time, swallow, hiccup, make a fist, and curl its toes that now have tiny nails. By the fourth month, the fetus starts to make even bigger movements. The mother now knows when her baby kicks its legs or stretches its arms.

During the fifth month, the fetus is about 20 cm long. Taste buds form on its tongue, and eyebrows form on its face. The fetus begins to hear sounds through the wall of its mother's uterus. Look at the timeline in **Figure 16** to review the changes that take place in the fetus.

Figure 16 Pregnancy Timeline

Weeks of the First Three Months	
1 and 2	The egg is fertilized by a sperm. The fertilized egg makes its way to the uterus, where it burrows into the lining. The fertilized egg is now called an embryo.
3 and 4	Most major organ systems have started to form. The heart starts to beat around day 22. The placenta is completely formed by the fourth week.
5 and 6	Facial features begin to take shape. The skeleton begins to form.
7 and 8	Muscle movement begins. The embryo is now called a fetus.
9 and 10	Arms, legs, hands, and feet have formed.
11 and 12	The internal organs have formed.

Weeks of the Second Three Months	
13 and 14	The circulatory system is working.
15 and 16	The mother may start to feel the fetus move.
17 and 18	The fetus responds to sound.
19 and 20	The fetus is now about 20 cm long.
21 and 22	
23 and 24	Eyelashes and eyebrows appear.

Weeks of the Third Three Months	
25 and 26	The eyes open.
27 and 28	The fetus can "practice breathe."
29 and 30	Layers of fat form beneath skin.
31 and 32	
33 and 34	Organs are fully functional.
35 and 36	The fetus responds to light.
Birth	The baby is born.

internet connect
TOPIC: Before Birth
GO TO: www.scilinks.org
*sci*LINKS NUMBER: HSTL635

IS THAT A FACT!

Between conception and birth, the developing fetus increases in size from a single cell to 6 trillion cells!

Seventh to Ninth Month The seventh month is when the fetus's memories begin to form. During this time, its lungs start to "practice breathe," moving up and down continuously as if breathing real air. If the fetus's mother smokes one cigarette during this stage, the fetus's lung movement will stop for up to an hour. The fetus in **Figure 17** is starting its first lung movement.

By the eighth month, the fetus's open eyes can perceive light through its mother's abdominal wall, and its sleeping pattern starts to be influenced by sunlight. When the fetus is asleep, it dreams. Can you imagine what its dreams might be about?

Figure 17 A 21-Week-Old Fetus

Actual size of hand

Birth

After about 9 months, the fetus is ready to live outside of its mother. The mother goes through a series of muscular contractions called *labor*. During labor, the fetus is usually squeezed headfirst through the vagina. There is little room to spare, and the fetus's head is temporarily squashed out of shape as the fetus passes through its mother's pelvis. Suddenly bright lights and cold air surround the newborn baby. It gasps, fills its lungs with air for the first time, and cries.

The baby in **Figure 18** is still connected to the placenta by its umbilical cord. The doctor or midwife assisting the mother ties and cuts the umbilical cord. The baby's navel is all that will remain of the point where the umbilical cord was attached. After the mother expels the placenta from her body, labor is complete.

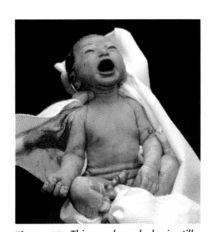

Figure 18 *This newborn baby is still attached to its umbilical cord. The average mass of a newborn baby is 3.3 kg. The average length is 50 cm.*

From Birth to Death

Of all the animals on this planet, humans have one of the longest life spans. Human infancy lasts 2 years—the same time it takes for most rabbits to be born, grow old, and die. Our childhood extends over a full decade, longer than many cats or dogs live. Humans can live for more than 100 years!

BRAIN FOOD

Sea horses are one of very few kinds of organisms in which the male carries the fertilized eggs. After courtship, the female deposits eggs into the pouch of her male mate. The male then fertilizes the eggs and carries them until they develop. After a gestation period of between 10 and 60 days, depending on the species, tiny, fully formed sea horses emerge from an opening in the pouch.

CONNECT TO EARTH SCIENCE

It is possible to create pictures using sound waves. Sonograms are pictures obtained by bouncing high-frequency sound waves off of an object. Doctors can use sonograms to "see" a human fetus while it is still in the womb. Sonar, which uses the same principle, is used for navigating and determining an object's position.

Use the following Teaching Transparency to illustrate how sonar works.

Teaching Transparency 154 "How Sonar Works" LINK TO EARTH SCIENCE

3) Extend

RETEACHING

Divide the class into small groups, and challenge each group to create a board game. Provide each group with a piece of poster board, plain index cards, and markers. Direct them to create a game board that leads players through prenatal development. The first player who is "born" wins the game. Have them use the index cards to write clues that direct players' movements through "gestation." For example, they might write, "Advance to 4 months if you can describe my abilities at 13 weeks." (At 13 weeks, the fetus can blink its eyes, swallow, hiccup, make a fist, and curl its toes.)

Have students create written rules. Then have them exchange games and play. **Sheltered English**

 Teaching Transparency 95 "Growth Chart"

 Reinforcement Worksheet "The Beginning of a Life"

 Critical Thinking Worksheet "One to Grow On!"

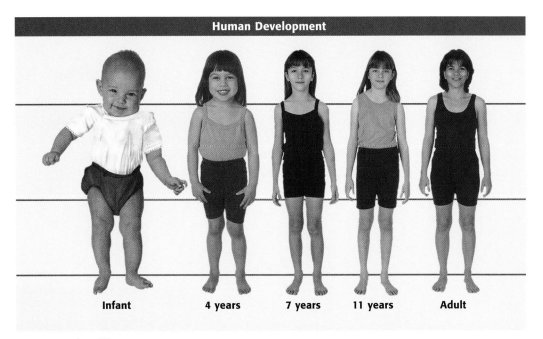

Figure 19 *Five different stages of development are shown the same size so you can see how body proportions change as a person develops.*

Activity

Create a poster or timeline illustrating the different stages of human growth. **TRY at HOME**

Infancy What life stages have you gone through since you were born? You have probably gone through most of the stages shown in **Figure 19.** You were an infant from birth to 2 years of age. During this time, you grew rapidly. Your teeth began to appear. You also became more coordinated as your nervous system developed. This enabled you to begin to walk.

Childhood Your childhood extends from 2 years to puberty. This is also a period of rapid growth. Your first set of teeth were slowly shed and replaced by permanent teeth. Your muscles became more coordinated, allowing you to do activities such as riding a bicycle and jumping rope. Your intellectual abilities also developed during this time.

Adolescence You are considered an adolescent from puberty to adulthood. During puberty, the reproductive systems of young males and females become mature. Puberty occurs in most boys sometime between the ages of 11 and 16. The young male body becomes more muscular, the voice becomes deeper, and body and facial hair appear. In most girls, puberty occurs between the ages of 9 and 14. During puberty in females, the amount of fat in the hips and thighs increases, the breasts enlarge, and body hair appears in areas such as the armpits. At this time, the young female also begins to menstruate.

IS THAT A FACT!

Girls attain three-quarters of their adult height by the age of $7 \frac{1}{2}$. Boys attain three-quarters of their adult height by the age of 9.

Adulthood From about age 20 to age 40, you will be considered a young adult. You will be at the peak of your physical development. Beginning around age 30, certain changes associated with aging begin. The changes will be gradual and slightly different for everyone. Some of the early signs of aging include decreasing muscle flexibility, deteriorating eyesight, increasing body fat, and increasing hair loss.

The aging process will continue in a middle-aged adult (someone between 40 and 65). During this period, hair may become gray, athletic abilities will decline, and skin will wrinkle. Any person over 65 years old is considered an older adult. Although aging persists during this period of an individual's life, older adults can still lead active lives. Some of this country's most productive citizens are older adults, as can be seen in **Figure 20.**

Figure 20 John Glenn, the first American to orbit Earth, returned to space at the age of 77.

QuickLab

Life Grows On

Use Figure 19 on the previous page to complete this activity.

1. Use a **ruler** to measure the infant's head height. Then measure the infant's entire body height, including the head.
2. Calculate the percentage of the infant's head height to the infant's total height.
3. Repeat these measurements and calculations for the other stages shown in the figure.

Answer the following question in your ScienceLog:

As a baby grows into an adult, does the head grow faster or slower than the rest of the body? Why do you think this is so?

TRY at HOME

SECTION REVIEW

1. What is the difference between an embryo and a fetus?
2. Why does a membrane enclose an egg once a sperm has entered?
3. What developmental changes take place from birth to puberty?
4. **Applying Concepts** When astronauts work in space, they are sometimes attached to the spacecraft by a line called an umbilical. Why do you think the line has been given this name?

internet connect

SCLINKS
NSTA

TOPIC: Before Birth, Growth and Development
GO TO: www.scilinks.org
sciLINKS NUMBER: HSTL635, HSTL645

4) Close

QuickLab
MATERIALS

FOR EACH STUDENT:
• ruler
• calculator

Answer to QuickLab
Slower; student explanations should include the following information: babies must be born with large heads to hold large brains, which enable them to learn quickly, and as babies grow older, their bodies begin to catch up in size.

Quiz

1. What is implantation? (It is the process by which an embryo embeds itself in the uterus.)
2. What functions does the placenta serve? (It is a two-way exchange organ that allows oxygen and nutrients to travel to the fetus from the mother and allows wastes to travel from the fetus to the mother.)

ALTERNATIVE ASSESSMENT

Writing Ask students to imagine that they have not yet been born. Have them write first-person stories describing their time in utero. Encourage creativity, but direct students to include the stages of development they went through as a fetus. Allow time for students to share their stories with the class.

▼ **Answers to Section Review**

1. An embryo is less developed than a fetus.
2. The membrane keeps other sperm from entering. This is important because it helps maintain the characteristic chromosome number.
3. When a person is first born, he or she is an infant. During infancy, the teeth and nervous system develop. During childhood, beginning at age 2, physical and mental capabilities develop. During adolescence, sexual characteristics develop.
4. The astronaut's umbilical cord supplies life support for the astronaut.

Skill Builder Lab

It's a Comfy, Safe World!
Teacher's Notes

Time Required

Two 45-minute class periods

Lab Ratings

TEACHER PREP 🝆🝆
STUDENT SET-UP 🝆
CONCEPT LEVEL 🝆
CLEAN UP 🝆🝆🝆

MATERIALS

This lab may require some larger plastic bags, a meterstick, and various other materials, depending on the students' designs. Soft-boiled eggs will simplify the cleanup. Students may wear gloves.

Safety Caution

Remind students to review all safety cautions and icons before beginning this lab activity. Students should wash their hands after handling eggs.

Lab Notes

This lab should be done over a large plastic sheet. You may also want to do this lab outside.

Skill Builder Lab

It's a Comfy, Safe World!

Before human babies are born, they lead a comfy life. By the seventh month, they just lie around sucking their thumb, blinking their eyes, and perhaps even dreaming. Most mammal babies develop within their mother's uterus, where they are surrounded by fluid and a placenta. Baby birds live inside a hard, protective shell until the baby has used up all the food supply. Is the internal environment in a placental mammal safer than a baby bird's environment? In this activity, you will create a model of a placental mammal's uterus to see how well it protects a fetus.

MATERIALS

- sealable plastic bags
- water
- mineral oil, cooking oil, syrup, or other thick liquid to represent fluid surrounding the fetus
- cotton, soft fabric, or other soft materials
- 3–5 soft-boiled eggs
- protective gloves

Procedure

1. Brainstorm several ideas about how you will construct and test your model. A peeled, soft-boiled egg will represent the fetus in your mammalian model. Review the structure of a uterus and amnion. Then build your model.

2. Using a computer or graph paper, make a data table similar to the "First Model Test" table below. Test your model, examine the egg for damage, and record your results.

3. If your soft-boiled egg broke during the first round of tests, you may need to modify your design. Remember to use your knowledge of a human uterus and amnion to help you improve your model. Repeat step 2, and record your results.

4. When you are satisfied with the design of your model, obtain another peeled, soft-boiled egg and an egg in the shell.

First Model Test	
Original model	**Modified model**

DO NOT WRITE IN BOOK

116

 Datasheets for LabBook

Randy Christian
Stovall Junior High School
Houston, Texas

116 Chapter 5 • Skill Builder Lab

Final Model Test	
	Test results
Model	
Egg in shell	

5. Make another data table similar to the "Final Model Test" table above. Repeat step 2 with your new eggs. Record your results in your data table. Use your table to organize, examine, and evaluate your data.

Analysis

6. Explain any differences in the test results for the mammalian model and the egg in the shell.

7. Use the direct evidence you gathered to evaluate which modification to your model was the most effective at protecting your fetus.

8. How well did your model represent a natural uterus? Identify the limitations of your model. Make recommendations for improving it.

Going Further
Compare the development of placental mammals with the development of marsupial mammals and monotremes.

Skill Builder Lab

Answers

6. Answers will vary, but an egg inside a viscous liquid in a plastic bag, wrapped in soft cotton and all inside another bag, should not be damaged when dropped from a height of 1 m. An egg protected only by a shell should break.

7–8. The answers will vary according to students' observations. Most students will observe that the soft wrapping might be thicker to protect the egg better.

Going Further
Students may want to research in the library or on the Internet about marsupials and monotremes.

Chapter Highlights

VOCABULARY DEFINITIONS

SECTION 1

asexual reproduction reproduction in which a single parent produces offspring that are identical to the parent

sexual reproduction reproduction in which two sex cells join to form a unique individual

egg sex cell produced by a female

sperm sex cell produced by a male

zygote a fertilized egg

external fertilization fertilization of an egg by sperm that occurs outside the body of the female

internal fertilization fertilization of an egg by sperm that occurs inside the body of the female

SECTION 2

testes organs in the male reproductive system that make sperm and testosterone

puberty the time of life when the sex organs become mature

vas deferens the tube in males where sperm is mixed with fluids to make semen

semen a mixture of sperm and fluids

penis the male reproductive organ that transfers semen into the female's body during sexual intercourse

Chapter Highlights

SECTION 1

Vocabulary
- asexual reproduction (p. 102)
- sexual reproduction (p. 103)
- egg (p. 103)
- sperm (p. 103)
- zygote (p. 103)
- external fertilization (p. 104)
- internal fertilization (p. 104)

Section Notes
- During asexual reproduction, a single parent produces offspring that are genetically identical to the parent. Budding and fragmentation are examples of asexual reproduction.
- During sexual reproduction, there is a union of an egg and a sperm.
- Each egg and sperm is the product of meiosis and contains half the usual number of chromosomes. The usual number of chromosomes is restored in the zygote.
- Sperm fertilize eggs outside the female's body in external fertilization. Sperm fertilize eggs inside the female's body in internal fertilization.
- Monotremes are egg-laying mammals. Marsupials are mammals that give birth to partially developed young. Placentals are mammals that give birth to well-developed young.

SECTION 2

Vocabulary
- testes (p. 106)
- puberty (p. 106)
- vas deferens (p. 106)
- semen (p. 106)
- penis (p. 106)
- ovaries (p. 107)
- fallopian tube (p. 107)
- uterus (p. 107)
- vagina (p. 107)
- infertile (p. 109)

☑ Skills Check

Math Concepts

EGGS IN EXILE A woman does not ovulate while she is pregnant. Therefore, if a woman has three children, she will release at least 27 fewer eggs from her ovaries than she would if she never became pregnant.

3 children × 9 months of pregnancy = 27 eggs

Visual Concepts

MALE AND FEMALE REPRODUCTIVE SYSTEMS The diagrams on pp. 106 and 107 show the male and female reproductive systems. Take another look at them, and make sure you recognize all the structures. Also note the similarities between the two systems. For instance, the ovaries have a similar function to the testes, and the fallopian tubes have a similar function to the vas deferens.

Lab and Activity Highlights

It's a Comfy, Safe World! **PG 116**

My, How You've Grown! **PG 182**

 Datasheets for LabBook (blackline masters for these labs)

118 Chapter 5 • Reproduction and Development

SECTION 2

Section Notes

- The male reproductive system produces sperm and delivers it to the female reproductive system. Sperm are produced in the seminiferous tubules and stored in the epididymis. Sperm leave the body through the urethra.

- The female reproductive system produces eggs, nourishes the developing embryo, and gives birth. An egg leaves one of two ovaries each month and travels to the uterus. If the egg is not fertilized, it disintegrates and menstruation occurs.

- Reproductive system disorders include infertility, cancer, and sexually transmitted diseases.

SECTION 3

Vocabulary
embryo (p. 110)
implantation (p. 110)
placenta (p. 111)
umbilical cord (p. 111)
fetus (p. 112)

Section Notes

- Fertilization occurs in a fallopian tube. From there, the zygote travels to the uterus and implants itself in the uterus's wall.

- After implantation, the placenta develops. The umbilical cord connects the embryo to the placenta. The amnion surrounds and protects the embryo.

- The embryo grows, developing limbs, nostrils, eyelids, and other features. By the eighth week, the embryo is developed enough to be called a fetus.

- Human life stages are infant (birth to 2 years), child (2 years to puberty), adolescent (puberty to 20 years), young adult (20 to 40 years), middle-aged adult (40 to 65 years), and older adult (older than 65 years).

Labs
My, How You've Grown! (p. 182)

VOCABULARY DEFINITIONS, continued

ovaries in animals, organs in the female reproductive system that produce eggs

fallopian tube the tube that leads from an ovary to the uterus

uterus an organ in the female reproductive system where a zygote grows and develops

vagina the passageway in the female reproductive system that receives sperm during sexual intercourse

infertile the state of being unable to have children

SECTION 3

embryo an organism in the earliest stage of development

implantation the process in which an embryo embeds itself in the lining of the uterus

placenta an organ that provides a developing fetus with nutrients and oxygen from the mother

umbilical cord a cord that connects the embryo to the placenta

fetus an embryo during the later stages of development within the uterus

---

internet connect

GO TO: go.hrw.com

Visit the **HRW** Web site for a variety of learning tools related to this chapter. Just type in the keyword:

KEYWORD: HSTBD5

GO TO: www.scilinks.org

Visit the **National Science Teachers Association** on-line Web site for Internet resources related to this chapter. Just type in the *sci*LINKS number for more information about the topic:

TOPIC: Reproduction — *sci*LINKS NUMBER: HSTL630
TOPIC: Before Birth — *sci*LINKS NUMBER: HSTL635
TOPIC: Reproductive System Irregularities or Disorders — *sci*LINKS NUMBER: HSTL640
TOPIC: Growth and Development — *sci*LINKS NUMBER: HSTL645

Vocabulary Review Worksheet

Blackline masters of these Chapter Highlights can be found in the **Study Guide**.

Lab and Activity Highlights

LabBank

Long-Term Projects & Research Ideas,
Get a Whiff of This!

Chapter Review Answers

Using Vocabulary
1. internal fertilization
2. seminiferous tubules
3. semen
4. ovulation
5. placenta

Understanding Concepts

Multiple Choice
6. a
7. d
8. c
9. d
10. c
11. d

Short Answer
12. The testes produce sperm, and the ovaries produce eggs.
13. the placenta
14. (must list 4) infancy, childhood, adolescence, young adulthood, middle-aged adulthood, older adulthood
15. an egg and a sperm
16. Budding is a type of asexual reproduction in which a young organism develops off a small part of the parent. Fragmentation is a type of asexual reproduction in which an organism develops from a separated piece of the parent.

Chapter Review

USING VOCABULARY

To complete the following sentences, choose the correct term from each pair of terms listed below:

1. Reptiles, birds, and mammals reproduce sexually by __?__. (*internal fertilization* or *external fertilization*)

2. Sperm are produced in the __?__ within the testes. (*epididymis* or *seminiferous tubules*)

3. The sperm-containing fluid that exits the male body is known as __?__. (*semen* or *amniotic fluid*)

4. The release of an egg from the ovary occurs once each month and is called __?__. (*ovulation* or *menstruation*)

5. The organ of exchange between the developing embryo and the mother is the __?__. (*amnion* or *placenta*)

UNDERSTANDING CONCEPTS

Multiple Choice

6. The sea star can reproduce asexually by
   a. fragmentation.
   b. budding.
   c. external fertilization.
   d. internal fertilization.

7. The correct path of sperm through the male reproductive system is
   a. testes → epididymis → urethra → vas deferens.
   b. epididymis → urethra → testes → vas deferens.
   c. testes → vas deferens → epididymis → urethra.
   d. testes → epididymis → vas deferens → urethra.

8. If the first day of the menstrual cycle is the onset of menstruation, on what day does ovulation typically occur?
   a. 2nd day   c. 14th day
   b. 5th day   d. 28th day

9. Monotremes are different from placental mammals because they
   a. are mammals.
   b. have hair.
   c. nurture their young with milk.
   d. lay eggs.

10. All of the following are sexually transmitted diseases *except*
    a. chlamydia.   c. infertility.
    b. AIDS.   d. genital herpes.

11. Fertilization occurs in the __?__, and implantation occurs in the __?__.
    a. uterus, fallopian tube
    b. fallopian tube, vagina
    c. uterus, vagina
    d. fallopian tube, uterus

Short Answer

12. What human reproductive organs produce sperm? egg cells?

13. Through what structure does oxygen from the mother pass into the fetus's body?

14. What are four stages of human life following birth?

15. What two cells combine to make a zygote?

16. What is the difference between budding and fragmentation?

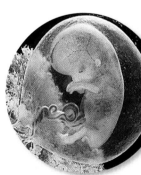

Concept Mapping

17. Use the following terms to create a concept map: asexual reproduction, budding, external fertilization, fragmentation, reproduction, internal fertilization, sexual reproduction.

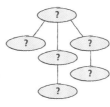

CRITICAL THINKING AND PROBLEM SOLVING

Write one or two sentences to answer the following questions:

18. Explain why the testes are found in the scrotum instead of inside the male body.

19. What is the function of the uterus? How is its function related to the menstrual cycle?

20. How is meiosis important to human reproduction?

MATH IN SCIENCE

21. Hardy Junior High School has 2,750 students. If 1 pair of identical twins is born for every 250 births, about how many pairs of identical twins will be attending the school?

22. Mrs. Schmidt had a baby April 30th. Her baby developed inside her uterus for 9 months. What month was her egg fertilized?

23. In the United States, seven infants die before their first birthday for every 1,000 births. Convert this figure to a percentage. Is your answer greater than or less than 1 percent?

24. In Haiti, a small country in the Caribbean, 74 infants die before their first birthday for every 1,000 births. Convert this figure to a percentage. Is your answer greater or less than 1 percent? Why do you think there is such a difference between the United States and Haiti?

INTERPRETING GRAPHICS

The following graph illustrates the cycles of the male hormone, testosterone, and the female hormone, estrogen. The blue line shows the estrogen level in a female over a period of 28 days. The red line shows the testosterone level in a male over a period of 28 days.

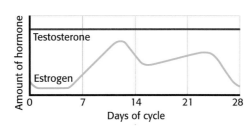

Hormone Cycles

25. What is the major difference between the two hormone levels over the 28-day period?

26. What cycle do you think estrogen affects?

27. Why might the level of testosterone stay the same?

Reading Check-up: Take a minute to review your answers to the Pre-Reading Questions found at the bottom of page 100. Have your answers changed? If necessary, revise your answers based on what you have learned since you began this chapter.

Concept Mapping

17. An answer to this exercise can be found at the front of this book.

CRITICAL THINKING AND PROBLEM SOLVING

18. Sperm develop best at a temperature three degrees lower than normal body temperature. Therefore, the testes are suspended away from the body to keep them cooler.

19. The uterus is the organ in the female reproductive system in which an embryo can develop into a fetus. Every month it builds up tissue that can help nourish a developing embryo. If no embryo is present, the tissue will discharge and cause menstruation.

20. Meiosis ensures that the characteristic human chromosome number remains the same. Since two cells must combine to form a zygote, each sex cell needs only half the 46 chromosomes that all other human body cells have.

MATH IN SCIENCE

21. 11
22. August
23. Approximately 0.7 percent of American infants die before their first birthday. This is less than 1 percent.
24. Approximately 7.4 percent of Haitian infants die before their first birthday. This is more than 1 percent. This is higher than the American percentage because Americans generally have access to better health care than Haitians.

INTERPRETING GRAPHICS

25. Estrogen levels fluctuate, but testosterone stays at the same level throughout the month.
26. Estrogen affects the menstrual cycle.
27. Testosterone levels stay the same because men continually produce sperm.

Concept Mapping Transparency 26

Blackline masters of this Chapter Review can be found in the **Study Guide.**

Across the Sciences
Acne

Background

This feature will provide students with a look at some of the changes affecting their body. Puberty is a dynamic time because hormone levels fluctuate from day to day. During puberty, acne will be more severe at times. Students should realize that the physical changes that cause acne are a necessary part of maturation and that acne is not necessarily caused by poor hygiene. Students may be reassured to learn that acne generally clears up after puberty, when hormone levels fluctuate less.

Most importantly, students should know that acne is a treatable condition and should understand how common treatments, such as over-the-counter topical creams and antibiotics, work.

Students may be interested to find out that the male sex hormones, called *androgens*, cause acne in both men and women. Indeed, men and women share the same hormones, only in different quantities and levels of activity.

ACROSS THE SCIENCES

LIFE SCIENCE • CHEMISTRY

Acne

If you are a teenager, you probably have some firsthand experience with acne. If you don't, you probably will. And contrary to what you may have heard, acne is not caused by greasy foods and candy, though these foods can aggravate the problem. The hormonal fluctuations that occur as young people mature into adults often cause acne.

What Are Pimples?

Skin contains thousands of tiny pores. Each pore contains sebaceous (suh BAY shuhs) glands that produce sebum, the oil you may have noticed on the surface of your skin. This oil is necessary to maintain healthy skin. The production and release of sebum is stimulated by androgens, the male sex hormones, which become active in both girls and boys during puberty.

Sebum usually escapes from the pores without a problem. But sometimes skin cells do not shed properly, and they clog the pores. The sebum that collects in the pores causes lesions, commonly called pimples.

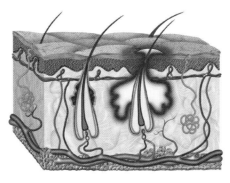

▲ *Acne is caused by the buildup of sebum and dead cells in the pores of the skin.*

Learn Your Lesions

There are two kinds of lesions—noninflamed lesions and inflamed lesions. Noninflamed lesions include blackheads and whiteheads. Some people think blackheads are pores filled with dirt. The dark color of these lesions is actually the result of dark skin pigments or oil trapped in the pores. Whiteheads are white because their contents are hidden under the skin's surface. Inflamed lesions are caused by bacteria and are often red and swollen. Bacteria live in healthy pores, and when pores become clogged, the bacteria are trapped and can cause irritation and infection.

Heredity

Family history appears to be a factor in the development of acne. Unfortunately, if your parents or brothers and sisters had acne, you are likely to have acne too. The causes of hereditary acne remain unclear. Your skin may be genetically programmed to produce more sebum than is produced in other teenagers.

Is There Hope?

Certain over-the-counter products can clean the dead skin cells and sebum out of the pores. Many medications inhibit the production of sebum or encourage the shedding of skin cells. Sometimes doctors prescribe antibiotics, such as tetracycline or erythromycin, to treat severe cases of acne. Most acne clears up as people become adults.

On Your Own

▶ Find out what the active ingredient is in an over-the-counter acne medication. Do some research on this ingredient to find out how it works. Report your findings to the class.

Answer to On Your Own

The primary ingredient in over-the-counter acne medications is benzoyl peroxide, a strong oxidizing agent that kills bacteria. Once absorbed into the skin, benzoyl peroxide is metabolized into benzoic acid, which then exits the body as benzoate through the urine. Some of the side effects of using medications with this ingredient can include different types of skin irritations, such as burning, blistering, crusting, itching, or severe redness.

Science, Technology, and Society

Technology in Its Infant Stages

Every year thousands of babies are born with life-threatening diseases or severe birth defects. What if medical treatments were available to these babies before they were born? Doctors at San Francisco, Harvard, and Vanderbilt Universities are performing experimental fetal surgery with encouraging results.

When Is Fetal Surgery an Option?

To date, approximately 100 fetal operations have been performed across the country. Corrective treatments can take place between the 18th and 30th weeks of pregnancy. Many factors determine whether fetal surgery is appropriate. Surgery is considered to be an option only if the condition is life threatening. However, fetuses with several defects or chromosomal abnormalities are not eligible for surgery.

Successful surgeries have been performed on fetal patients with spina bifida, diaphragmatic hernias, malformations of the lungs, and urinary tract obstructions. Spina bifida is a defect that leaves the spine exposed. A diaphragmatic hernia is a hole in the diaphragm. This condition causes severe breathing difficulties.

Surgery on a Small Scale

Fetal surgery can fall into one of three categories. The least traumatic type of treatment uses a laser scalpel or an endoscope. The scalpel is used to remove chest tumors. An *endoscope* is a video-guided tool that combines a camera lens and scissors that are less than 0.2 cm wide. The doctor guides the scissors through a tiny cut in the abdominal and uterine walls. The doctor is unable to see the fetus directly during this surgery because the cut is so small. Therefore, he or she must watch the video images provided by the endoscope during the operation.

A more traumatic option is open fetal surgery. In this treatment, the mother's abdomen and uterus are opened, and the fetus is partially exposed.

The third, and relatively new, option is called fetal stem cell transplant. This treatment is essentially a bone marrow transplant for the fetus. It is used to treat genetic diseases and diseases of the immune system.

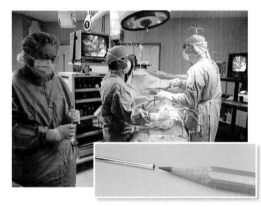

▲ *The endoscope shown here is used to perform fetal surgery.*

What the Future Holds

Each fetal surgery results in the improvement of techniques and treatments, as well as in the expansion of the types of defects and diseases that can be treated. As the number of fetal surgeries increases, fetal surgery will become much more routine.

Going Further

▶ The endoscopes used in fetal surgery use a technology called fiber optics. Research what items around your home also use fiber optics.

Chapter Organizer

CHAPTER ORGANIZATION	TIME MINUTES	OBJECTIVES	LABS, INVESTIGATIONS, AND DEMONSTRATIONS
Chapter Opener p. 124	45	National Standards: UCP 2, SAI 1, 2, HNS 3	**Start-Up Activity,** Invisible Invaders, p. 125
Section 1 Disease	90	▶ Explain the difference between infectious diseases and noninfectious diseases. ▶ Identify five ways that you might come into contact with a pathogen. ▶ Discuss four methods that have helped reduce the spread of disease. SAI 1, 2, ST 2, SPSP 1, 5, HNS 1, 3, LS 1f; Labs UCP 2, SAI 1	**Discovery Lab,** Passing the Cold, p. 184 **Datasheets for LabBook,** Passing the Cold
Section 2 Your Body's Defenses	90	▶ Describe how your body keeps out pathogens. ▶ Explain how the immune system works. ▶ Discuss the purpose of a fever. UCP 1–5, SAI 1, HNS 3, LS 1b, 1d–1f, LS 3a, 3b; Labs UCP 2, SAI 1	**QuickLab,** It's Only Skin Deep, p. 131 **Making Models,** Antibodies to the Rescue, p. 138 **Datasheets for LabBook,** Antibodies to the Rescue
Section 3 Challenges to the Immune System	90	▶ Explain the difference between allergies and autoimmune diseases. ▶ Discuss what cancer is. ▶ Describe how HIV affects the immune system. UCP 3, 4, LS 1d–1f, 3b	**Long-Term Projects & Research Ideas,** A Chuckle a Day Keeps the Doctor Away

*See page **T23** for a complete correlation of this book with the*

NATIONAL SCIENCE EDUCATION STANDARDS.

TECHNOLOGY RESOURCES

 Guided Reading Audio CD English or Spanish, Chapter 6

 One-Stop Planner CD-ROM with Test Generator

 Science Discovery Videodisc Science Sleuths: The NatureHouse Malady

 CNN Scientists in Action, In Search of Nature's Cures, Segment 29

Science, Technology, & Society, Computer Healing, Segment 34

Chapter 6 • Body Defenses and Disease

Chapter 6 • Body Defenses and Disease

CLASSROOM WORKSHEETS, TRANSPARENCIES, AND RESOURCES	SCIENCE INTEGRATION AND CONNECTIONS	REVIEW AND ASSESSMENT
Directed Reading Worksheet **Science Puzzlers, Twisters & Teasers**		
Directed Reading Worksheet, Section 1 **Critical Thinking Worksheet,** Vaccine for Super Bug Found! **Transparency 233,** Thermal Energy	**Physics Connection,** p. 128 **Connect to Physical Science,** p. 128 in ATE **Apply,** p. 129 **MathBreak,** p. 129 **Careers:** Naturopathic Physician, p. 144	**Self-Check,** p. 127 **Section Review,** p. 129 **Quiz,** p. 129 in ATE **Alternative Assessment,** p. 129 in ATE
Directed Reading Worksheet, Section 2 **Transparency 96,** An Antibody's Shape Matches a Pathogen **Transparency 97,** The Five Steps of the Immune System: A **Transparency 98,** The Five Steps of the Immune System: B **Reinforcement Worksheet,** Immunity Teamwork	**Multicultural Connection,** p. 131 in ATE **Scientific Debate:** Frogs in the Medicine Cabinet?, p. 145	**Section Review,** p. 134 **Quiz,** p. 134 in ATE **Alternative Assessment,** p. 134 in ATE
Directed Reading Worksheet, Section 3	**Real-World Connection,** p. 136 in ATE	**Homework,** p. 136 in ATE **Section Review,** p. 137 **Quiz,** p. 137 in ATE **Alternative Assessment,** p. 137 in ATE

END-OF-CHAPTER REVIEW AND ASSESSMENT

Chapter Review in Study Guide
Vocabulary and Notes in Study Guide
Chapter Tests with Performance-Based Assessment, Chapter 6 Test
Chapter Tests with Performance-Based Assessment, Performance-Based Assessment 6
Concept Mapping Transparency 27

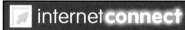

 Holt, Rinehart and Winston On-line Resources
 go.hrw.com

For worksheets and other teaching aids related to this chapter, visit the HRW Web site and type in the keyword: **HSTBD6**

 National Science Teachers Association
 www.scilinks.org

Encourage students to use the *sci*LINKS numbers listed in the internet connect boxes to access information and resources on the **NSTA** Web site.

Chapter 6 • Chapter Organizer

Chapter Resources & Worksheets

Visual Resources

TEACHING TRANSPARENCIES

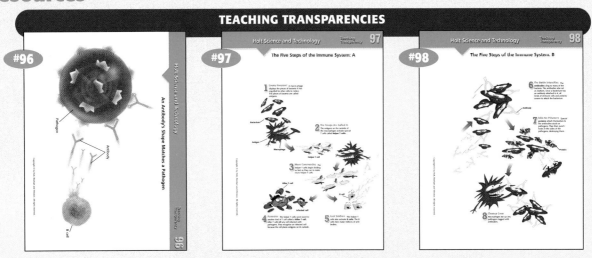

TEACHING TRANSPARENCIES / CONCEPT MAPPING TRANSPARENCY

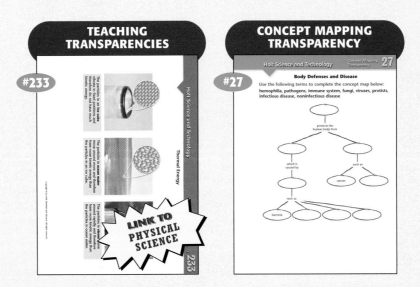

Meeting Individual Needs

DIRECTED READING

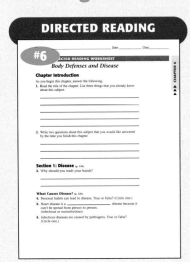

REINFORCEMENT & VOCABULARY REVIEW

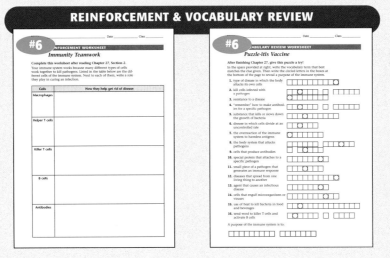

SCIENCE PUZZLERS, TWISTERS & TEASERS

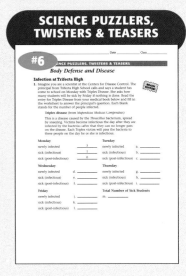

Chapter 6 • Body Defenses and Disease

Chapter 6 • Body Defenses and Disease

Review & Assessment

STUDY GUIDE

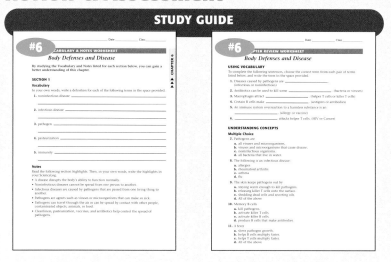

CHAPTER TESTS WITH PERFORMANCE-BASED ASSESSMENT

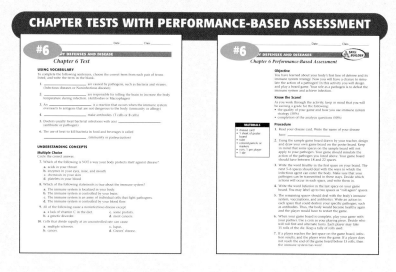

Lab Worksheets

LONG-TERM PROJECTS & RESEARCH IDEAS

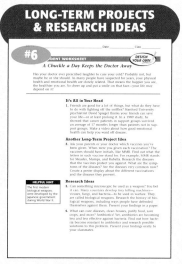

DATASHEETS FOR LABBOOK

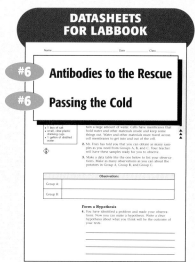

Applications & Extensions

CRITICAL THINKING & PROBLEM SOLVING

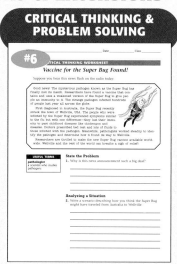

SCIENCE TECHNOLOGY

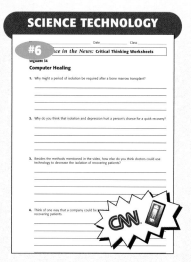

SCIENTISTS IN ACTION

Chapter 6 • Chapter Resources & Worksheets   123D

Chapter Background

SECTION 1

Disease

▶ Pathogens and the Diseases They Cause

Pathogens are agents that cause disease. Pathogens can be living or nonliving. Living pathogens include bacteria, protists, fungi, worms, and insects. Nonliving pathogens include viruses, viroids, and prions.

- Bacterial diseases include bubonic plague, cholera, dental caries, lyme disease, pneumonia, and typhoid fever.

- Viral diseases include colds, influenza, chickenpox, measles, rubella, mumps, smallpox, infectious hepatitis, polio, and AIDS.

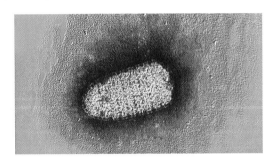

- Protists cause malaria, Chagas' disease, toxoplasmosis, and giardiasis. Common fungal diseases include ringworm, athlete's foot, vaginal yeast infections, jock itch, and histoplasmosis.

IS THAT A FACT!

- Bacteria were the first organisms to inhabit Earth. They are thought to have been the only organisms on Earth for about 1 billion years.

- All other life-forms are thought to have evolved from ancestral forms of bacteria.

- Some species of bacteria can reproduce every 20 minutes. Given unlimited resources, one bacterium could produce 1 million kilograms of bacteria in a day!

▶ Emerging Viruses

The news is full of instances in which new viruses have suddenly begun to infect people. Examples include HIV, which became common in the 1980s, hantavirus, which caused an outbreak in 1993, and new strains of Ebola, which were announced in the 1990s. Such viruses are called emerging viruses.

- Emerging viruses can evolve or mutate from an existing virus that was not infectious. They can also spread from one species to another. Another way that a virus can suddenly emerge is through globalization. When people build roads through a previously isolated area or travel to new places, viruses can spread.

▶ Types of Vaccines

Vaccines are harmless forms of a pathogen that are introduced into the body, helping the immune system develop antibodies against the pathogen should the disease-causing form enter the body in the future.

- There are several ways that vaccines are prepared today. One way to make a vaccine is to inactivate the viruses or to kill the bacteria that cause the disease. Another way is to use a similar strain of pathogen or an attenuated or weakened version of the pathogen. The latter kind of vaccine is sometimes called a live attenuated vaccine.

IS THAT A FACT!

- Not all vaccines are injections. One form of the polio vaccine is given orally, in a sugar cube.

- Researchers at the Boyce Thompson Institute for Plant Research at Cornell University are working to make a vaccine that subjects receive by eating a genetically engineered banana.

SECTION 2

Your Body's Defenses

▶ The Body's Defenses Against Pathogens

The body has two main kinds of defenses against disease—specific and nonspecific defenses.

- The body's so-called first and second lines of defense are its nonspecific defense mechanisms. The first line

Chapter 6 • Body Defenses and Disease

of defense is the skin, the mucous membranes, and the secretions of the mucous membranes. Oil, sweat, tears, mucus, and saliva wash pathogens away and contain proteins that digest the cell walls of microbes.

- The body's second line of defense includes white blood cells that ingest or destroy foreign agents that they detect in the body. The second line of defense also includes proteins that help cells avoid infection by foreign agents. Antimicrobial proteins include complement proteins and interferons. Also part of the body's second line of defense is its inflammatory response, which occurs in response to cuts or other incisions through the skin.

▶ Fever
A fever occurs when the body's temperature rises above 98.6°F (37°C) when measured orally.

- Fevers, which commonly accompany infectious diseases, help the body thwart invading pathogens. However, fevers may also accompany noninfectious conditions, such as dehydration and heart attack.

IS THAT A FACT!
- A very high fever can cause a coma, seizures, or brain damage.

SECTION 3
Challenges to the Immune System

▶ AIDS
AIDS is the final stage in an HIV infection. Upon infection, the viruses first multiply quickly and are also quickly fought off by immune system cells. Some viruses remain in the body and replicate slowly over time. Immune system cells continue the fight, sometimes for many years. HIV attacks T4 helper cells, and when the number of helper T cells falls below a certain level, a person is said to have AIDS.

- HIV is spread through the exchange of blood and other body fluids. It is not spread through casual contact, however.

- There is no cure for AIDS at this time. There are only drug regimens that may prolong the life of some patients.

- The best way to avoid contracting AIDS is to avoid the behaviors known to put one at risk for acquiring HIV. These behaviors include sharing needles and engaging in unprotected sexual contact.

▶ Allergic Reactions
The first time a person is exposed to an allergen, such as pollen, he or she usually shows no allergic response. As when a vaccine is administered, the body produces antibodies against the allergen. Thus prepared, the body develops the allergic response upon subsequent exposures to the allergen.

IS THAT A FACT!
- Allergic responses may be a defense mechanism left over from the body's defenses against parasitic worms. This hypothesis comes from the observation that the body's method of combating parasitic worms is very similar to the allergic response seen in hay fever and asthma.

- Large, colorful flowers are not usually the culprits for pollen allergies. Pollen allergies are typically the result of small, plain flowers whose pollen is dispersed via wind.

For background information about teaching strategies and issues, refer to the *Professional Reference for Teachers.*

CHAPTER 6

Body Defenses and Disease

 Pre-Reading Questions

Students may not know the answers to these questions before reading the chapter, so accept any reasonable response.

Suggested Answers

1. A pathogen, more specifically a virus, infects your body.
2. A fever helps cells in your immune system reproduce faster and also inhibits the growth of many pathogens.

Body Defenses and Disease

Sections

1. Disease 126
   - Physics Connection . . 128
   - MathBreak 129
   - Apply 129
   - Internet Connect 129
2. Your Body's Defenses . 130
   - QuickLab 131
   - Internet Connect 134
3. Challenges to the Immune System 135
   - Internet Connect 137

Chapter Lab 138
Chapter Review 142
Feature Articles 144, 145
LabBook 184–185

1. When you "catch a cold," what is it that infects your body?
2. How does a fever help you get well?

Alien Invaders

No, this photo is not from a sci-fi movie. It is not an alien insect soldier. This is, in fact, a greatly enlarged image of a house dust mite that is tinier than the dot of an *i*. Huge numbers of these creatures live in carpets, beds, and sofas in every home. Dust mites often cause problems for people who have asthma or allergies. In this chapter, you will learn how the body's immune system fights diseases and alien factors, such as dust mites, that cause allergies. You will also get some tips on controlling the spread of disease.

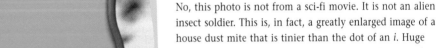

 HRW On-line Resources
go.hrw.com
For worksheets and other teaching aids, visit the HRW Web site and type in the keyword: **HSTBD6**

www.scilinks.com
Use the *sci*LINKS numbers at the end of each chapter for additional resources on the **NSTA** Web site.

www.si.edu/hrw
Visit the Smithsonian Institution Web site for related on-line resources.

www.cnnfyi.com
Visit the CNN Web site for current events coverage and classroom resources.

124 Chapter 6 • Body Defenses and Disease

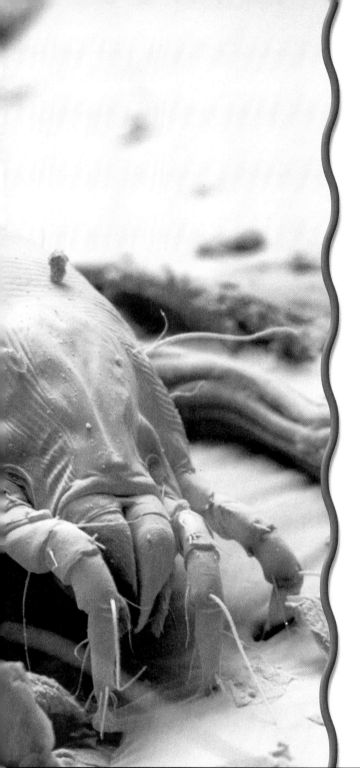

INVISIBLE INVADERS

In this activity, you will use a technique that makes "invisible" lifeforms become visible.

Procedure

1. Obtain two **Petri dishes containing nutrient agar.** Label them "Washed" and "Unwashed."

2. Rub two **marbles** between the palms of your hands. Observe the appearance of the marbles.

3. Roll one marble in the Petri dish labeled "Unwashed."

4. Put on a pair of **disposable gloves.** Wash the other marble with **soap** and **warm water** for 4 minutes. Does the appearance of the marble change after it is washed?

5. Roll the washed marble in the Petri dish labeled "Washed."

6. Secure the lids of the Petri dishes with **transparent tape**. Place the dishes in a warm, dark place. **Caution:** Do not open the Petri dishes after they are sealed.

7. Observe the Petri dishes each day for a week. Record your observations in your ScienceLog.

Analysis

8. How did the washed and unwashed marbles compare? How did the Petri dishes differ after several days?

9. Why is it important to wash your hands before eating?

INVISIBLE INVADERS

MATERIALS

FOR EACH GROUP:
- 2 Petri dishes
- nutrient agar
- 2 marbles
- sterile disposable gloves
- soap
- warm water
- transparent tape

Safety Caution

Remind students to review all safety cautions and icons before beginning this lab activity. Tell students not to open the Petri dishes once they are sealed. Treat all growth in the Petri dishes as pathogenic, and dispose of the dishes as you would any other biohazard.

Teacher's Notes

Keep the lids on the Petri dishes except when rolling the marble on the agar. This helps keep outside contamination to a minimum. In step 6, it might be helpful to use an incubator set at 37°C.

Answers to START-UP Activity

8. Descriptions can vary. The Petri dish labeled "Unwashed" should have the most bacterial growth.

9. It is important to wash to help decrease the number of microorganisms you put in your mouth.

Chapter 6 • Body Defenses and Disease

SECTION 1

Focus

Disease

This section introduces the difference between noninfectious diseases and infectious diseases. Students will learn ways they can come into contact with pathogens and how cleanliness, pasteurization, vaccines, and antibiotics can help reduce the spread of pathogens.

Bellringer

Writing Ask students to list as many different diseases as they can. After students have completed their list, you may want to make a master list on the board or on a transparency. The list might include physical illnesses as well as mental illnesses. Explain that in this chapter, students will learn how illnesses are caused by pathogens and how they can be prevented.

Directed Reading Worksheet Section 1

---

internetconnect

sciLINKS NSTA

TOPIC: What Causes Diseases?
GO TO: www.scilinks.org
sciLINKS NUMBER: HSTL655

TOPIC: Pathogens
GO TO: www.scilinks.org
sciLINKS NUMBER: HSTL660

Terms to Learn

noninfectious disease
infectious disease
pathogen
immunity

What You'll Do

- Explain the difference between infectious diseases and non-infectious diseases.
- Identify five ways that you might come into contact with a pathogen.
- Discuss four methods that have helped reduce the spread of disease.

Disease

You've probably heard it before: "Cover your mouth when you sneeze!" "Wash your hands!" "Don't put that in your mouth!" What is all the fuss about? When people say these things to you, they are concerned about the spread of disease.

What Causes Disease?

When you have a *disease*, your normal body functions are disrupted. Some diseases, such as most cancers and heart disease, are not spread from one person to another. They are called **noninfectious diseases.**

Noninfectious diseases can be caused by a variety of factors. For example, a genetic disorder causes the disease hemophilia, in which a person's blood does not clot properly. The disease scurvy is caused by a lack of vitamin C in the diet. Smoking, lack of physical activity, and a high-fat diet can greatly increase a person's chances of getting certain noninfectious diseases. Avoiding harmful habits may help you avoid noninfectious diseases.

A disease that can be passed from one living thing to another is an **infectious disease.** Infectious diseases are caused by agents called **pathogens.** Viruses and some bacteria, fungi, protists, and worms may all cause diseases. **Figure 1** shows some common pathogens.

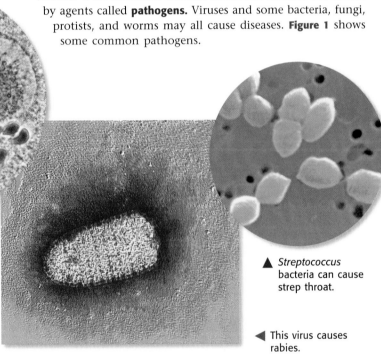

Figure 1 Pathogens, such as these, are often referred to as germs.

▲ This protist, *Entamoeba*, can infect the intestines of humans, causing a disease called dysentery.

▲ *Streptococcus* bacteria can cause strep throat.

◀ This virus causes rabies.

MISCONCEPTION ALERT

Students may think that all bacteria are pathogens. Point out that very few bacteria are pathogenic. Bacteria serve many useful, and in some cases essential, functions. They fix nitrogen. They assist in digestion. They are also used to make foods such as yogurt and cheese, to clean up oil spills, and to recycle wastes. In addition, they can be genetically engineered to produce insulin and other human proteins that are needed to treat disease.

126 Chapter 6 • Body Defenses and Disease

Pathways to Pathogens

There are many ways pathogens can be passed from one person to another. Being aware of them can help you stay healthy.

Through the Air Some pathogens travel through the air. For example, a single sneeze, like the one shown in **Figure 2,** releases thousands of tiny droplets of moisture that can carry pathogens.

Contaminated Objects A person who is sick may leave bacteria or viruses on objects such as doorknobs, keyboards, drinking glasses, towels, or combs. If you drink from a glass that an infected person has just used, you could become infected with a pathogen.

Person to Person Some pathogens are spread by direct person-to-person contact. You can become infected with some illnesses by kissing, shaking hands, or touching the sores of an infected person.

Animals Some pathogens are carried by animals. For example, humans can get a fungus called ringworm from handling an infected dog or cat. Also, ticks may carry bacteria that cause Lyme disease or Rocky Mountain spotted fever.

Food and Water Drinking water in the United States is generally safe, but water lines can break, or treatment plants can become flooded, allowing microorganisms to enter the public water supply. Bacteria growing in foods and beverages can cause illness too. Refrigerating foods can slow the growth of many of these pathogens, but meat, fish, and eggs that are not cooked enough can still contain dangerous bacteria or parasites. Leaving food out at room temperature can give bacteria such as *Salmonella* time to grow and produce toxins in the food. For these reasons, it is important to wash all used cooking tools.

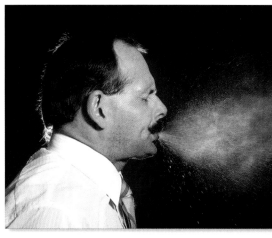

Figure 2 *A sneeze can force thousands of pathogen-carrying droplets out of your body at up to 160 km per hour.*

In developing countries, 80 percent of diseases are related to contaminated drinking water.

Self-Check

Jackie cut up raw meat on her kitchen counter. If her brother makes a sandwich on the same counter later, how could he come in contact with a pathogen? *(See page 212 to check your answer.)*

Answer to Self-Check

If Jackie did not wash the counter after cutting up the meat, bacteria could grow on the counter where the meat was. This bacteria could contaminate her brother's sandwich.

3) Extend

GOING FURTHER

Writing Certain strains of bacteria that were first noticed in the 1950s have genes that make them resistant to antibiotics. Because bacteria can share genes with one another and reproduce rapidly, the number of bacteria that are resistant to antibiotics is increasing. The misuse of antibiotics, such as taking antibiotics for viral infections, can cause the number of antibiotic-resistant bacteria to increase. Have students research and report on the health hazards of these resistant bacteria.

CONNECT TO PHYSICAL SCIENCE

Pasteurization is the process of killing harmful bacteria using thermal energy. Milk is pasteurized one of two ways: by heating it to 66°C for 30 minutes or by heating it to 75°C for 16 seconds. The organisms that can survive pasteurization will eventually spoil the milk, but they are not generally harmful to people. Boiling the milk would kill even more of the bacteria in it—not just the harmful ones—but it would also change the milk. Use Teaching Transparency 233 to illustrate how increasing the temperature of a substance increases its thermal energy.

Teaching Transparency 233 "Thermal Energy" LINK TO PHYSICAL SCIENCE

Critical Thinking Worksheet "Vaccine for Super Bug Found"

Physics CONNECTION

Hospitals use a machine called an autoclave to kill bacteria on surgical instruments. An autoclave works by increasing the pressure of steam as its temperature increases. The combined effect of pressure and temperature kills bacteria at a lower temperature than would normally be needed.

Putting Pathogens in Their Place

Until the twentieth century, surgery patients often died of bacterial infections. But as doctors learned more about disease, it became clear that simple cleanliness could help prevent the spread of some diseases. Today, hospitals and clinics use a variety of technologies to prevent the spread of pathogens. For example, ultraviolet radiation, boiling water, and chemicals are used in health facilities to kill pathogens.

Pasteurization During the mid-1800s, Louis Pasteur, a French scientist, discovered that microorganisms caused wine to spoil. The uninvited microorganisms were bacteria. Pasteur devised a method of using heat to kill most of the bacteria in the wine. This method is called *pasteurization*, and it is still used today. The milk that the girl in **Figure 3** is drinking has been pasteurized.

Vaccines and Immunity In the late 1700s, no one knew what a pathogen was. During this time, British physician Edward Jenner studied a disease called smallpox. He observed that people who had been infected with cowpox seemed to have protection against smallpox. This protection, or resistance to a disease, is called **immunity**. Jenner's work led to the first modern *vaccine*. A vaccine is a substance that helps your body develop immunity to a disease.

Today vaccines are used all over the world to prevent many serious diseases. Modern vaccines contain pathogens that are killed or specially treated so that they can't make you very sick. The vaccine is enough like the pathogen to allow your body to develop a defense against the disease.

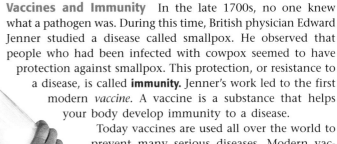

Figure 3 Today pasteurization is used to kill pathogens in many different types of food, including dairy products, eggs, meats, and juices.

WEIRD SCIENCE

During the Civil War some surgeons began to find links between unclean conditions and certain infections, such as gangrene and blood poisoning. By 1867, a British surgeon named Joseph Lister insisted on cleanliness during surgery. So it came as a shock in 1872 when an American surgeon, Addinell Hewson, claimed that the best way to treat ulcers, burns, and gunshot wounds was with dirt. Although Hewson was much criticized at the time, 93 of his severely wounded patients recovered. It appears he may have stumbled onto natural antibiotics found in the soil, more than 50 years before penicillin was discovered.

Chapter 6 • Body Defenses and Disease

Antibiotics Bacterial infections can be a serious threat to your health. Fortunately, doctors can usually treat these kinds of infections with antibiotics. An *antibiotic* is a substance that can kill bacteria or slow the growth of bacteria. Antibiotics may also be used to treat infections caused by other microorganisms, like fungi. If you take an antibiotic when you are sick, it is important that you take it according to your doctor's instructions to ensure that all the pathogens are killed.

Viruses, such as those that cause colds, are not affected by antibiotics. The only way to destroy viruses in your body is to locate and kill the cells they have invaded. In the next section, you'll see how a healthy immune system does just that.

MATH BREAK

Epidemic!

You catch a cold and return to your school while sick. Your friends don't have immunity to your cold. On the first day, you expose five friends to your cold. The next day, each of those friends passes the virus to five more people. If this pattern continues for 5 more days, how many people will be exposed to the virus?

Cold Calamity

Frank caught a bad cold just before the opening night of his school play. He visited his doctor and asked her to prescribe antibiotics for his cold. The doctor politely refused and suggested that Frank stay home and get plenty of rest. Why do you think the doctor refused to give Frank antibiotics? Explain your answer.

SECTION REVIEW

1. How is an infectious disease different from a noninfectious disease?
2. List five ways that you might come into contact with a pathogen.
3. How does a vaccine work?
4. **Inferring Relationships** Why might the risk of infectious disease be high in a community that has no water-treatment facility?

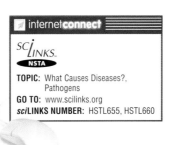

internet connect

SCILINKS NSTA

TOPIC: What Causes Diseases?, Pathogens
GO TO: www.scilinks.org
*sci*LINKS NUMBER: HSTL655, HSTL660

---

4) Close

Answers to MATHBREAK

1st day: five people exposed
2nd day: 25 (5^2) people exposed
7th day: 78,125 (5^7) people exposed

Quiz

Ask students whether these statements are true or false.

1. All diseases are caused by pathogens. (false)
2. You can become infected with some pathogens by shaking hands with an infected person. (true)
3. Leaving food out at room temperature helps kill bacteria. (false)
4. Vaccines kill or inhibit the growth of bacteria. (false)
5. Pasteurization uses very cold temperatures to kill bacteria. (false)

ALTERNATIVE ASSESSMENT

Writing Have students develop a disease-prevention pamphlet. The pamphlet should include information about what pathogens are, how they are spread, and how they can be avoided. Encourage students to include illustrations. If possible, allow students to use a desktop publishing program to produce their pamphlet.

Answers to APPLY

The doctor refused to give Frank antibiotics because colds are caused by viruses. Antibiotics have no effect on viruses.

Answers to Section Review

1. An infectious disease can be transmitted to other organisms. A noninfectious disease cannot be transmitted to another organism.
2. Answers can vary, but may include being sneezed upon by an infected person, touching an infected person, having sexual contact with an infected person, eating infected food, or breathing in the pathogen.
3. A vaccine is enough like a pathogen to help a body develop a defense against that pathogen.
4. Without water treatment, a community's water supply can become easily infected with pathogens. These pathogens can be ingested or can contaminate objects washed in the water.

SECTION 2

Focus

Your Body's Defenses

This section introduces the means by which the body protects itself from pathogens. Students will learn how the skin keeps pathogens out of the body and how the immune system works. Students will also learn the purpose of a fever.

Bellringer

Have students make a list in their ScienceLog of the ways that pathogens might enter the body. (Examples include through the mouth, ears, nose, and cuts in the skin. Pathogens can travel in the water, air, and food.)

1 Motivate

ACTIVITY

Reactions to Illness Ask students to think of a time when they were ill. Have students make a list of the ways their body reacted to the illness. (Answers might include having fever, chills, sweating, running nose, sore throat, a rash, and throbbing pain.)

Encourage students to share their lists with a partner. Then have students skim this lesson and try to link their body's reactions with the reactions of the immune system to pathogenic invasions.

 Directed Reading Worksheet Section 2

Terms to Learn

immune system   B cell
macrophage   antibody
T cell   memory B cell

What You'll Do

- Describe how your body keeps out pathogens.
- Explain how the immune system works.
- Discuss the purpose of a fever.

Your Body's Defenses

Although you probably don't realize it, your body must constantly protect itself against pathogens that are trying to invade it. But how does your body do that? Luckily, your body has its own built-in defense system.

Your Suit of Armor

For a pathogen to harm you, it must attack a part of your body. Usually, though, only a small percentage of the pathogens around you ever make it past your first lines of defense.

Eyes, Nose, and Mouth Many organisms that try to enter your eyes or mouth are destroyed by special enzymes. Pathogens that enter your nose are washed down the back of your throat by mucus. The mucus carries the pathogens to your stomach, where most are quickly digested.

Skin Your skin is made of many layers of flat cells. The outermost layers are dead. As a result, any pathogen that lands on your skin cannot find a live cell to infect. As **Figure 4** shows, the dead skin cells are constantly dropping off of your body as new skin cells grow from beneath. As the dead skin cells flake off, they carry away viruses, bacteria, and other microorganisms. In addition, glands secrete oil onto your skin's surface. The oil contains chemicals that kill many pathogens.

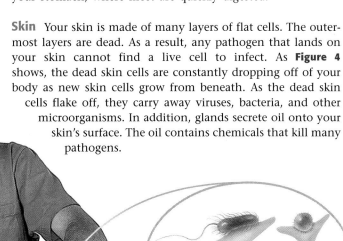

Figure 4 *Your body loses and replaces approximately 1 million skin cells every 40 minutes. In the process, countless pathogens are sloughed off.*

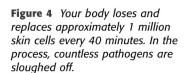

Earwax is one of your body's defenses against foreign invaders. Wax collects dirt, bacteria, fungi, and other foreign matter that could cause an ear infection.

A Forced Entry

Sometimes skin is burned, cut, or punctured. When this happens, pathogens can enter the body. The body acts quickly to keep out as many pathogens as possible. Blood flow to the injured area increases. Cell parts in the blood called *platelets* help seal the open wound so that no more pathogens can enter.

The increased blood flow also brings cells that belong to the **immune system,** the body system that fights pathogens. The immune system is not localized in any one place in your body, nor is it controlled by any one organ, such as the brain. Instead, it is an army of individual cells, tissues, and organs that work together to combat invading pathogens.

Soldiers of the Immune System

The immune system consists mainly of three kinds of cells. One kind is the **macrophage** (MAK roh FAYJ). Macrophages engulf, or eat, any microorganisms or viruses floating around. If only a few microorganisms and viruses have entered the wound, the macrophages can easily stop them.

The other two main types of immune-system cells are **T cells** and **B cells.** T cells play an important role in coordinating the immune system. Many B cells make **antibodies,** which are proteins that attach to specific pathogens. Your body is capable of making billions of different antibodies, but each antibody usually attaches to only one type of pathogen, as illustrated in **Figure 5.**

It's Only Skin Deep

Cut an **apple** in half. Place **plastic wrap** over both halves. The plastic wrap will act as skin. Use **scissors** to cut the plastic wrap on one of the apple halves, then use an **eyedropper** to drip **food coloring** on each apple half. The food coloring represents pathogens coming into contact with your body. Now, answer the following questions:

1. What happened to each apple half?
2. How is the plastic wrap similar to skin?
3. How is the plastic wrap different from skin?

Figure 5 *An antibody's shape is very specialized. It matches a pathogen like a key fits a lock.*

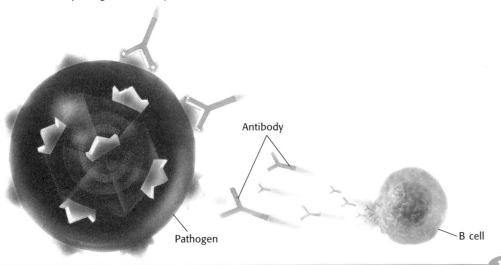

 **Multicultural CONNECTION**

In the sixteenth century, Hernando Cortes and the Spanish conquistadors overwhelmed and conquered the flourishing cultures of what is now Mexico. One reason the Europeans were able to overtake the Native Americans so quickly is that the Europeans incidentally brought with them diseases such as smallpox. These diseases were new to the populations in the Americas, and they devastated entire nations. In the span of just two generations, an estimated 12,000,000 to 25,000,000 Native Americans died as a result of European diseases.

2) Teach, continued

MEETING INDIVIDUAL NEEDS

Learners Having Difficulty Have students trace the proper sequence of the pictures across these two pages. Point out that each picture has its own caption that begins with a bold heading. Then have students outline the information on these two pages by listing the bold headings in order on a sheet of paper. Have pairs of students compare their outlines. **Sheltered English**

USING THE FIGURE

As students look at the pictures on these pages, point out that the pictures are numbered. After students have studied the sequence of events portrayed in the pictures, have them point to each kind of white blood cell shown in the pictures on these two pages as you call out their names: helper T cell, killer T cell, B cell, macrophage. Then ask students to describe the role of each kind of white blood cell. **Sheltered English**

Teaching Transparency 97
"The Five Steps of the Immune System: A"

internet connect

TOPIC: Body Defenses
GO TO: www.scilinks.org
sciLINKS NUMBER: HSTL665

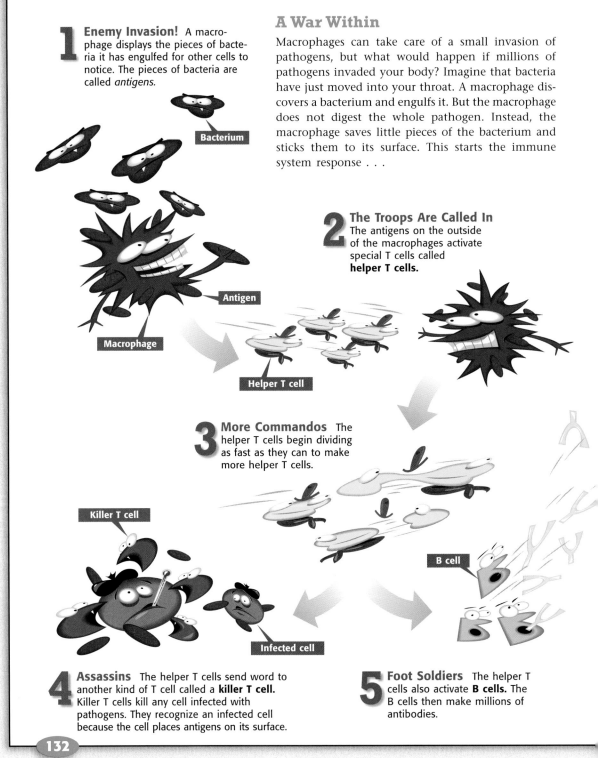

1 Enemy Invasion! A macrophage displays the pieces of bacteria it has engulfed for other cells to notice. The pieces of bacteria are called *antigens*.

2 The Troops Are Called In The antigens on the outside of the macrophages activate special T cells called **helper T cells.**

3 More Commandos The helper T cells begin dividing as fast as they can to make more helper T cells.

4 Assassins The helper T cells send word to another kind of T cell called a **killer T cell.** Killer T cells kill any cell infected with pathogens. They recognize an infected cell because the cell places antigens on its surface.

5 Foot Soldiers The helper T cells also activate **B cells.** The B cells then make millions of antibodies.

A War Within

Macrophages can take care of a small invasion of pathogens, but what would happen if millions of pathogens invaded your body? Imagine that bacteria have just moved into your throat. A macrophage discovers a bacterium and engulfs it. But the macrophage does not digest the whole pathogen. Instead, the macrophage saves little pieces of the bacterium and sticks them to its surface. This starts the immune system response . . .

IS THAT A FACT!

In 1918, a strain of flu viruses called the Spanish flu killed at least 20 million people. That's more people than were killed in combat in World War I. Other flu pandemics include the Asian flu, which broke out in 1957, and the Hong Kong flu, which occurred in 1968. All three of these serious flu outbreaks were caused by the A form of the influenza virus. Type A influenza viruses are unstable and can mutate into different, infectious forms fairly quickly. Thus, new vaccines are developed before each flu season.

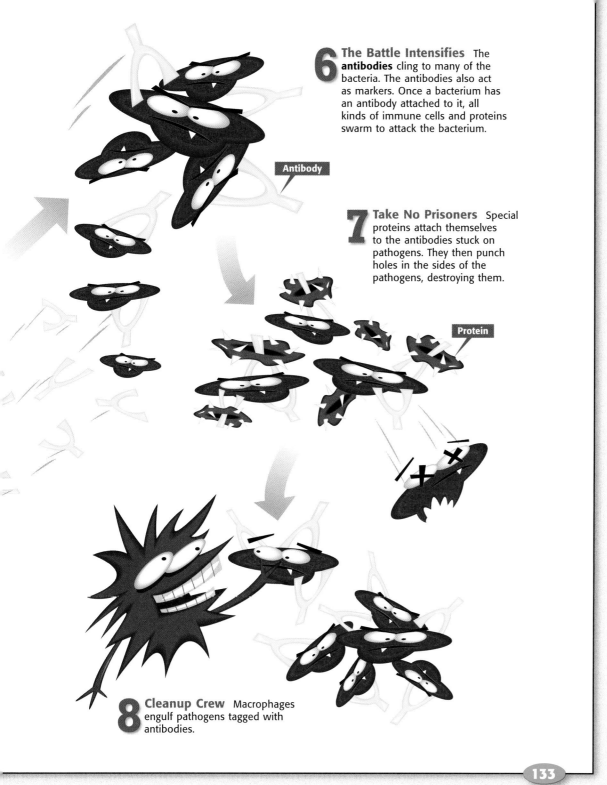

6 The Battle Intensifies The **antibodies** cling to many of the bacteria. The antibodies also act as markers. Once a bacterium has an antibody attached to it, all kinds of immune cells and proteins swarm to attack the bacterium.

7 Take No Prisoners Special proteins attach themselves to the antibodies stuck on pathogens. They then punch holes in the sides of the pathogens, destroying them.

8 Cleanup Crew Macrophages engulf pathogens tagged with antibodies.

3) Extend

Research

Writing Have students research the immune problems of "bubble" children, such as David Vetter (the "Bubble Boy"). These children suffer from severe combined immune deficiency (SCID). (Students should find that children afflicted with this disease lack T cells and B cells and have a very weak immune system. Childhood illnesses that most children can shrug off can kill a child with SCID. People with SCID must live in a germ-free environment, such as in a sterile plastic bubble or cubicle in which the air is cleaned.)

Activity

Immunity Skit Have students perform a skit that demonstrates the immune system's response to an invasion of bacteria. Assign a role to each student, and have students study the text to learn about their role. Allow students time to prepare their skit and any props they feel are necessary. For example, the students playing B cells may wish to make antibodies, and the students playing bacteria may wish to construct simple body cells to infect.

 Teaching Transparency 98 "The Five Steps of the Immune System: B"

WEIRD SCIENCE

In some cases, a pregnant woman may form antibodies against the blood of the baby she is carrying. The situation is likely to occur when the mother's blood is Rh-negative, meaning it lacks the Rh antigen, and the baby's blood is Rh-positive, meaning it carries the Rh antigen. The mother's immune system recognizes the Rh antigen as foreign and begins to attack the baby's blood cells.

4 Close

Quiz

Ask students whether these statements are true or false.

1. As dead skin cells drop off your body, some pathogens may enter your skin. (false)
2. Oil secreted by oil glands in your skin have chemicals that kill pathogens. (true)
3. Antibodies are specific to certain pathogens. (true)
4. Fevers are dangerous because they can increase the growth of pathogens in your body and make you sicker. (false)

ALTERNATIVE ASSESSMENT

Making Models Provide students with an assortment of craft materials, including colored modeling clay, construction paper, pipe cleaners, plastic foam, and other materials. Have students use the materials to construct a "battlefield" that shows the fight among immune system cells and invading pathogens. Allow students to work with a partner to portray the battle. Students should identify the pathogens, the disease they cause, and the following aspects of the immune system: B cells, helper T cells, killer T cells, macrophages, and antibodies. **Sheltered English**

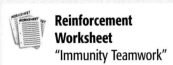
Reinforcement Worksheet
"Immunity Teamwork"

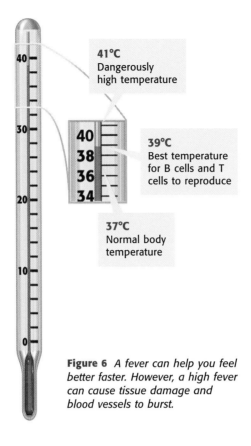

Figure 6 *A fever can help you feel better faster. However, a high fever can cause tissue damage and blood vessels to burst.*

- 41°C Dangerously high temperature
- 39°C Best temperature for B cells and T cells to reproduce
- 37°C Normal body temperature

internet connect
sciLINKS NSTA
TOPIC: Body Defenses
GO TO: www.scilinks.org
sciLINKS NUMBER: HSTL665

Heating Things Up

When macrophages activate the helper T cells, they also send a chemical signal that tells your brain to turn up the thermostat. In a few minutes, your body's temperature can rise several degrees. A moderate fever of one or two degrees actually helps you get well faster because it slows the growth of some pathogens. As is shown in **Figure 6**, a fever also helps B cells and T cells multiply faster than usual.

Haven't We Met Somewhere Before?

The immune system responds very quickly if your B cells recognize the invading pathogen and can produce antibodies for it. However, B cells must have had previous contact with a pathogen before they can make the correct antibodies. During the first encounter with a new pathogen, specialized B cells make antibodies that are effective against that particular invader. This process takes about 2 weeks, which is far too long to prevent an infection. Therefore, the first time you are infected, you usually get sick.

A few of the B cells become **memory B cells** that "remember" how to make an antibody for a particular pathogen. If the pathogen shows up again, the memory B cells produce B cells that make enough antibodies to protect you in just 3 or 4 days.

SECTION REVIEW

1. List three ways your body defends itself against pathogens.
2. Name three different cells in the immune system, and describe how they respond to pathogens.
3. Can you make antibodies for diseases you have never come in contact with? Why or why not?
4. **Applying Concepts** If you had chickenpox at age 7, what would be your chances of getting chickenpox again if your memory B cells lived only 2 months?

▼ **Answers to Section Review**

1. Sample answer: Skin keeps most pathogens out. Chemical defenses are in your eyes, stomach, and mouth. If you have a cut in your skin, blood platelets help to close the wound so more microorganisms cannot enter. Any microorganisms that do enter encounter immune system cells.
2. Sample answer: **macrophages**—they engulf pathogens and stick antigens on their outer membranes; **helper T cells**—they activate killer T cells and B cells; **killer T cells**—they kill any body cell infected with pathogens
3. No; B cells can't make antibodies for a disease until they have encountered that disease at least once before.
4. You could get chickenpox every other month, because there would be no memory B cells to "remember" the pathogen.

Chapter 6 • Body Defenses and Disease

SECTION 3

Terms to Learn
allergy
autoimmune disease
cancer

What You'll Do
- Explain the difference between allergies and autoimmune diseases.
- Discuss what cancer is.
- Describe how HIV affects the immune system.

Challenges to the Immune System

The immune system is a very effective body-defense system, but it is not invincible. There are some diseases that the immune system is unable to deal with. There are also conditions in which the immune system does not work properly.

Ragweed pollen

A-a-achoo!

Sometimes the immune system overreacts to antigens that are not dangerous to the body. This inappropriate reaction is called an **allergy**. Allergies may be caused by many things, including certain foods and medicines. Some of the culprits behind allergic reactions are shown in **Figure 7**. Symptoms can range from a runny nose and itchy eyes to more serious conditions, such as asthma.

Doctors are not sure why the immune system overreacts in some people. Scientists think allergies might be useful because the mucus draining from your nose carries away pollen, dust, and microorganisms.

Figure 7 Things That Cause Allergies

Pollen

Dust mite

Animal hair and dander (skin flakes)

Cigarette smoke

IS THAT A FACT!

Peanuts are dangerous, even potentially deadly, to people who are allergic to them. A person nearly died from eating a tuna sandwich because the knife used to cut the sandwich had just been used to make a peanut butter sandwich.

internet connect

TOPIC: Allergies
GO TO: www.scilinks.org
sciLINKS NUMBER: HSTL670

2 Teach

REAL-WORLD CONNECTION

Allergies are usually treated two ways once the allergen is identified. An allergist may try to desensitize a patient to the allergen with small, periodic doses of the allergen. In addition, an allergist may prescribe drugs to minimize the body's reactions to allergens.

Homework

✏️ Writing | Ask students to research and report on the topic anaphylactic shock. (Students should find that anaphylactic shock is a serious allergic reaction that can result in death. A small amount of an allergen can produce anaphylactic shock. People who are prone to such an allergic reaction may carry with them a syringe containing the hormone epinephrine to combat the onset of anaphylactic shock.)

3 Extend

GOING FURTHER

✏️ Writing | Have students find an article or news report about one of the immune disorders discussed in this lesson. Have students work in small groups to summarize their article and link it to the concepts presented in this chapter, either in writing or orally.

internetconnect

SCI_{LINKS} | **TOPIC:** Cancer and HIV
GO TO: www.scilinks.org
NSTA | **sciLINKS NUMBER:** HSTL675

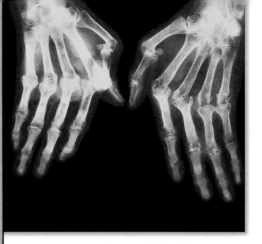

Figure 8 *In rheumatoid arthritis, immune-system cells cause joint-tissue swelling, which can lead to joint deformities.*

Autoimmune Diseases

An **autoimmune disease** is a disease in which the immune system attacks the body's own cells. This happens when immune-system cells are not able to tell the difference between pathogens and particular body cells. One autoimmune disease is rheumatoid arthritis, in which the immune system attacks the joints. The most common location for rheumatoid arthritis is the joints of the hands, as shown in **Figure 8**. Other autoimmune diseases include type 1 diabetes, Graves' disease, multiple sclerosis, and lupus.

Cancer

Healthy cells divide at a carefully regulated rate. Occasionally, though, a cell doesn't respond to the body's regulation and begins dividing at an uncontrolled rate. As can be seen in **Figure 9,** killer T cells destroy this type of cell. But sometimes division of these cells gets out of the control of the immune system. This causes a condition known as **cancer.**

Many cancers will invade nearby tissues. They can also enter the cardiovascular system or lymphatic system. This way, cancers can be transported to other places in the body. Cancers disrupt the normal activities of organs they have invaded, often leading to death. Today, though, there are many treatments for cancer. Radiation and certain drugs can be used to kill cancer cells or slow their division.

Figure 9 The Destruction of an Unregulated Cell

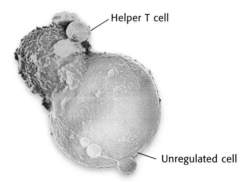

Helper T cell

Unregulated cell

❶ A killer T cell attacks an unregulated cell.

❷ The cell's membrane ruptures as the cell dies.

WEIRD SCIENCE

There is much evidence to suggest that emotional stress affects a person's immune system. Nearly 2,000 years ago, the Greek physician Galen (A.D. 129–c. 199) observed that people who were mentally depressed tended to be more susceptible to certain diseases than those who were mentally healthy. Today, it is known that hormones secreted in response to stressful situations can affect the number of white blood cells in the blood and can weaken the immune system.

136 Chapter 6 • Body Defenses and Disease

AIDS

The human immunodeficiency virus (HIV) causes the acquired immune deficiency syndrome (AIDS). Most viruses infect cells in the nose, mouth, lungs, or intestines, but HIV is different. As you can see in **Figure 10,** HIV infects the immune system itself, using helper T cells as factories to produce more viruses. The helper T cells are destroyed in the process. Remember that the helper T cells put the B cells and killer T cells to work.

People with AIDS have very few helper T cells, so nothing activates the B cells and killer T cells. Therefore, the immune system cannot attack HIV or any other pathogen. People with AIDS don't usually die of AIDS itself. They die of other diseases that they are unable to fight off.

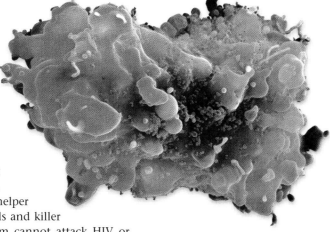

Figure 10 The blue particles on this helper T cell are human immunodeficiency viruses. They were made inside the cell and can now go and infect other cells.

SECTION REVIEW

1. What is the difference between allergies and autoimmune diseases?
2. Why is it important for immune-system cells to be able to recognize all of the body's own cells?
3. What characterizes a cancerous cell?
4. **Interpreting Graphs** Over time, people with AIDS become very sick and are unable to fight off infection. Use the information in the graph below to explain why this occurs.

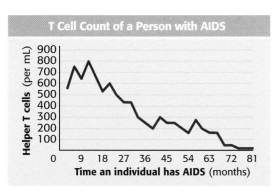

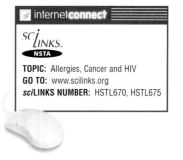

TOPIC: Allergies, Cancer and HIV
GO TO: www.scilinks.org
sciLINKS NUMBER: HSTL670, HSTL675

Making Models Lab

Antibodies to the Rescue
Teacher's Notes

Time Required
One 45-minute class period

Lab Ratings

- TEACHER PREP 🧪🧪
- STUDENT SET-UP 🧪🧪
- CONCEPT LEVEL 🧪🧪
- CLEAN UP 🧪🧪

MATERIALS

You will be able to expand this activity from very simple to a grand art project! Have students bring craft supplies from home. You may want to include poster paper in the supply list.

Safety Caution
Remind students to review all safety cautions and icons before beginning this lab activity.

Preparation Notes
Encourage students to be imaginative in this exercise. Have them come up with interesting shapes for the viruses and the antibodies that fit them. Remind students that the main lesson of this exercise is that the antibodies can fit only a specific pathogen and in only one specific way, just as a key fits only one lock.

Making Models Lab

Antibodies to the Rescue

Some cells of the immune system, called B cells, make antibodies that attack and kill invading viruses and microorganisms. These antibodies help make your body immune to disease. Have you ever had chickenpox? If you have, your body has built up antibodies that can recognize that particular virus. Antibodies will attach themselves to the virus, tagging it for destruction. If you are exposed to the same disease again, the antibodies remember that virus. They will attack the virus even quicker and in greater number than they did the first time. That is why you will probably never have chickenpox more than once.

In this activity, you will construct simple models of viruses and their antibodies. You will see how antibodies are specific for a particular virus.

MATERIALS

- craft materials, such as buttons, fabric scraps, pipecleaners, and recycled materials
- scissors
- tape or glue
- colored paper

Procedure

1. Draw the virus patterns shown on this page on a separate piece of paper, or design your own virus models from the craft supplies. Remember to design different receptors on each of your virus models.

2. In your ScienceLog, write a few sentences describing how your viruses are different.

3. Cut out the viruses, and attach them to a piece of colored paper with tape or glue.

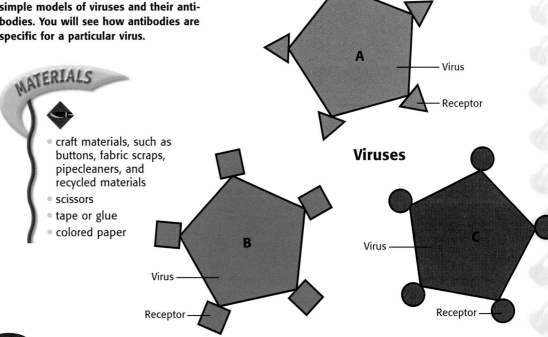

Viruses

Datasheets for LabBook

Martha Kisiah
Fairview Middle School
Tallahassee, Florida

138 Chapter 6 • Making Models Lab

Antibodies

4 Select the antibodies drawn above or design your own antibodies that will exactly fit on the receptors on your virus models. Draw or create each antibody enough times to attach one to each receptor site on the virus.

5 Cut out the antibodies you have drawn. Arrange the antibodies so that they bind to the virus at the appropriate receptor. Attach them to the virus with tape or glue.

Analysis

6 Explain how an antibody "recognizes" a particular virus.

7 After the attachment of antibodies to the receptors, what would be the next step in the immune response?

8 Many vaccines use weakened copies of the virus to protect the body. Use the model of a virus and its specific antibody to explain how vaccines work.

9 Use your model of a virus to demonstrate to the class how a receptor might change or mutate so that a vaccine would no longer be effective.

Going Further
Research in the library or on the Internet to find information about the discovery of the Salk vaccine for polio. Include information on how polio affects people today.

Research in the library or on the Internet to find information about filoviruses. What do they look like? What diseases do they cause? Why are they especially dangerous? Is there an effective vaccine against any filovirus? Explain.

Making Models Lab

Going Further
Jonas Edward Salk (1914–1995) was born in New York City. His work on an anti-influenza vaccine led him and his colleagues to develop an inactivated vaccine against polio in 1952. After successful wide-scale testing in 1954, the vaccine was distributed nationally, greatly reducing the incidence of the disease. With the Salk vaccine, the cases of polio have drastically gone down year after year; there hasn't been a single case of polio in the Western Hemisphere since 1994, and there were only 4,000 polio cases reported worldwide in 1996. Because we are nearing full eradication of the disease, the World Health Organization plans to end routine polio vaccination sometime around the year 2005.

Filoviruses belong to the family Filoviridae, one of several groups of viruses that can cause hemorrhagic fever, such as Ebola, in animals and humans. When magnified many thousands of times by an electron microscope, filoviruses have the appearance of long filaments or threads. Because filoviruses can be extremely hazardous, laboratory studies of these viruses must be conducted in special maximum-containment facilities. The reservoir and natural history of filoviruses remain unknown. Filoviruses have the highest fatality rates (as high as 90 percent for epidemics of hemorrhagic fever caused by Ebola-Zaire virus). No vaccine exists to protect against filovirus infection, and no specific treatment is available for diseases caused by these viruses.

Answers

6. Sample answer: Antibodies recognize and bind to specific pathogens because they are shaped to match the specific three-dimensional shape of the antigen.

7. Sample answer: Antibodies bind to specific pathogens and either inactivate the pathogen or trigger its destruction by macrophages.

8. Sample answer: Vaccines produce immunity because they contain antigens that stimulate an immune response producing memory cells. Students can use their models to demonstrate or give examples.

9. Sample answer: The antibodies specific for a pathogen are ineffective when the pathogen changes these proteins. Students can show this by slightly changing their model so that the virus and the antibody no longer fit like a lock and key.

Chapter Highlights

VOCABULARY DEFINITIONS

SECTION 1

noninfectious disease a disease that cannot spread from one person to another

infectious disease a disease caused by a pathogen

pathogen an agent that causes disease

pasteurization a method of heating food and beverages to kill bacteria

immunity resistance to a disease

SECTION 2

immune system a collection of cells that fight pathogens; these cells include macrophages, B cells, memory B cells, helper T cells, and killer T cells

macrophage an immune-system cell that engulfs pathogens

T cell an immune-system cell that matures in the thymus

B cell an immune-system cell that matures in bones and makes antibodies

antibody a special protein that can recognize specific pathogens

memory B cell an immune-system cell that "remembers" how to make a specialized antibody for a particular pathogen

Chapter Highlights

SECTION 1

Vocabulary
noninfectious disease (p. 126)
infectious disease (p. 126)
pathogen (p. 126)
immunity (p. 128)

Section Notes
- A disease disrupts the body's ability to function normally.

- Noninfectious diseases cannot be spread from one person to another.
- Infectious diseases are caused by pathogens that are passed from one living thing to another.
- Pathogens are agents such as viruses or microorganisms that can make us sick.
- Pathogens can travel through the air or can be spread by contact with other people, contaminated objects, animals, or food.
- Cleanliness, pasteurization, vaccines, and antibiotics help control the spread of pathogens.

Labs
Passing the Cold (p. 184)

SECTION 2

Vocabulary
immune system (p. 131)
macrophage (p. 131)
T cell (p. 131)
B cell (p. 131)
antibody (p. 131)
memory B cell (p. 134)

Section Notes
- Enzymes in your eyes, nose, and mouth kill most pathogens that try to enter. Other pathogens are washed down the throat and destroyed in the stomach.
- Dead skin cells and oil help to keep germs out of the body.

☑ Skills Check

Math Concepts

SPREAD OF DISEASES It is easy to infect a large group of people with a disease. For instance, suppose a man with the flu gets onto an empty train. The train stops at 10 different towns. If five people get off and five people get on at every stop, how many people could the man expose to his illness?

10 stops × 5 people = 50 people exposed

Visual Understanding

IMMUNE RESPONSE Look at the immune response illustration on pp. 132–133. Review each step to make sure you understand how your immune system works. Think about how many different cells the immune system uses to destroy pathogens. Also, notice that each cell has a special job.

Lab and Activity Highlights

Antibodies to the Rescue PG 138

Passing the Cold PG 184

 Datasheets for LabBook (blackline masters for these labs)

140   Chapter 6 • Body Defenses and Disease

SECTION 2

- When pathogens get into your blood or tissues, the immune system reacts.
- Macrophages engulf pathogens. Macrophages then display parts of the pathogens, called antigens, on their surface.
- Macrophages activate helper T cells. The helper T cells put the killer T cells and B cells to work.
- Killer T cells kill infected cells. B cells make antibodies.
- Antibodies cling to antigens and attract macrophages and other cells. Special proteins kill pathogens with antibodies stuck to them.
- Macrophages cause fever, which speeds the division of T cells and B cells.
- Memory B cells stand ready to produce more B cells that make antibodies if the pathogen appears again.

SECTION 3

Vocabulary
allergy *(p. 135)*
autoimmune disease *(p. 136)*
cancer *(p. 136)*

Section Notes
- The immune system can overreact to a harmless antigen. This reaction is called an allergy.
- Autoimmune diseases are diseases in which the immune system attacks the body's healthy tissue.
- Cancer cells can enter the body's circulatory systems and infect other areas of the body.
- HIV attacks helper T cells, preventing the immune system from functioning properly.

VOCABULARY DEFINITIONS, *continued*

SECTION 3

allergy an inappropriate immune-system reaction to a harmless antigen

autoimmune disease a disease in which the immune system attacks the cells of the body it is meant to protect

cancer a condition in which certain body cells begin dividing at an uncontrolled rate

 Vocabulary Review Worksheet

 Blackline masters of these Chapter Highlights can be found in the **Study Guide.**

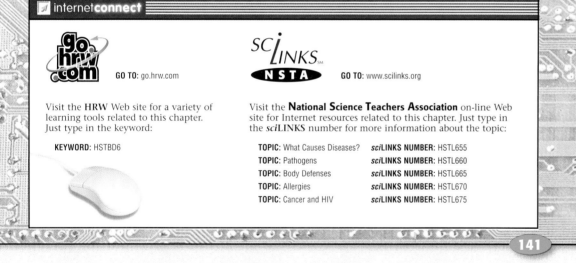

internet connect

GO TO: go.hrw.com

Visit the **HRW** Web site for a variety of learning tools related to this chapter. Just type in the keyword:

KEYWORD: HSTBD6

GO TO: www.scilinks.org

Visit the **National Science Teachers Association** on-line Web site for Internet resources related to this chapter. Just type in the *sci*LINKS number for more information about the topic:

TOPIC	sciLINKS NUMBER
What Causes Diseases?	HSTL655
Pathogens	HSTL660
Body Defenses	HSTL665
Allergies	HSTL670
Cancer and HIV	HSTL675

Lab and Activity Highlights

LabBank

 Long-Term Projects & Research Ideas,
A Chuckle a Day Keeps the Doctor Away

Chapter Review

Chapter Review Answers

USING VOCABULARY

1. infectious
2. bacteria
3. helper T cells
4. antibodies
5. allergy
6. HIV

UNDERSTANDING CONCEPTS

Multiple Choice

7. b
8. d
9. c
10. d
11. d
12. d

Short Answer

13. When a macrophage engulfs a pathogen, it places pieces of the pathogen called antigens on its outer membrane. The antigens attract helper T cells.
14. Helper T cells activate B cells and killer T cells.

Concept Mapping Transparency 27

Blackline masters of these Chapter Highlights can be found in the **Study Guide.**

Chapter Review

USING VOCABULARY

To complete the following sentences, choose the correct term from each pair of terms listed below:

1. Diseases caused by pathogens are __?__. (*infectious* or *noninfectious*)

2. Antibiotics can be used to kill some __?__. (*bacteria* or *viruses*)

3. Macrophages attract __?__. (*helper T cells* or *killer T cells*)

4. Certain B cells make __?__. (*antigens* or *antibodies*)

5. An immune system overreaction to a harmless substance is a(n) __?__. (*allergy* or *vaccine*)

6. __?__ attacks helper T cells. (*HIV* or *Cancer*)

UNDERSTANDING CONCEPTS

Multiple Choice

7. Pathogens are
   a. all viruses and microorganisms.
   b. viruses and microorganisms that cause disease.
   c. noninfectious organisms.
   d. all bacteria that live in water.

8. The following is an infectious disease:
   a. allergies
   b. rheumatoid arthritis
   c. asthma
   d. common cold

9. The skin keeps pathogens out by
   a. staying warm enough to kill pathogens.
   b. releasing killer T cells onto the surface.
   c. shedding dead cells and secreting oils.
   d. All of the above

10. Memory B cells
    a. kill pathogens.
    b. activate killer T cells.
    c. activate killer B cells.
    d. produce B cells that make antibodies.

11. A fever
    a. slows pathogen growth.
    b. helps B cells multiply faster.
    c. helps T cells multiply faster.
    d. All of the above

12. Macrophages
    a. make antibodies.
    b. release helper T cells.
    c. live in the gut.
    d. engulf pathogens.

Short Answer

13. Explain how macrophages start an immune response.

14. Describe the role of helper T cells in responding to an infection.

142 Chapter 6 • Body Defenses and Disease

Concept Mapping

15. Use the following terms to create a concept map: macrophages, helper T cells, B cells, antibodies, antigens, killer T cells, memory B cells.

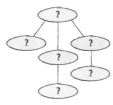

CRITICAL THINKING AND PROBLEM SOLVING

Write one or two sentences to answer the following questions:

16. Why does the disappearance of helper T cells in AIDS patients damage the immune system?

17. Many people take fever-reducing drugs as soon as their temperature exceeds 37°C. Why might it not be a good idea to immediately reduce a fever with drugs? What are the benefits of taking fever-reducing drugs?

18. The risk of dying from a whooping cough vaccine is about one in 1 million. In contrast, the risk of dying from whooping cough itself is about one in 500. Discuss the pros and cons of this vaccination.

MATH IN SCIENCE

19. Suppose you have 50,000 flu viruses on your fingers and you rub your eyes. Only 20,000 viruses make it into your eyes, 10,000 dissolve in chemicals, and 10,000 are washed down into your nose. Of those, you sneeze out 2,000. How many viruses are left to wash down the back of your throat and start an infection?

INTERPRETING GRAPHICS

Immune Response

The graph above compares the concentration of antibodies in the blood the first time you are exposed to a pathogen with the concentration of antibodies the next time you are exposed to the pathogen.

20. Are there more antibodies present during the first week of the first exposure or the first week of the second exposure? Why do you think this is so?

21. What is the difference in recovery time between the first exposure and second exposure? Why?

Take a minute to review your answers to the Pre-Reading Questions found at the bottom of page 124. Have your answers changed? If necessary, revise your answers based on what you have learned since you began this chapter.

Concept Mapping

15. An answer to this exercise can be found at the front of this book.

CRITICAL THINKING AND PROBLEM SOLVING

16. Helper T cells are needed to activate B cells and killer T cells. Without B cells, the immune system cannot produce antibodies. Without killer T cells, the immune system cannot destroy infected cells.

17. Fevers can help T cells and B cells multiply faster. Fevers can also slow the growth of some pathogens. High fevers can be dangerous, though. If a fever exceeds 40.6°C, fever-reducing drugs can help bring the body's temperature down to a safer temperature.

18. The vaccine gives you an immunity to whooping cough, which is a dangerous disease. However, you may never come into contact with whooping cough and, therefore, may not need the vaccine. If you receive the vaccine anyway, you will have a small risk of dying from the vaccine. If no one is vaccinated, however, the risk of contracting the disease rises.

MATH IN SCIENCE

19. 8,000 viruses

INTERPRETING GRAPHICS

20. There are more antibodies present in the second exposure because memory B cells recognize the pathogen and immediately start the production of antibody-producing B cells.

21. Recovery time is much shorter for the second exposure because your body can produce more antibodies in less time than during the first exposure.

CAREERS

Naturopathic Physician—Stacey Kargman

Background

One kind of sensitivity involves what are called the IgE antibodies. These antibodies engender an immediate reaction. For example, when a person who is allergic to strawberries eats a strawberry, and a few minutes later that person breaks into hives or notices that he or she has a swollen, irritated tongue, that is an IgE response.

To help establish which foods are problematic for someone, naturopaths such as Kargman rely on specialized blood tests. After taking blood, the naturopath will send it to a lab, where it is exposed to different substances and tested for reactions. Once a sensitivity has been discovered, a patient can avoid the foods that have a negative impact on his or her immune system.

NATUROPATHIC PHYSICIAN

Dr. Stacey Kargman of Tucson, Arizona, is a doctor of naturopathic medicine (NMD), commonly referred to as a naturopath. An NMD has similar training to an MD but is less likely than a traditionally trained doctor to use prescription drugs or surgery to treat a patient's symptoms. Naturopaths tend to look for a natural way to treat a patient, using drugs or surgery as a last resort. Dr. Kargman tries to strengthen her patients' immune systems by focusing on things like nutrition.

Dr. Kargman attended the Southwest College of Naturopathic Medicine, where she studied all the sciences a medical doctor would study—like biochemistry, anatomy, pharmacology, and physiology. Beyond the standard medical school sciences, naturopaths spend an additional four years studying subjects like botanical medicines, homeopathy, acupuncture, counseling, and nutrition. "Naturopathy is a way of looking at the person as a whole," says Kargman.

The Keystone to Good Health

Many naturopaths believe that nutrition is the keystone to good health. "Most MDs don't talk to their patients about their diets," Kargman explains. "I'm in a position to talk to them about what they eat and how it may be affecting their health. Food allergies can cause an immune reaction in the body—anything from depression to skin problems to migraine headaches. Even though I can prescribe prescription medications, I usually defer to MDs when it comes to prescription medications."

Dr. Kargman treats many HIV and AIDS patients. She encourages these patients and others who need prescription medications to work with their medical doctor and their naturopath at the same time. That way, patients get the best care.

A Fulfilling Career

Dr. Kargman says the best part of her work is making people feel better. "Someone might come to me and say they have terrible migraines that they can no longer live with and that they've seen every doctor. After examining them, I might be able to tell them something as simple as, 'Stop eating wheat.' The simplest thing can change someone's life . . . It's not like putting a bandage on it. It's fixing the cause of the problem."

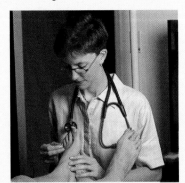

▲ Stacey Kargman, NMD, tries to treat the patient as a whole.

On Your Own

▶ Do some research about naturopaths. Find out how an NMD's training and practice differ from the training and practice of an MD.

Answer to On Your Own

Answers will vary. Students might contact the American Naturopathic Medical Association to learn more about how naturopaths are trained and how they practice.

Health Watch

Frogs in the Medicine Cabinet?

Frog skin, mouse intestines, cow lungs, and shark stomachs—sounds like the ingredients for a witch's brew, doesn't it? Actually, these animal parts are being tested in an effort to create more effective medicines to combat harmful bacteria.

Leapin' Lily Pads—It's Infection Protection

In 1896, a biologist named Michael Zasloff was studying African clawed frogs. He noticed that cuts in the frogs' skin healed quickly and never became infected. Zasloff decided to investigate further. He found that when a frog was cut, its skin released a liquid antibiotic that killed invading bacteria.

Scientists have found other animals whose bodies contain similar infection fighters. For example, the stomach and tissues of sand sharks (also called dogfish) contain chemicals that kill bacteria and other microorganisms. These useful antibiotics are also in moths, pigs, mice, cows, and even the small intestines of humans!

What's in Dog Spit?

A healthy dog licks cuts, scrapes, and minor wounds to clean them. A mother cat licks her kittens clean. Have you ever wondered why animals do that? Well, dogs, cats, humans, and some other animals have an antibacterial enzyme in their saliva. When animals lick a wound, the enzymes kill the bacteria and help the wound heal.

▲ *African clawed frogs produce a natural antibiotic.*

POW! Punching Holes in Bacteria

Bacteria are becoming resistant to many man-made antibiotics, which means that the drugs no longer affect the bacteria. Scientists now face the challenge of developing new antibiotics that can overcome the resistant strains of bacteria.

Antibiotics from animals pack a different punch than some man-made antibiotics. These substances bore holes through the membranes that surround bacterial cells, causing the cells to disintegrate and die. Bacterial membranes don't mutate often, so they are less likely to become resistant to the animal antibiotics.

Getting Well

▶ When your doctor prescribes antibiotics, you are usually reminded to finish the entire bottle even if you start to feel better. Call or visit a local pharmacy to investigate why this is so important when you are taking antibiotics.

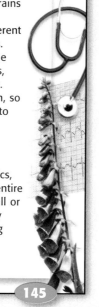

Answer to Getting Well

Not finishing the bottle might enable the remaining bacteria to begin another infection.

Scientific Debate

Frogs in the Medicine Cabinet

Background

Antibiotics are drugs used to treat diseases caused by microorganisms. Antibiotics interfere with the disease-causing (pathogenic) bacteria by either damaging the cell membrane or disrupting chemical processes in the cell.

Antibiotics must be used properly to be effective. Some possible side effects from antibiotics include allergic reactions, damage to organs and tissues, and destruction of helpful microorganisms.

Modern scientific study of antibiotics began in the 1800s. Louis Pasteur discovered that bacteria spread infectious diseases, and Robert Koch developed methods for isolating and growing different kinds of bacteria. A breakthrough in treating bacterial diseases came in the early 1900s when Alexander Fleming discovered penicillin, an antibiotic formed from mold. Streptomycin, a fungal antibiotic, was discovered by Selman A. Waksman in 1943. Doctors now use antibiotics to treat diseases such as strep throat, bacterial meningitis, and tuberculosis.

Chapter Organizer

CHAPTER ORGANIZATION	TIME MINUTES	OBJECTIVES	LABS, INVESTIGATIONS, AND DEMONSTRATIONS
Chapter Opener pp. 146–147	45	National Standards: SAI 1	**Start-Up Activity,** Conduct a Survey, p. 147
Section 1 What We Put into Our Bodies	90	▶ Identify the six groups of nutrients and explain their importance to good health. ▶ Use dietary guidelines and the food pyramid to plan a healthy diet. ▶ Understand nutrition information labels. ▶ Explain the dangers of various nutritional disorders. UCP 1, 4, SAI 1, SPSP 1, 4; Labs SPSP 1	**QuickLab,** A Healthy Diet, p. 154 **Discovery Lab,** To Diet or Not to Diet, p. 166 **Datasheets for LabBook,** To Diet or Not to Diet **Labs You Can Eat,** Snack Attack
Section 2 Risks of Alcohol and Other Drugs	90	▶ Distinguish between the positive and negative uses of drugs. ▶ Explain the hazards of tobacco, alcohol, and illegal drugs. SAI 1, SPSP 1, 4, 5	**Demonstration,** The Cost of Smoking, p. 157 in ATE
Section 3 Healthy Habits	90	▶ Describe four important aspects of good hygiene. ▶ Explain why exercise and sleep are important to good health. ▶ Describe methods of handling stress. ▶ List ways to stay safe at home, on the road, and outdoors. SPSP 1, 3–5; Labs UCP 2, SAI 1, SPSP 1	**Discovery Lab,** Keep it Clean, p. 167 **Datasheets for LabBook,** Keep It Clean **Inquiry Lab,** Consumer Challenge **Long-Term Projects and Research Ideas,** Breakfast, Lunch, and Dinner of Champions

See page **T23** for a complete correlation of this book with the

NATIONAL SCIENCE EDUCATION STANDARDS.

TECHNOLOGY RESOURCES

 Guided Reading Audio CD
English or Spanish, Chapter 7

 One-Stop Planner CD-ROM with Test Generator

 CNN. **Multicultural Connections,** Muslims and Medicine, Segment 13
A Healthy Mexican Tradition, Segment 14

 Science Discovery Videodisc
Image and Activity Bank with Lesson Plans: Fountain of Youth

145A Chapter 7 • Staying Healthy

Chapter 7 • Staying Healthy

CLASSROOM WORKSHEETS, TRANSPARENCIES, AND RESOURCES	SCIENCE INTEGRATION AND CONNECTIONS	REVIEW AND ASSESSMENT
Directed Reading Worksheet Science Puzzlers, Twisters & Teasers		
Directed Reading Worksheet, Section 1 Transparency 253, Covalent Bonds in a Water Molecule Transparency 99, The Food Pyramid Transparency 100, How to Read a Food Label Critical Thinking Worksheet, A Daily Routine Reinforcement Worksheet, To Eat, or Not to Eat . . .	Oceanography Connection, p. 150 Connect to Physical Science, p. 150 in ATE Multicultural Connection, p. 151 in ATE Connect to Chemistry, p. 151 in ATE Math and More, p. 152 in ATE MathBreak, What Percentage? p. 153 Math and More, p. 153 in ATE Multicultural Connection, p. 153 in ATE Health Watch: Meatless Munching, p. 173	Homework, pp. 149, 150 in ATE Section Review, p. 151 Self-Check, p. 153 Section Review, p. 154 Quiz, p. 154 in ATE Alternative Assessment, p. 154 in ATE
Directed Reading Worksheet, Section 2	Multicultural Connection, p. 156 in ATE MathBreak, Deadly Averages, p. 158 Real-World Connection, p. 158 in ATE Cross-Disciplinary Focus, p. 159 in ATE	Self-Check, p. 156 Homework, pp. 156, 159 in ATE Self-Check, p. 159 Section Review, p. 160 Quiz, p. 160 in ATE Alternative Assessment, p. 160 in ATE
Directed Reading Worksheet, Section 3	Connect to Earth Science, p. 161 in ATE Cross-Disciplinary Focus, p. 162 in ATE Apply, p. 163 Real-World Connection, p. 163 Science, Technology, and Society: Bacteria at Your Service, p. 172	Homework, pp. 162, 163, 164 in ATE Section Review, p. 165 Quiz, p. 165 in ATE Alternative Assessment, p. 165 in ATE

END-OF-CHAPTER REVIEW AND ASSESSMENT

Chapter Review in Study Guide
Vocabulary and Notes in Study Guide
Chapter Tests with Performance-Based Assessment, Chapter 7 Test
Chapter Tests with Performance-Based Assessment, Performance-Based Assessment 7
Concept Mapping Transparency 28

 Holt, Rinehart and Winston On-line Resources
go.hrw.com

For worksheets and other teaching aids related to this chapter, visit the HRW Web site and type in the keyword: **HSTBD7**

 National Science Teachers Association
www.scilinks.org

Encourage students to use the sciLINKS numbers listed in the internet connect boxes to access information and resources on the **NSTA** Web site.

Chapter 7 • Chapter Organizer   **145B**

Chapter Resources & Worksheets

Visual Resources

TEACHING TRANSPARENCIES

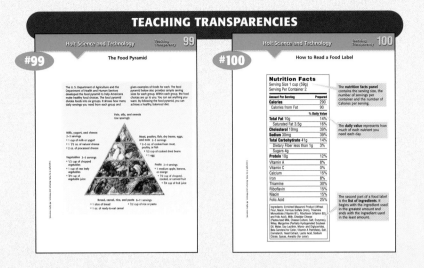

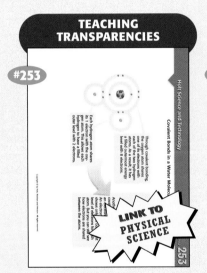

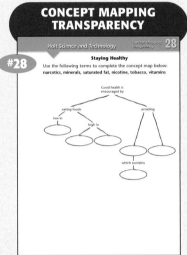

Meeting Individual Needs

DIRECTED READING

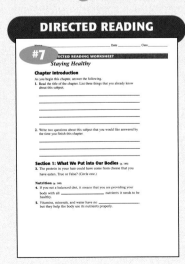

REINFORCEMENT & VOCABULARY REVIEW

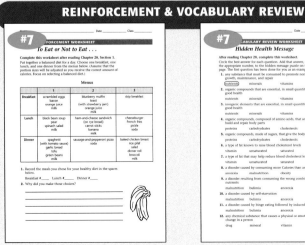

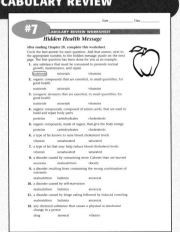

SCIENCE PUZZLERS, TWISTERS & TEASERS

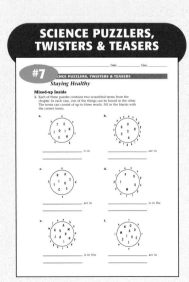

Chapter 7 • Staying Healthy

Chapter 7 • Staying Healthy

Review & Assessment

STUDY GUIDE

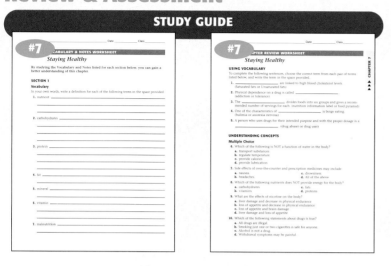

CHAPTER TESTS WITH PERFORMANCE-BASED ASSESSMENT

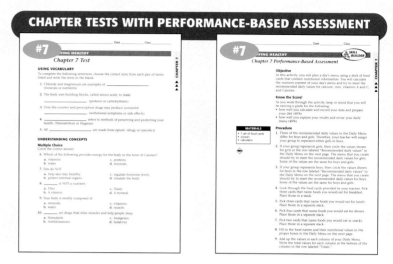

Lab Worksheets

INQUIRY LABS

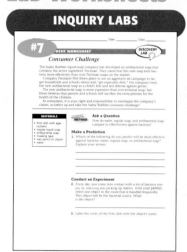

LABS YOU CAN EAT

LONG-TERM PROJECTS & RESEARCH IDEAS

DATASHEETS FOR LABBOOK

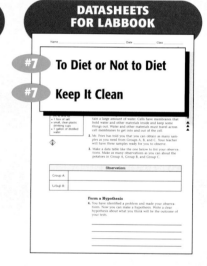

Applications & Extensions

CRITICAL THINKING & PROBLEM SOLVING

MULTICULTURAL CONNECTIONS

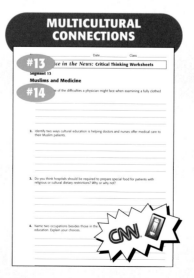

Chapter 7 • Chapter Resources & Worksheets

Chapter Background

SECTION 1

What We Put into Our Bodies

▶ Nutrient Needs in Adolescence

Puberty is a period of rapid growth, and its changes affect every organ of the body. Adolescents need extra nutrients to meet their needs during the growth spurt that accompanies puberty.

- The onset of menstruation in girls and the increase in lean body mass in boys increases the body's need for iron. The Recommended Daily Allowance (RDA) for iron in adolescence is 12–15 mg.

- The increase in skeletal mass that occurs during adolescence boosts the body's calcium needs. Seventy-five to 80 percent of the skeleton is built during this period, increasing the RDA for calcium to 1,200 mg.

- The increased growth rate of adolescents increases the body's caloric needs. Energy needs vary depending on growth rate, body composition, and activity level. In general, boys have a larger proportion of lean body mass to fat and require more calories than girls do.

▶ Fiber

Although fiber contains no vitamins or minerals, it is essential to good health. Insoluble fiber (obtainable from whole-grain foods), along with fluids, helps the colon move waste (fecal matter) out of the bowels. A lack of insoluble fiber can increase a person's risk for constipation, colon and bowel cancer, and diverticulosis.

- Soluble fiber—obtainable from fruit, beans, peas, and other legumes—can reduce the risk of heart disease by lowering cholesterol. Legume-based fiber also aids in the regulation of blood glucose levels.

▶ The Biopsychosocial Nature of Nutrition in Youth

The physical, emotional, and social changes of adolescence have a great impact on teen nutrition. Adolescence is typically a period of increased autonomy. With increased independence comes increased opportunity to make food choices. But rather than making food choices based on long-term health, teens tend to be influenced by social pressures to reach cultural ideals of thinness, gain the acceptance of peers, and assert their independence from their parents.

- The characteristic adolescent preoccupation with body image extends to nutrition. If sound nutritional principles are taught and reinforced during adolescence, long-term health can benefit.

IS THAT A FACT!

▬ Just before astronauts Neil Armstrong (1930–) and Edwin Aldrin Jr. (1930–1999) embarked on the first moonwalk on July 20, 1969, each astronaut ate four bacon squares, three sugar cookies, and peaches and drank pineapple-grapefruit juice and coffee.

SECTION 2

Risks of Alcohol and Other Drugs

▶ Inhalant Abuse

One of the most disturbing trends in substance abuse today is the increase in inhalant abuse. According to national surveys, one in five eighth graders has used inhalants for the purpose of getting high. Nevertheless, most parents are unaware of the popularity and hazards of inhalant use.

Chapter 7 • Staying Healthy

- More than 1,000 common products are used as inhalants. These seemingly benign, everyday substances can prove deadly when inhaled. Initially, the user may feel slightly stimulated or less inhibited, but nearly all inhalants produce effects similar to anesthesia. The body functions slow down, and loss of consciousness may result.

- A frightening risk associated with inhalant use is the threat of sudden sniffing death syndrome, which may occur any time a person uses an inhalant. Other physical consequences of inhalant use are damage to the brain, heart, liver, lung, and bone marrow.

▶ Preventing Accidental Drug Poisonings
Beginning in 1973, the federal government required that all drugs and medications be packaged in child-proof containers. Between 1973 and 1976, there was a 50 percent decrease in childhood poisoning deaths. The development of single-dose packets, product alterations to reduce toxicity, and the creation of poison control centers have also contributed to this decline.

▶ Smokeless Tobacco
In recent years, the use of smokeless tobacco among adolescents has increased. Many teens believe that smokeless tobacco is safer than cigarettes are. Unfortunately, ignorance has a high cost: Smokeless tobacco use is linked with many health problems, including mouth ulcers and oral cancer. Furthermore, teens who use smokeless tobacco are more likely to begin to smoke.

 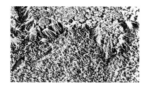

IS THAT A FACT!

- In the 1800s, scientists learned how to isolate drug compounds from their natural state in plants. Morphine, cocaine, and heroin became readily available and were welcomed as safe, powerful pain relievers because, at first, no one knew of their addictive properties. By the early 1900s, there was an epidemic of drug abuse in the United States.

SECTION 3
Healthy Habits

▶ Safety Notes
In the United States, injuries are the leading cause of death for children and adolescents; in fact, they surpass all major diseases as the cause of death in children between the ages of 1 and 19. Adolescents are particularly prone to accidents, injuries, and death because of their risk-taking behavior, desire to exert autonomy, and sense of invincibility.

- In 1994, motor-vehicle accidents claimed the lives of 5,768 Americans between the ages of 13 and 19. Tragically, almost half of these deaths were easily preventable. Research has shown that the use of seat belts reduces the risk of fatal injury by 45 percent and the risk of critical injury by 50 percent.

▶ Excuse Me?
Preventable hearing loss among adolescents is increasing at an alarming rate. Results of hearing tests in a 1995 report showed that 11 percent of ninth graders and 10.6 percent of twelfth graders could not hear high-frequency sounds. Experts attribute this trend to increased exposure to damaging levels of noise, including loud music.

- When sounds enter the ear, they cause the eardrum to vibrate. Tiny hair cells in the inner ear respond to these vibrations by sending nerve impulses to the brain. Ordinarily, these hair cells slowly die as a person ages. When exposed to loud noises, however, more hair cells die than usual. Fewer impulses are sent to the brain. Once damage occurs, it is irreversible. Fortunately, prevention is as simple as inserting a pair of earplugs or turning down the volume.

For background information about teaching strategies and issues, refer to the *Professional Reference for Teachers.*

Staying Healthy

Pre-Reading Questions

Students may not know the answers to these questions before reading the chapter, so accept any reasonable response.

Suggested Answers

1. You need to eat foods that will provide all the necessary nutrients. These foods are shown on the food pyramid. Limit your intake of sweets and fats.

2. Drugs are helpful when they are used to cure symptoms or illness. They are harmful when they lead to addiction and health problems.

3. Everyday habits that help keep you healthy include exercise, getting plenty of rest, eating a balanced diet, washing your hands, and taking care of your skin and teeth.

CHAPTER 7

Staying Healthy

Sections

1. **What We Put into Our Bodies** 148
   - Oceanography Connection 150
   - Internet Connect 151
   - MathBreak 153
   - QuickLab 154
   - Internet Connect 154

2. **Risks of Alcohol and Other Drugs** 155
   - MathBreak 158
   - Internet Connect 160

3. **Healthy Habits** 161
   - Apply 163

Chapter Labs 166, 167
Chapter Review 170
Feature Articles 172, 173

Pre-Reading Questions

1. What types of foods should you eat daily to have a healthy diet? What foods should you avoid?
2. How are drugs helpful? harmful?
3. What everyday habits can help to keep you healthy?

THE GLOW OF HEALTH

What do you notice most about this photo? Sure, you can see five boys and girls facing the camera. Besides that, though, what else does the picture tell you? The bright eyes, happy smiles, shiny hair, and a certain glow from the faces give off a feeling of radiant health. Having a clear mind, high energy, and a long, active life all depend on having a healthy body. In this chapter, you will learn some basic steps for maintaining the body—your personal "tool kit" for thriving in the world.

---

internet connect

HRW On-line Resources
go.hrw.com
For worksheets and other teaching aids, visit the HRW Web site and type in the keyword: **HSTBD7**

www.scilinks.com
Use the *sci*LINKS numbers at the end of each chapter for additional resources on the **NSTA** Web site.

www.si.edu/hrw
Visit the Smithsonian Institution Web site for related on-line resources.

www.cnnfyi.com
Visit the CNN Web site for current events coverage and classroom resources.

START-UP Activity

CONDUCT A SURVEY

How healthy is your class? Collect data and see for yourself.

Procedure

1. Copy the questionnaire below.
2. Circle *yes* or *no* to answer the questions. Do not put your name on the survey.

Analysis

3. Record the data from your survey and the surveys of your classmates in a chart. Count the number of students who answered *yes* to each question. For each question, calculate the percentage of your class that answered *yes*.
4. What things does your class do well? What health habits can be improved?

---

1. Do you exercise at least three times a week?   Yes   No
2. Do you wear a seat belt every time you ride in a car?   Yes   No
3. Do you eat five or more servings of fruits and vegetables every day?   Yes   No
4. Do you use sunscreen to protect your skin when you are outdoors?   Yes   No
5. Do you eat a lot of high-fat foods?   Yes   No

147

START-UP Activity

CONDUCT A SURVEY

MATERIALS

FOR EACH GROUP:
- questionnaire
- paper
- pen or pencil

Teacher's Notes

Tell students to use this formula to find the percentages in step 3:

$$\frac{\text{\# of } yes \text{ answers}}{\text{total \# of students}} \times 100$$

Answers to START-UP Activity

3–4. Answers will vary.

Chapter 7 • Staying Healthy   **147**

SECTION 1

Focus

What We Put into Our Bodies

In this section, students learn what the six essential nutrients are and why they are important for good health. Dietary guidelines and the food pyramid are used to illustrate the principles of sound nutrition, and students learn how to use these tools to create a healthy diet. Finally, students learn about the dangers of nutritional disorders.

 Bellringer

Post the following on the board or an overhead projector in two columns:

Terms	Descriptions
1. nutrients	a. energy in nutrients
2. calories	b. fats found in meats
3. carbohydrates	c. overweight
4. proteins	d. essential substances
5. saturated fats	e. inorganic elements
6. minerals	f. tissue builders
7. obese	g. main source of energy

Challenge students to recall information from previous chapters and match the terms in column A with the descriptions in column B. (1. d; 2. a; 3. g; 4. f; 5. b; 6. e; 7. c)

Terms to Learn

nutrient   mineral
carbohydrate vitamin
protein   malnutrition
fat

What You'll Do

- Identify the six groups of nutrients, and explain their importance to good health.
- Use dietary guidelines and the food pyramid to plan a healthy diet.
- Understand nutrition information labels.
- Explain the dangers of various nutritional disorders.

What We Put into Our Bodies

"You are what you eat." Does this familiar saying mean that you are pizza or candy? Of course not! But, the substances in the pizza and candy enter your body. The protein in the cheese may become part of your hair; the carbohydrates in the crust can give you energy to run your next race. The sugar in the candy can give you a quick energy boost but make you tired later when your blood sugar level drops.

Nutrition

Are you more likely to have potato chips or broccoli for a snack? a candy bar or a banana? If you lean toward foods that are high in sugar and fat, such as potato chips and candy, your food choices probably are not as healthy as they could be. Does that mean you have to cut out all of your favorite foods to eat healthy? No! Broccoli *is* a healthier food than potato chips. But eating only broccoli every day, like the person in **Figure 1,** is not much better than eating only potato chips! Either way, you do not get a balanced diet.

Balancing Act In order to stay healthy, you need to take in more than 40 different substances every day. These substances, or **nutrients,** nourish your body and are essential to life. To get them all, you must eat a wide variety of foods. Nutrients are grouped into six categories: *carbohydrates, proteins, fats, vitamins, minerals,* and *water.* Three of these—carbohydrates, proteins, and fats—provide energy for the body. The energy in these nutrients is measured in units called Calories. The other three nutrients—vitamins, minerals, and water—do not provide energy in Calories but help the body use all of the nutrients properly.

Figure 1 Eating only one food, even a healthy food, will not give you all the substances your body needs.

 **Directed Reading Worksheet** Section 1

IS THAT A FACT!

Frozen vegetables are often just as nutritious as fresh vegetables.

TOPIC: Nutrition
GO TO: www.scilinks.org
*sci*LINKS NUMBER: HSTL680

148 Chapter 7 • Staying Healthy

Body Fuel A **carbohydrate** is a chemical composed of one or more simple sugars. Carbohydrates are your body's main source of energy. They help digest fats, lubricate joints, and keep skin, bones, and nails healthy. Plant foods are the major source of carbohydrates.

There are two basic types of carbohydrates: simple and complex. *Simple carbohydrates* are sugars. They are easily digested and give your body quick energy. *Complex carbohydrates* are made of many sugar molecules linked together. They are digested more slowly than simple carbohydrates and give your body long-lasting energy.

Body Builders Protein is found in body fluids, muscle, bone, skin, and all other tissues. **Proteins** are nutrients used to build and repair body parts. Your body makes the proteins it needs, but it must have the necessary building blocks, called *amino acids*, to make them. Your digestive system breaks down the protein in food into individual amino acids that are then used to make new proteins. If your body does not get enough carbohydrates, it can also use proteins for energy.

Some foods, such as poultry, fish, milk, and eggs provide all of the essential amino acids. These food sources are called complete proteins. Incomplete proteins contain only some of the essential amino acids. Most plants are incomplete sources of protein, but eating a variety of plant foods each day will provide all of the amino acids your body needs.

Energy Storage **Fats** are energy-storage nutrients that help the body store some vitamins. Too much fat in the diet has been linked to weight gain, heart disease, and some kinds of cancer. But fats are essential to a balanced diet. They are needed to transport vitamins, produce hormones, keep skin healthy, protect vital organs, and provide insulation. Fats provide more than twice as much energy as proteins and carbohydrates per unit mass.

Simple carbohydrates

Complex carbohydrates

Proteins

Fats

Figure 2 Energy-Producing Nutrients

2 Teach, continued

BRAIN FOOD

Point out to students that many vitamin manufacturers promote their products by proudly proclaiming that their vitamins are "natural" or "organic," rather than synthetic (manufactured). Vitamins touted as "natural" are almost always more expensive than their synthetic counterparts. Point out that most experts agree that organic vitamins and synthetic vitamins are identical in both structure and function.

Homework

Writing Have students research and write a report about the physical effects of vitamin deficiencies for at least three of the essential vitamins.

CONNECT TO PHYSICAL SCIENCE

Water is a very simple molecule made up of one oxygen atom and two hydrogen atoms. Water is vital to all living organisms. Remind students that water makes up nearly 70 percent of the human body. Illustrate the water molecule using the following Teaching Transparency.

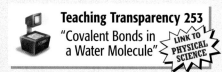

Teaching Transparency 253 "Covalent Bonds in a Water Molecule" **LINK TO PHYSICAL SCIENCE**

Oceanography CONNECTION

Kelp, a type of brown seaweed, is a nutritious food that is grown on special farms in oceans off the coasts of China and Japan.

Figure 3 *You need to drink about eight glasses of water a day. When you exercise, you need even more.*

There are two types of fats, saturated and unsaturated. *Saturated fats* are found in meats, dairy products, coconut oil, and palm oil. Saturated fats are known to raise blood cholesterol levels. *Cholesterol* is a fatlike substance found naturally in the body. Although cholesterol is important to the body, high levels in the blood increase the risk of heart disease. *Unsaturated fats* may help reduce blood cholesterol levels. Your body can make its own saturated fats, but it cannot make certain unsaturated fats. You must get these from your diet. Vegetable oils and fish contain unsaturated fats.

Flushing the System A human cannot survive for more than a few days without water. Your body is about 70 percent water. Water is in every cell and every kind of tissue. Water's three main functions are to transport substances, regulate temperature, and provide lubrication. You should drink 8 to 10 glasses of water daily, as shown in **Figure 3.** You also get water from the other liquids you drink and the foods you eat. Fresh fruits and vegetables, juices, soups, and milk contain large amounts of water.

Small Necessities **Minerals** are elements that are essential for good health. Six minerals are needed in large amounts: calcium, chloride, magnesium, phosphorus, potassium, and sodium. There are at least 12 minerals that are required in very small amounts. These include fluorine, iodine, iron, and zinc. If you eat a balanced diet, you should get all of the minerals you need. Calcium and magnesium are necessary for strong bones and teeth. Magnesium and sodium help the body use proteins. Potassium is needed to regulate your heartbeat and produce muscle movement, and iron is necessary for red blood cell production.

WEIRD SCIENCE

You are what you eat, as the saying goes, and in some cases, you can even begin to look like it. For example, consuming a lot of carrots or carrot juice can give the skin a yellowish color, a harmless condition that results from consuming the beta carotene found in some vegetables. In 1960, a doctor was puzzled by a patient whose skin looked orange. It turned out that the patient not only was eating large quantities of carrots but also was eating a lot of tomatoes, which can give the skin a reddish color. The two colors mixed, and *voilà*, he had orange skin!

Body Controllers **Vitamins** are organic compounds that control many body functions. Most vitamins cannot be made by the body, so you have to get them from food. The following table provides information about the 13 essential vitamins.

The Essential Vitamins

Vitamin	What it does	Where you get it
A	keeps skin and eyes healthy; builds strong bones and teeth	yellow and orange fruits and vegetables; dark, leafy greens; meat; and milk
B_1 (thiamine)	helps body use carbohydrates; helps nerves and heart function	meats, whole grains, beans, peas, nuts, and seafood
B_2 (riboflavin)	helps cells use carbohydrates and oxygen; keeps skin and eyes healthy	dairy products, fruits, whole grains, eggs, leafy vegetables, and poultry
B_3 (niacin)	helps body use carbohydrates; helps cells use oxygen; helps digestion	meats, peanuts, whole grains, peas, and beans
B_6	helps body use proteins, carbohydrates, and fats	poultry, fish, meat, eggs, potatoes, avocados, and bananas
B_{12}	keeps blood and nerves healthy	meats, poultry, eggs, fish, and milk
Folic acid (a B vitamin)	helps red blood cell formation	leafy greens, peas, beans, nuts, whole grains, liver, and oranges
Pantothenic acid (a B vitamin)	helps body use protein, carbohydrates, and fat; keeps body tissues healthy	meats, fish, whole grains, beans, peas, eggs, and corn
Biotin (a B vitamin)	helps body use protein, carbohydrates, some B vitamins, and fat	eggs, milk, meats, nuts, peas, beans, and whole grains
C	strengthens blood vessels and connective tissue; helps the body absorb iron; helps the body fight disease	citrus fruits; dark, leafy greens; broccoli; peppers; cabbage; tomatoes; potatoes; and strawberries
D	builds strong bones and teeth; helps the body use calcium and phosphorus	sunlight, enriched milk, eggs, and fish
E	protects red blood cells from destruction; needed for some enzymes to work	oils, fats, eggs, whole grains, wheat germ, liver, and leafy greens
K	assists with blood clotting	leafy greens, tomatoes, and potatoes

SECTION REVIEW

1. Name the six groups of nutrients, and explain why each is important to the body.
2. If vitamins and minerals do not supply energy, why are they important to a healthy diet?
3. **Applying Concepts** Name some of the nutrients that can be found in a glass of milk.

TOPIC: Nutrition, Vitamins
GO TO: www.scilinks.org
sciLINKS NUMBER: HSTL680, HSTL685

CONNECT TO CHEMISTRY

Vitamins are important because they enable enzymes to function. Enzymes perform their functions by fitting together with other body chemicals, and either combining them or breaking the chemicals down. In order for this to happen, vitamins bind to these enzyme proteins, activating them by changing the proteins' shape so that they fit properly with the chemicals they act upon. Without vitamins, many enzymes cannot do their jobs.

In many parts of the world, especially the tropics, insects are part of people's diet. Besides being plentiful and high in protein, insects can be quite tasty. Termites taste like pineapple, and baked bees have a nutty taste, similar to breakfast cereal.

MEETING INDIVIDUAL NEEDS

Writing **Advanced Learners**
Encourage students to use library or Internet resources to learn about vitamin D. (Vitamin D is produced by the body when the skin is exposed to sunlight.)

Have students share their findings with the class.

TOPIC: Vitamins
GO TO: www.scilinks.org
sciLINKS NUMBER: HSTL685

▼ Answers to Section Review

1. Sample answer: Carbohydrates provide energy to the body. Proteins provide amino acids so the body can make other proteins. Fats are an energy source, and they provide insulation to the body. Vitamins and minerals control many body functions. Water transports substances, regulates body temperature, and provides lubrication.

2. Vitamins and minerals help the body use other nutrients properly, and they control many body functions.

3. Milk contains carbohydrates; protein; saturated fat; vitamins A, B_2 (riboflavin), B_{12}, and D; biotin; calcium; magnesium; potassium; and water.

Section 1 • What We Put into Our Bodies

2) Teach, continued

MATH and MORE

Have students keep track of all the food they eat in a day. Have them determine how many servings from each of the food groups they consume. Then have them determine the percentage of recommended servings these numbers represent. For example, two glasses of milk and a grilled cheese sandwich represents 100 percent of the recommended dairy servings. One apple is 25–50 percent of the recommended fruit servings. Encourage students to identify how they could change their diet to make it more healthful.

Reinforcement Worksheet
"To Eat, or Not to Eat . . ."

Critical Thinking Worksheet
"A Daily Routine"

Teaching Transparency 99
"The Food Pyramid"

internet connect

SCLINKS NSTA
TOPIC: Food Pyramids
GO TO: www.scilinks.org
sciLINKS NUMBER: HSTL690

Eating for Good Health

Now you know what nutrients you need for good health. But how can you be sure to get them all in the right amounts? To begin, keep in mind that most teenage girls need about 2,200 Calories a day and most boys need 2,800 Calories. Since different foods contain different nutrients, *where* you get your Calories is as important as *how many* you get. The food pyramid below can help you make good food choices.

The Food Pyramid

The U.S. Department of Agriculture and the Department of Health and Human Services developed the food pyramid to help Americans make healthy food choices. The food pyramid divides foods into six groups. It shows how many daily servings you need from each group and gives examples of foods for each. This food pyramid also provides sample serving sizes for each group. Within each group, the food choices are up to you. You can eat anything you want. By following the food pyramid, you can achieve a healthy, balanced diet.

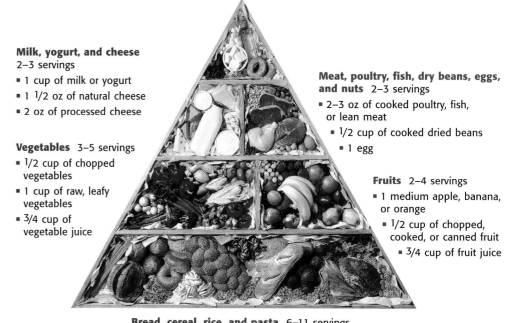

Fats, oils, and sweets
Use sparingly.

Milk, yogurt, and cheese 2–3 servings
- 1 cup of milk or yogurt
- 1 1/2 oz of natural cheese
- 2 oz of processed cheese

Vegetables 3–5 servings
- 1/2 cup of chopped vegetables
- 1 cup of raw, leafy vegetables
- 3/4 cup of vegetable juice

Meat, poultry, fish, dry beans, eggs, and nuts 2–3 servings
- 2–3 oz of cooked poultry, fish, or lean meat
- 1/2 cup of cooked dried beans
- 1 egg

Fruits 2–4 servings
- 1 medium apple, banana, or orange
- 1/2 cup of chopped, cooked, or canned fruit
- 3/4 cup of fruit juice

Bread, cereal, rice, and pasta 6–11 servings
- 1 slice of bread
- 1 oz of ready-to-eat cereal
- 1/2 cup of rice or pasta
- 1/2 cup of cooked cereal

MISCONCEPTION ALERT

Snacking is a well-established eating pattern among teenagers, and popular wisdom has long held that it should be discouraged. This is not necessarily true, however. Snacks, when balanced with healthy food choices for the rest of the day, can be an important source of nutrients and calories.

IS THAT A FACT!

It's not a lack of food, per se, that makes us hungry. Lack of nutrients in the bloodstream to keep our vital organs functioning causes a message to be sent to the brain, which then stimulates the stomach and the intestines—also known as "stomach growling"!

How to Read a Food Label

Packaged foods are required by law to have nutrition information labels. The illustration below shows a nutrition information label for a box of macaroni and cheese. Reading labels will help you make healthy eating choices.

Some of the nutrients found in each serving are listed on the label. The daily values shown are based on a 2,000-Calorie-per-day diet. You will need to calculate your own daily values based on your personal Calorie needs.

The **nutrition facts panel** contains the serving size, the number of servings per container, and the number of Calories per serving.

The **daily value** represents how much of each nutrient you need each day.

Nutrition Facts
Serving Size 1 cup (59g)
Serving Per Container 2

Amount Per Serving	Prepared
Calories	290
Calories from Fat	90
	% Daily Value
Total Fat 10g	14%
Saturated Fat 3.5g	16%
Cholesterol 10mg	39%
Sodium 30mg	39%
Total Carbohydrate 41g	14%
Dietary Fiber less than 1g	3%
Sugars 4g	
Protein 10g	12%
Vitamin A	8%
Vitamin C	0%
Calcium	15%
Iron	8%
Thiamine	30%
Riboflavin	15%
Niacin	15%
Folic Acid	25%

Ingredients: Enriched Macaroni Product (Wheat Flour, Niacin, Ferrous Sulfate (Iron), Thiamine Mononitrate (Vitamin B1), Riboflavin (Vitamin B2), and Folic Acid), Milk, Cheddar Cheese (Pasteurized Milk, Cheese Culture, Salt, Enzymes), Whey, Margarine (Partially Hydogenated Soybean Oil, Water, Soy Lecithin, Mono- and Diglycerides, Beta Carotene for Color, Vitamin A Palmitate), Salt, Cornstarch, Yeast Extract, Lactic Acid, Sodium Citrate, Spices, Annatto (for color).

The second part of a food label is the **list of ingredients.** It begins with the ingredient used in the greatest amount and ends with the ingredient used in the least amount.

MATH BREAK

What Percentage?

Use the nutrition information label on this page to answer the following questions for yourself, based on the Calorie needs of teenagers described on the previous page.

1. What percentage of your daily Calorie needs does one serving of macaroni and cheese provide?

2. The recommended daily value of fat is 72 g for teenage girls and 90 g for teenage boys. What percentage of the daily recommended fat value is provided by one serving of packaged macaroni and cheese?

Self-Check

For breakfast, you eat 1 cup of hot cereal with a banana and 1 cup of milk. What servings from the food pyramid have you eaten? *(See page 212 to check your answer.)*

Multicultural CONNECTION

In Japan, the incidence of breast cancer is very low. After moving to United States, however, Japanese women develop the disease nearly as often as American women do. Scientists attribute this to cultural dietary differences. They believe that the high-fat diet of many Americans plays a pivotal role in the development of breast cancer.

3) Extend

Answers to MATHBREAK

1. One serving will give girls 13 percent of their daily Calories and will give boys 10 percent.

2. One serving provides 14 percent for girls and 11 percent for boys.

GOING FURTHER

Have students check out a popular diet book from the library. Have them evaluate the diet in terms of its ability to meet the nutrient and energy needs of the body, and encourage them to identify any necessary changes, if any, to make the diet healthful. Have them prepare brief reports, and allow them to share their findings with the class.

MATH and MORE

If there are 3,312 Cal in a pound of butter, and a single serving of butter is 14.2 g (1tbsp), how many Calories are in a single serving of butter?

$$\frac{3{,}312 \text{ Cal}}{1 \text{ lb}} = \frac{x \text{ Cal}}{2.2 \text{ lbs/kg}} = 7{,}286 \text{ Cal/kg}$$

$$\frac{7{,}286 \text{ Cal}}{1{,}000 \text{ g}} = \frac{x \text{ Cal}}{14.2 \text{ g}} = 103 \text{ Cal in } 14.2\text{g}$$

Answer to Self-Check

You have eaten two servings from the bread, cereal, rice, and pasta group; one serving from the fruit group; and one serving from the milk, yogurt, and cheese group.

 Teaching Transparency 100 "How to Read a Food Label"

Section 1 • What We Put into Our Bodies

4) Close

QuickLab

MATERIALS

For Each Student:
- food pyramid
- daily-servings table
- paper
- pen or pencil

Answers to QuickLab

3. Answers will vary. Girls should base their menu on a 2,200 Cal diet, and boys should base their menu on a 2,800 Cal diet. Answers will vary as the students compare this diet with their normal diet.

Quiz

1. What needs must a healthy diet meet? (It must provide enough essential nutrients and meet the body's energy needs.)

2. Why should people strive to limit the fat in their diet? How might a very low-fat diet affect the body? (Too much fat in the diet has been linked to weight gain, heart disease, and some kinds of cancer. Too little fat, however, might result in hormone imbalances, dry skin, and vitamin deficiencies.)

Alternative Assessment

Concept Mapping Have students create a concept map using the new terms of the section and any additional words that are necessary. Have them begin their map with the concept, "What we put into our bodies."

QuickLab

A Healthy Diet

1. Use the recommended daily Calories for teenagers and the **table** below to estimate how many servings from each group in the food pyramid you should eat daily.

2. Create a menu for 2 days that includes the correct number of servings from each group in the food pyramid.

3. Compare this diet with your normal diet. How different are they? What can you do to improve your diet?

Daily Number of Servings

Food group	2,200 Calories	2,800 Calories
Bread	9	11
Fruit	3	4
Vegetable	4	5
Dairy	3	3
Meats	6 oz	7 oz

TRY at HOME

TOPIC: Food Pyramids, Nutritional Disorders
GO TO: www.scilinks.org
*sci*LINKS NUMBER: HSTL690, HSTL695

154

Nutritional Disorders

Unhealthy eating habits can cause nutritional disorders. For example, malnutrition can result from consuming too few Calories or too few of the right nutrients. Eating too many Calories or too many of the wrong nutrients can also cause malnutrition. **Malnutrition** occurs when you do not consume the right combination of nutrients.

Anorexia Nervosa and Bulimia Anorexia nervosa and bulimia can lead to malnutrition. Many of the people who suffer from anorexia nervosa and bulimia are teenage girls. *Anorexia nervosa* is an eating disorder characterized by self-starvation and an intense fear of gaining weight. This can cause weak bones, low blood pressure, and heart problems. *Bulimia* is a disorder characterized by binge eating followed by induced vomiting. Sometimes people suffering from bulimia also use laxatives and diuretics to rid the body of food and water. Bulimia can damage the teeth and digestive system and can also lead to kidney or heart failure. These disorders can both be fatal, but they can be cured with medical help.

Obesity *Obesity* is a condition characterized by an extremely high percentage of body fat. Eating too many foods from the top of the food pyramid and having an inactive lifestyle that involves little exercise can contribute to obesity. Obesity increases the risk of high blood pressure, heart disease, and diabetes.

SECTION REVIEW

1. What information is found on a nutrition information label?
2. How do anorexia nervosa and bulimia differ?
3. **Applying Concepts** How can someone who is obese suffer from malnutrition?

Answers to Section Review

1. The nutrition facts panel includes the serving size, the number of servings per container, the number of Calories per serving, and the percent daily value of each type of nutrient per serving.

2. Anorexia nervosa involves self-starvation. Bulimia involves binge eating followed by induced vomiting or the use of laxatives.

3. An overweight person can suffer from malnutrition by eating too many of the wrong nutrients and too few of the right nutrients.

SECTION 2

Reading Warm-Up

Terms to Learn
- drug
- addiction
- nicotine
- alcoholism
- narcotics

What You'll Do
- Distinguish between the positive and negative uses of drugs.
- Explain the hazards of tobacco, alcohol, and illegal drugs.

Risks of Alcohol and Other Drugs

You see them in movies and on television. You read about them in magazines. You hear about them in music and in your school. You are exposed to information, and misinformation, about tobacco, alcohol, and other drugs almost every day.

What Is a Drug?

A **drug** is any chemical substance that causes a physical or emotional change in a person. Drugs come in many forms, as shown in **Figure 4.** They can be pills, powders, fumes, liquids, or creams. Some drugs enter the body through the skin, and others are swallowed, inhaled, or injected.

When used safely and correctly, legal drugs can help your body heal a variety of ailments from athlete's foot to pneumonia. They can also provide relief from pain, congestion, and other symptoms. When used illegally or improperly, however, drugs can become killers.

What Effects Do Drugs Have?

Different drugs have different effects. One way to classify drugs is by function. *Analgesics,* like aspirin, are pain-relieving drugs. Drugs that relax muscles and help people sleep are *sedatives.* *Antibiotics* fight bacterial infections, and *antihistamines* help control symptoms of allergies, asthma, and colds.

Another way to classify drugs is by their effect on the central nervous system. *Stimulants* speed up the action of the central nervous system and may cause a person to feel more alert. *Depressants* have the opposite effect. They slow down body functions and may reduce a person's alertness.

Figure 4 All of these products contain drugs.

WEIRD SCIENCE

Humans may not be the only organisms that use drugs. Chimpanzees have been observed swallowing entire leaves of a plant that is known to be effective against infections and parasites. In Ethiopia, baboons actively seek out a fruit that fights parasites. It seems that muriqui monkeys of Brazil are able to control their fertility with a plant eaten only certain times of the year, and howler monkeys may even select the sex of their offspring by eating different plants.

SECTION 2

Focus

Risks of Alcohol and Other Drugs

In this section, students learn that drugs can have positive and negative effects on the body. They will learn that herbal medicines, tobacco, and alcohol are drugs. Finally, they will learn about the causes, effects, and treatments of drug abuse and addiction.

Bellringer

Pose this question to students:
> Are drugs good or bad for you?

Explain that a drug is not inherently good or bad. Any chemical substance that causes a physical or emotional change in a person is a drug.

1) Motivate

ACTIVITY

Poster Project Have students cut out advertisements for prescription drugs, over-the-counter drugs, alcohol, and tobacco from magazines. Each student can prepare a poster with the ads. Students should identify the type of drug and the type of person that would find the advertisement appealing. They should also calculate the percentage of advertisements for drugs in the magazine.

 Directed Reading Worksheet Section 2

Section 2 • Risks of Alcohol and Other Drugs   **155**

2) Teach

Answer to Self-Check
Regular use of some drugs may cause tolerance or addiction. Withdrawal symptoms occur when the body does not receive a drug that it is addicted to.

USING THE FIGURE
Refer students to **Figure 5**. Encourage them to compare the label of a prescription drug with that of the over-the-counter equivalent. Have them identify the information that both labels contain. (Both OTC and prescription labels list the name of the drug, the means of administration, the dosage, and the frequency with which the drug should be taken. In addition, both labels list possible side effects of the drug.)

Why is it important to read the label before taking an over-the-counter or prescription drug? (to make sure that you take the correct amount in the correct way and so that you are aware of possible side effects)

Homework
Encourage students to visit a pharmacy to make a list of 25 drugs that are available without a prescription. Challenge them to include drugs that have different modes of administration. Have them prepare a table that categorizes the drugs by their active ingredients and by their uses.

> ✓ **Self-Check**
>
> How are tolerance, addiction, and withdrawal symptoms related? *(See page 212 to check your answer.)*

Dependence and Addiction Regular use of some drugs can cause the body to develop *tolerance*. This means larger and larger doses of the drug are needed to get the same effect. **Addiction** is physical dependence on a drug. When a person is addicted to a drug, the body has a chemical need for the drug. If the body doesn't receive the drug, withdrawal symptoms may occur. These include nausea, vomiting, pain, and other physical symptoms. Once addicted, it is very difficult to stop taking a drug.

Sometimes dependence on a drug is not due to a physical need. Some people form *psychological dependence* on a drug and feel powerful cravings for it.

Types of Drugs
There are many kinds of drugs and many ways to use them. Some drugs are obtained from plants, and some are made synthetically. You can buy some drugs off the shelf, while others must be taken under the supervision of a doctor. Some drugs are illegal to buy, sell, or even possess.

Over-the-Counter and Prescription Drugs An over-the-counter drug can be purchased legally without a doctor's prescription. A prescription is a note written by a doctor to allow a patient to buy a medicine. It specifies the drug, directions for use, and the amount of the drug to be used.

Many over-the-counter and prescription drugs are powerful healing agents. However, some of these drugs also produce unwanted side effects. Side effects are uncomfortable symptoms such as nausea, headaches, drowsiness, or more serious problems caused by a drug.

Whether purchased with or without a prescription, all drugs must be used with care. Each year about 75,000 people in the United States become ill or die from the misuse of drugs. Information on proper use can be found on the label. **Figure 5** shows an example of a prescription drug label. The table below gives some drug safety tips.

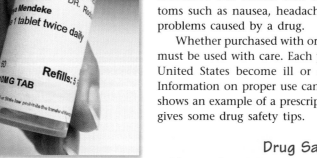

Figure 5 Prescription drug labels provide instructions for use and list possible side effects.

Drug Safety Tips
- Never take another person's prescription medicine.
- Read the label before each use. Always follow the instructions on the label and those provided by your doctor or pharmacist.
- Do not take more or less medication than prescribed.
- Consult a doctor if you have any side effects.
- Throw away leftover and out-of-date medicines.

Multicultural CONNECTION

The Chickasaw Indians relieved pain and fever by chewing on the bark of the willow tree. Today many people relieve their minor ills with a chemically related substance—aspirin. Derived from willow, aspirin is the most widely used drug in the world. We have the Chickasaw to thank for the "wonder drug" that relieves pain and fever and prevents blood clotting.

Herbal Medicines Information about medicinal herbs has been handed down by word-of-mouth for centuries. Some plants contain chemicals with important healing properties. However, these herbs are drugs and should be used carefully. **Figure 6** shows some medicinal herbs.

Tobacco About 50 million people in the United States—one-third of all adults—smoke. Smoking has serious health risks, and the nicotine in cigarettes is addictive. **Nicotine** is a chemical stimulant in tobacco that increases heart rate and blood pressure. Many smokers also experience a loss of appetite and a decrease in physical endurance. Smoking increases the chances of lung cancer by 10 times, and it has also been linked to other cancers, emphysema, chronic bronchitis, and heart disease. About 400,000 people die from smoking-related illnesses each year. **Figure 7** shows one of the effects of smoking.

Smokeless tobacco is also a health hazard. Nicotine is absorbed through the lining of the mouth, and the amount of nicotine that reaches the blood can be the same as for a smoker. Smokeless tobacco increases the risk of several types of cancer, including mouth and throat cancer. It also causes gum disease and discoloration of the teeth.

Figure 6 Some herbs can be purchased in health food stores. Medicinal herbs should always be used with care.

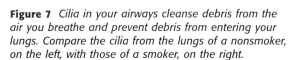

Figure 7 Cilia in your airways cleanse debris from the air you breathe and prevent debris from entering your lungs. Compare the cilia from the lungs of a nonsmoker, on the left, with those of a smoker, on the right.

 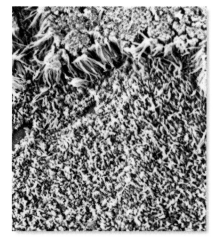

IS THAT A FACT!

The likelihood that a first-time cigarette smoker will become addicted is 90 percent. Fewer than half are able to quit, which puts their health at great risk. About 90 percent of people who get lung cancer are smokers or former smokers. Cigarette smokers are also at greater risk of getting cancers of the larynx, lip, esophagus, bladder, pancreas, and kidneys.

READING STRATEGY

Prediction Guide Before students read this page, have them answer the following question:

Which of the following can be caused by smoking?
1. nicotine addiction
2. increased blood pressure
3. emphysema
4. all of the above

(4)

Have students evaluate their answer after they read this section.

DEMONSTRATION

The Cost of Smoking Write the following equation on the board:

$2 \times 3 \times 365 \times 30 = 65{,}700$

Ask students if they know what this might refer to. (Accept all reasonable responses. This equation demonstrates the financial cost of smoking two packs of cigarettes per day, at $3 per pack, for 30 years.

$2 \frac{\text{packs}}{\text{day}} \times \frac{\$3}{\text{pack}} \times 365 \frac{\text{days}}{\text{year}} \times 30 \text{ years} = \$67{,}500$)

DISCUSSION

Teenage Views on Smoking Remind students that the harmful effects of smoking are well documented, and encourage them to speculate about why people begin to smoke knowing the hazards. Remind students that 87 percent of young people do not smoke; in fact, most teenagers consider smoking to be anything *but* cool. In a survey by the Centers for Disease Control, 86 percent of teens said that they would rather date people who don't smoke, and more than half said that they strongly dislike being around people who smoke.

2) Teach, continued

DISCUSSION

The Advertising Age Tell students that beer, liquor, and wine companies in the United States spend more than $2 billion annually on advertising and promotion. The result of this advertising is increased alcohol consumption. Display for students a selection of advertisements for beer, liquor, or wine taken from magazines. Encourage students to identify advertising tactics that seem to target youth. (Students might note that the ads often show groups of young, beautiful, active people in outdoor settings. These ads imply that people cannot have fun without alcohol.)

REAL-WORLD CONNECTION

Current studies suggest that young people are making their first choices about alcohol between the ages of 12 and 14, while only a generation ago those choices were made between the ages of 16 and 18. By ninth grade, 56 percent of students have tried alcohol. By high school, the effects of alcohol abuse have already taken a toll. Furthermore, a recent poll indicated that 66 percent of teenagers believed that drugs and alcohol were the biggest problems facing them. And alcohol-related deaths are the primary killer of Americans between the ages of 15 and 24.

Answers to MATHBREAK

About six people between the ages of 16 and 20 die every day in alcohol-related car crashes. This means that these crashes kill one person every 4 hours.

Figure 8 On average, there is one alcohol-related fatality every 31 minutes.

MATH BREAK

Deadly Averages

Approximately 2,200 people between the ages of 16 and 20 die in alcohol-related crashes each year. On average, how many die every day?

Alcohol Alcohol depresses the central nervous system and causes relaxation and memory loss. Excessive use of alcohol can damage the liver, pancreas, brain, nerves, and cardiovascular system. In very large quantities, alcohol can cause respiratory failure and even death. In addition, alcohol is a factor in more than half of all accidental deaths, suicides, and murders. **Figure 8** shows one example of an alcohol-related accident. It is illegal in the United States for people under the age of 21 to use alcohol.

About 10 million alcohol users are alcoholics. They suffer from **alcoholism,** which means that they are physically and psychologically dependent on alcohol. Alcoholism is considered a disease, and genetic factors are thought to influence a person's tendency to become an alcoholic.

Marijuana Marijuana is an illegal drug made from the Indian hemp plant. Marijuana produces a mind-altering effect that varies from user to user. It may cause a relaxed feeling or may increase anxiety. Marijuana slows reaction time, impairs thinking, and causes a loss of coordination.

Cocaine Cocaine and its more purified form, crack, are stimulants made from the South American coca plant. Both drugs are illegal and highly addictive, and users can become dependent on them in a very short time. Cocaine and crack produce feelings of intense excitement followed by anxiety and depression. Both drugs increase heart rate and blood pressure and can cause heart attacks even among first-time users.

Hallucinogens Hallucinogens (huh LOO si nuh juhnz) distort the senses and cause changes in mood and thought processes. Users have hallucinations (huh LOO si NAY shuhnz), which means they see and hear things that are not real. Some hallucinations are extremely frightening and can cause people to respond violently. LSD and PCP, or "angel dust," are two powerful and illegal hallucinogens. Sniffing some glues and solvents also causes hallucinations and can cause serious brain damage.

MISCONCEPTION ALERT

One of the most common alcohol-related myths is that beer is not as "strong" as liquor. "It's just beer" is a common refrain heard from teenagers and adults alike. In fact, beer has the same effects as wine or liquor. Alcohol from any source is addictive. It can cause liver damage, memory loss, heart damage, stomach problems, and sexual impotence. It impairs judgment, contributes to traffic deaths, and plays a role in many suicides and homicides.

Narcotics Narcotics are drugs made from opium. Some narcotics are used to treat severe pain and are legal when prescribed by a doctor. But many narcotics are illegal.

One illegal narcotic is heroin. Heroin is one of the most addictive drugs known, and large doses can lead to death. Because it is so strongly addictive, users must continue taking the drug in order to prevent painful withdrawal symptoms. Heroin is usually injected, and users often share needles that are contaminated. Therefore, heroin users have a high risk of infecting themselves with diseases like hepatitis and AIDS.

Designer Drugs There are many other illegal drugs. Examples include inhalants, barbiturates (downers), amphetamines (uppers), and "designer" drugs, which are produced by making small changes to existing drugs. Some inhalants, barbiturates, and amphetamines are legal if prescribed by a doctor.

Drug Use and Drug Abuse

A drug user takes a drug to prevent or improve some medical condition. The drug user obtains the drug legally and uses the drug properly. A drug abuser does *not* take a drug to relieve a medical condition. The drug abuser may take drugs for the temporary good feelings they produce, to escape from problems, or to belong to a group. The drug is often obtained illegally, and it is often taken without knowledge of the strength or purity of the drug.

Drug Abuse: How Does It Start? Nicotine, alcohol, and marijuana are called *gateway drugs* because they are usually the first drugs a person tries. The abuse of other more dangerous drugs may follow the abuse of gateway drugs, as illustrated in **Figure 9**. Most teenagers who start smoking cigarettes, drinking alcohol, or using marijuana do not realize that these drugs are addictive or that they can seriously harm their health.

Peer pressure is the reason most young people begin to use drugs. Teenagers may feel the need to drink, smoke, or try marijuana in order to make friends or to avoid being ridiculed or threatened. Many young people feel that "everyone" is using drugs. Because drug abusers often stand out, the fact that many teenagers do not abuse drugs is sometimes hard to see.

Self-Check

1. What is a hallucination?
2. List three dangers of heroin use.

(See page 212 to check your answers.)

Figure 9 Nicotine, alcohol, and marijuana are called gateway drugs.

4) Close

Quiz

1. What can be done to prevent drug abuse? (Answers will vary but should mention that information about the risks associated with drug use is an effective weapon. Being well informed is the first step toward making informed choices.)

2. How are nicotine and alcohol alike? How do they differ? (Answers will vary but should reflect the understanding that both drugs are legal for adults and that both are addictive. The use of both nicotine and alcohol has been associated with heart disease. The primary physical consequences of smoking are respiratory and cardiac; alcohol mostly affects the central nervous system, the liver, and the digestive system. Of the two, only alcohol dangerously impairs judgment.)

ALTERNATIVE ASSESSMENT

Have students prepare tables comparing alcohol, tobacco, marijuana, cocaine, LSD, and heroin. Have them define each and describe their effects on the body. In addition, have them indicate whether the drug is legal or illegal, and whether or not it is addictive.

Answer to Activity

Answers will vary.

TOPIC: Drug and Alcohol Abuse
GO TO: www.scilinks.org
sciLINKS NUMBER: HSTL700

Many people who start using drugs do not recognize the dangers. Misinformation about drugs is everywhere. Several common drug myths are discussed below.

Drug Myths

Myth	Reality
"It's only alcohol, not drugs."	Alcohol is a mood-altering and mind-altering drug. It affects the central nervous system and is addictive.
"I won't get hooked on one or two cigarettes a day."	Addiction is not related to the amount of a drug used. Some people become addicted after using a drug once or twice.
"I can quit any time I want."	Addicts may quit and return to drug usage many times. Their inability to stay drug free shows how powerful the addiction is.

Activity

Write yourself a letter. Tell yourself why you should stay drug free. You may wish to refer to people, goals, activities, or values that are important to you. Put your letter in an envelope, and keep it in a safe place. If you ever find yourself thinking about using drugs, take out your letter and read it.

TRY at HOME

Signs of Use People who begin using drugs generally undergo emotional, physical, and behavioral changes. Teenagers may have problems with school or family. Changes in personality or physical appearance may occur. However, remember that many young people do not take drugs. These changes could be signs of other problems or normal age-related changes.

Getting Off Drugs The first step to quitting drugs is to admit to being a drug abuser and to decide to stop. When quitting drugs, it is important for the addicted person to get proper medical and psychological treatment. Getting off drugs can be extremely difficult. Withdrawal symptoms may be painful, and even after the symptoms are gone, powerful cravings for a drug continue. This is even true of smoking.

SECTION REVIEW

1. What is the difference between a prescription drug and an over-the-counter drug?
2. How does addiction occur?
3. **Analyzing Relationships** How are nicotine, alcohol, heroin, and cocaine similar? How are they different?

internet connect

TOPIC: Drug and Alcohol Abuse
GO TO: www.scilinks.org
sciLINKS NUMBER: HSTL700

▼ Answers to Section Review

1. A prescription drug can only be purchased after your doctor writes a prescription telling the pharmacist what drug and what dosage to give you. An over-the-counter drug can be purchased by anyone without a prescription.

2. Addiction occurs when a person takes an addictive drug over a period of time and becomes physically dependent on the drug.

3. Nicotine, alcohol, and cocaine are all addictive drugs. Nicotine and alcohol are legal for use by adults. Cocaine is illegal. Cocaine and nicotine are stimulants. Alcohol is a sedative.

SECTION 3

Reading Warm-Up

Terms to Learn
- hygiene
- stress
- aerobic exercise

What You'll Do
- Describe four important aspects of good hygiene.
- Explain why exercise and sleep are important to good health.
- Describe methods of handling stress.
- List ways to stay safe at home, on the road, and outdoors.

Healthy Habits

Do you like playing sports? acting in plays? going to the movies? swimming in the ocean? No matter what you enjoy, the better your health, the easier it will be to take part in the things you like to do. Keeping yourself healthy is a daily task. If you have healthy habits, you are likely to stay healthy for a long time.

Hygiene and Posture

Hygiene refers to methods of preserving and protecting your health. It sounds simple, but washing your hands is the best way to prevent the spread of disease and infection. You should always wash your hands after using the bathroom and before and after eating or preparing food.

Taking care of your skin, hair, and teeth are other important aspects of good hygiene. Using sunscreens can help prevent sunburn, wrinkles, and skin cancer. Shampoo your hair regularly. To prevent cavities and keep your teeth and gums healthy, eat a healthy diet, brush at least twice a day, and floss at least once daily. Get regular dental exams, and replace your toothbrush about every 3 months.

Posture Posture is also important to health. Good posture helps you look and feel good. Bad posture strains your muscles and ligaments and makes breathing difficult. To have good posture when you stand, imagine a vertical line passing through your ear, shoulder, hip, knee, and ankle, as shown in **Figure 10.** When working at a desk, pull your chair forward and plant your feet firmly on the floor.

Figure 10 A slumped posture strains your lower back.

CONNECT TO EARTH SCIENCE

Tell students that they may use fossils as part of their daily hygiene. Many toothpastes contain diatomaceous earth, which consists of fossilized diatoms—algae with silica shells. The silica in diatomaceous earth effectively polishes the surface of teeth.

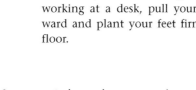 **Directed Reading Worksheet** Section 3

SECTION 3

Focus

Healthy Habits

In this section, students will learn how hygiene, exercise, rest, and play can enhance their well-being. Students learn how stress affects the body and how it can be reduced. Finally, students learn how injuries can be prevented.

Bellringer

Write this statement and question on the board or on an overhead projector:

In the 1800s and earlier, people who sustained simple cuts on their skin often ended up with serious infections. That rarely happens now. Why? (We now understand the importance of washing wounds with soap to keep them clean. We also have antiseptics today.)

1) Motivate

ACTIVITY

Don't Kick These Habits! Ask students to list all of their daily hygienic habits. (Answers may vary but will probably include washing their hands, bathing, brushing their teeth, using skin lotions, washing dishes, and using clean utensils to eat.)

Section 3 • Healthy Habits

2) Teach

READING STRATEGY

Prediction Guide Before students read this page, have them answer the following question:

Which pair of exercises is aerobic?
- **a.** walking and running
- **b.** swimming and stretching
- **c.** biking and yoga
- **d.** running and weight lifting

(a)

CROSS-DISCIPLINARY FOCUS

Language Arts Recite for students the adage, "An apple a day keeps the doctor away." Tell students to write slogans that encourage washing hands, brushing teeth, and good posture.

MEETING INDIVIDUAL NEEDS

Advanced Learners Encourage students to use library or Internet resources to investigate how weight-bearing exercise affects bone density. Have them prepare brief reports, and allow them to share their findings with the class.

Figure 11 *To stay with aerobic exercise, it is important to choose an activity you enjoy.*

Only half of a dolphin's brain sleeps at a time. A full, deep sleep would be a problem because dolphins must come to the surface for air every 20–30 seconds.

Exercise and Rest

Imagine that school starts at 7:30 A.M. You drag yourself out of bed at 6:00 every morning. You are tired, but somehow you make it through the school day. In the afternoon, you attend play rehearsal. When you get home, you take a break, eat dinner, and start your homework at 8:00 P.M. When you finish, you do not feel ready to sleep, so you watch TV, then read in bed, and finally, at 10:30, you drift off to sleep.

Keeping your body healthy requires giving it plenty of exercise and plenty of rest. If you have a schedule like the one above, you do not get enough of either.

Exercise Regular aerobic exercise at least three times a week is critical to good health. **Aerobic exercise** is vigorous, sustained exercise of the whole body for 20 minutes or more. Walking, running, swimming, and biking are all examples of aerobic exercise. For another example, see **Figure 11**.

Aerobic exercise increases the heart rate. As a result, more oxygen is taken in and distributed throughout the body. Over time, aerobic exercise strengthens the heart, lungs, and bones. It burns Calories, helps your body conserve nutrients, and aids digestion. It also gives you more energy and stamina. In other words, aerobic exercise protects your physical and mental health, and it's free! What a deal!

Sleep Believe it or not, teenagers actually need more sleep than younger children. Do you ever fall asleep in class, like the girl in **Figure 12**, or feel tired in the middle of the afternoon? If so, you may not be getting enough sleep. Scientists say that teenagers need about 9.5 hours of sleep each night.

At night, the body goes through several cycles of progressively deeper sleep, with periods of lighter sleep in between. If you do not sleep long enough, you will not enter the deepest, most restful period of sleep.

Figure 12 *If you fall asleep easily during the day, you are probably not getting enough sleep.*

Homework

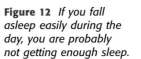 Have students list four benefits of aerobic exercise. (Answers could include the following: improves oxygen distribution; strengthens heart, lungs, and bones; burns calories; aids digestion; conserves proteins and minerals; improves energy and stamina; and reduces stress.)

Regular, intense exercise decreases body fat and therefore, in extreme cases, can restrict the release of estrogen in female athletes. When this occurs in prepubescent athletes, it delays the onset of puberty. Menstruation is delayed, and the period of bone growth is extended.

162 Chapter 7 • Staying Healthy

Coping with Stress

You have a big soccer game tomorrow. Are you excited and ready for action? You got a low grade on your math test. Are you upset or angry? The game and the test are causing you stress. **Stress** is the physical and mental response to pressure.

Some stress is a normal part of life, as shown in **Figure 13**. Stress stimulates your body to prepare for difficult or dangerous situations. However, sometimes you may have no outlet for the stress, and it builds up. Excess stress is harmful to your health and can decrease your ability to carry out your daily activities.

You may not even realize you are stressed until your body reacts. Perhaps you get a headache, have an upset stomach, or lie awake at night. You might feel tired all the time or begin an old nervous habit, such as nail-biting. You may become irritable or resentful. All of these things can be signs of too much stress.

Figure 13 *Working under stress can increase an athlete's ability to perform well.*

What to Do Different people are stressed by different things. Once you identify the source of the stress, you can find ways to deal with it. Here are some ideas for handling stress.

- Make a list of all the things you would like to get done, and rank them in order of importance. Do the most important things first.
- Exercise regularly, and get enough sleep.
- Pet a friendly animal.
- If you cannot remove a stressor, spend some quiet time alone or practice deep breathing or other relaxation techniques.
- Share your problems. Talk things over with someone you trust.

APPLY

Stress SOS
Your mother is out of town helping your grandfather move into a nursing home. Your little brother has been snooping in your room. You have a history project due in two days, an oral report due in English, and quizzes in both math and science! You are ready to scream! What are some healthy ways to handle all the stress you are feeling?

3) Extend

DISCUSSION

Injuries Tell students that injuries cause more deaths among children ages 1–19 than any of the major diseases. Remind them that many accidents can be prevented. Point out that bicycle injuries alone cause the death of more than 500 children and teens in the United States each year. Most of these deaths involve head trauma. Experts estimate that the simple act of wearing a bicycle helmet can reduce the risk of head injury by 85 percent; yet many children don't wear them. Ask students why they think people don't wear helmets. (Likely responses include: perhaps some people cannot afford helmets; the perception that helmets are "funny-looking" or "uncool.")

Point out that experts believe that one death per day and one head injury every 4 minutes would be prevented if every bike rider wore a helmet.

GROUP ACTIVITY

Poster Project
Divide the class into small groups, and provide each group with poster board and markers. Have them create posters designed to educate the public and motivate them to wear bicycle helmets. After students present their posters to the class, display them in the classroom.

Injury Prevention

Have you ever fallen off your bike or sprained your ankle? Maybe you avoided injury, or maybe you ended up in the emergency room. Accidents happen, and they can cause injury and even death. It is impossible to prevent all accidents, but you can decrease your risk by using your common sense and following basic safety rules.

Safety at Home Many accidents can be avoided. **Figure 14** below shows tips for safety around the house.

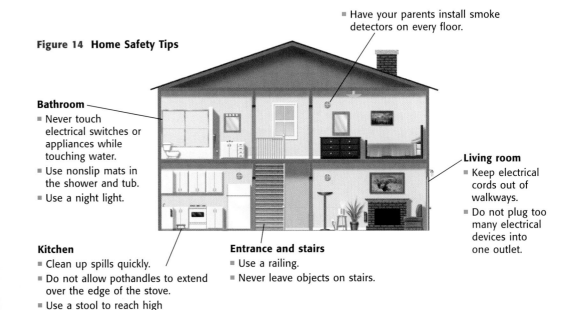

Figure 14 Home Safety Tips

Bathroom
- Never touch electrical switches or appliances while touching water.
- Use nonslip mats in the shower and tub.
- Use a night light.

Kitchen
- Clean up spills quickly.
- Do not allow pothandles to extend over the edge of the stove.
- Use a stool to reach high shelves.
- Keep grease and drippings away from open flames.

Entrance and stairs
- Use a railing.
- Never leave objects on stairs.

- Have your parents install smoke detectors on every floor.

Living room
- Keep electrical cords out of walkways.
- Do not plug too many electrical devices into one outlet.

Safety Outdoors Always dress appropriately for the weather and the activity. Never hike or camp alone. Tell someone where you are going and when you expect to return. If you do not bring water from home, be sure to purify any water you drink in the wilderness.

Learn how to swim. It could save your life! Never swim alone, and do not dive into shallow water or water of unknown depth. When in a boat, wear a life jacket. If a storm threatens, get out of the water and seek shelter.

Safety on the Road In the car, always wear a seat belt, even if you are traveling only a short distance. Most car accidents happen close to home, and seat belts save lives. When riding a bicycle, always wear a helmet. Ride with traffic, and obey all traffic rules. Be sure to signal when stopping or turning.

Homework

Writing Have students research and write a report about the physical symptoms and emergency treatments for hypothermia and heatstroke.

When Accidents Happen

No matter how well we practice safety measures, accidents can still happen. What should you do if a friend chokes on food and cannot breathe? What if he is stung by a bee and has a violent allergic reaction?

Call for Help Once you've checked for other dangers, call for medical help immediately, like the person in **Figure 15**. In most communities you can dial 911. Speak slowly and clearly. Give the complete address and a description of the location. Describe the accident, the number of people injured, and the types of injuries. Ask what to do and listen carefully to the instructions. Let the other person hang up first to be sure they have no more questions or instructions.

Figure 15 When calling 911, stay calm and listen carefully to what the dispatcher tells you.

Learn First Aid You may want to learn more about what to do in an emergency by taking a first-aid course or a CPR course, as shown in **Figure 16**. CPR is a lifesaving procedure designed to revive a person who is not breathing and has no heartbeat. If you are over 12 years old, you can become certified in both CPR and first aid. Some babysitting courses also provide basics in first aid and are a good idea for anyone who cares for young children. The American Red Cross, community organizations, and local hospitals offer these classes. You should not attempt any lifesaving procedure unless you have been trained.

Figure 16 These teenagers are taking a CPR course to prepare themselves for emergency situations.

SECTION REVIEW

1. What is aerobic exercise? Give three examples.
2. What should you do when calling for help in a medical emergency?
3. What is hygiene, and how does it help you stay healthy?
4. **Applying Concepts** What situations are causing excess stress in your life right now? What can you do to help relieve the stress you are feeling?

4) Close

Quiz

1. How are healthy habits related to stress? (Practicing healthy habits, such as exercising regularly and getting adequate rest, helps us cope with stress in a positive way.)
2. What do bicycle helmets and seat belts have in common? (They both decrease the risk of injury from accidents.)
3. When you call for emergency help, what should you tell the person who answers the phone? (Tell the person the location of the emergency, the nature of the emergency, the number of people injured, and the type of injuries involved. Always hang up last.)

ALTERNATIVE ASSESSMENT

Writing Have students write an analysis of their hygienic habits, exercise regimes, sleep patterns, and injury risks or safety measures during play and athletic activities. Tell them to explain whether their activities and habits are encouraging good health and safety or whether they need to make changes to their routines.

▼ Answers to Section Review

1. Aerobic exercise is vigorous, sustained exercise of the whole body for 20 minutes or more. Some examples of aerobic exercise include biking, in-line skating, jumping rope, and swimming.
2. Dial 911, speak slowly and clearly, and describe where you are, why you called, how many people were injured, and what types of injuries were sustained. Listen carefully to the instructions, and let the other person hang up first.
3. Hygiene includes methods of preserving and protecting your health. It keeps you healthy by keeping your body in top condition.
4. Answers will vary.

Discovery Lab

To Diet or Not to Diet
Teacher's Notes

Time Required
Two 45-minute class periods

Lab Ratings

Teacher Prep 🧪🧪
Student Set-Up 🧪
Concept Level 🧪🧪
Clean Up 🧪🧪

MATERIALS
A copy of a chart that fast-food restaurants post in their dining rooms outlining the nutritional information of the food they serve would be very helpful in analyzing a fast-food meal.

Answers
3. Answers will vary. Students may say they might try to eat fewer fatty foods. The nutritional value of foods is a subject that may be new to many students.

 Datasheets for LabBook

Ivora Washington
Hyattsville Middle School
Hyattsville, Maryland

Discovery Lab

To Diet or Not to Diet

There are six main classes of foods that we need in order to keep our bodies functioning properly: water, vitamins, minerals, carbohydrates, fats, and proteins. In this activity, you will investigate the importance of a well-balanced diet in maintaining a healthy body. Then you will create a poster or picture that illustrates the importance of one of the three energy-producing nutrients—carbohydrates, fats, and proteins.

MATERIALS

- nutrition reference books
- diet books
- white unlined paper
- crayons or colored markers
- fast-food menus (optional)

Procedure

1. In your ScienceLog, create a table like the one below. Research in the library, on nutrition labels, in nutrition or diet books, or on the Internet to find the information you need to fill out the chart.

2. Choose one of the foods you have learned about in your research, and create a poster or picture that describes its importance in a well-balanced diet.

Analysis

3. Based on what you have learned in this lab, how might you change your eating habits to have a well-balanced diet? Does the nutritional value of foods concern you? Why or why not? Write your answers in your ScienceLog, and explain your reasoning.

Nutrition Data Table			
	Fats	**Carbohydrates**	**Proteins**
Found in which foods			
Functions in the body			
Consequences of deficiency			

DO NOT WRITE IN BOOK

Discovery Lab

Keep It Clean

One of the best ways of preventing the spread of bacterial and viral infections is to wash your hands with soap and water frequently. Many companies advertise that their soap ingredients will destroy bacteria normally found on the body. In this activity, you will investigate how effective antibacterial soaps are at killing bacteria.

MATERIALS

- wax pencil
- new scrub brush
- 4 agar plates (Petri dishes filled with sterile nutrient agar)
- liquid antibacterial soap
- incubator
- transparent tape

Procedure

1. Label four agar plates with "Control," "Unwashed," "No soap," and "Soap." Open the "Control" plate for 1 minute. Cover the dish, and leave it closed for the rest of the experiment.

2. Carefully press different surfaces of your unwashed hands on the "Unwashed" plate. Replace the cover immediately.

3. Hold your right hand under running water for two minutes. During this time, ask your partner to scrub all surfaces of your hand with the scrub brush. When you are finished, your partner should turn off the water and open the plate marked "No soap." Carefully press the same surfaces of your right hand on the plate, as you did in step 2.

4. Repeat step 3, using liquid antibacterial soap on your left hand. Do not touch anything before pressing your left hand on the plate marked "Soap."

5. Tape the lid of each plate to its bottom. Incubate the plates upside down overnight at 37°C.

6. Remove the plates from the incubator and turn them right side up. Check each plate for the presence of bacterial colonies, and count the number of colonies present on each plate.
   Caution: Do not remove any of the lids.

Analysis

7. Which plate contained the most growth? Which contained the least?

8. Does water alone effectively kill bacteria? Explain.

Discovery Lab

Keep It Clean Teacher's Notes

Time Required
Two 45-minute class periods

Lab Ratings

EASY ——— HARD

TEACHER PREP 🧪🧪
STUDENT SET-UP 🧪
CONCEPT LEVEL 🧪🧪
CLEAN UP 🧪🧪

MATERIALS

You can reduce the number of scrub brushes needed by grouping students and having each group select a volunteer.

Safety Caution

Remind students to review all safety cautions and icons before beginning this lab activity. Check for known allergies to antibacterial or other soaps before beginning. Use plastic Petri dishes if possible.

 Datasheets for LabBook

Answers

7. Students should observe that the control dish shows the most bacterial growth. Ideally, the dish inoculated after scrubbing with antibacterial soap should show the least bacterial growth.

8. Water alone kills many bacteria because of the chlorine and other agents used in drinking water. Washing with antibacterial soap for a short time may be only slightly better than washing with water alone.

Elizabeth Rustad
Crane Junior High School
Yuma, Arizona

Chapter Highlights

VOCABULARY DEFINITIONS

SECTION 1

nutrient a substance that must be consumed or taken in by an organism to promote normal growth, maintenance, and repair

carbohydrate a biochemical composed of one or more simple sugars bonded together; used to provide and store energy

protein a biochemical composed of amino acids; its functions include regulating chemical reactions, transporting and storing materials, and providing support

fat an energy-storage nutrient that helps the body store some vitamins

mineral an element that is essential for good health

vitamin an organic compound that controls many body functions, including cell growth and hormone production

malnutrition a disorder resulting from consuming too few calories or too few nutrients

SECTION 2

drug any chemical substance that causes a physical or emotional change in a person

addiction a physical dependence on a drug

Chapter Highlights

SECTION 1

Vocabulary
- nutrient (p. 148)
- carbohydrate (p. 149)
- protein (p. 149)
- fat (p. 149)
- mineral (p. 150)
- vitamin (p. 151)
- malnutrition (p. 154)

Section Notes
- Carbohydrates, proteins, fats, vitamins, minerals, and water are the six types of nutrients that are essential to life.
- Teenage girls need about 2,200 Calories a day. Boys need about 2,800. It is important to obtain Calories from a variety of foods.
- The food pyramid gives information for eating a balanced diet.
- A nutrition information label on a packaged food lists serving size, Calorie and nutrient content, and the ingredients.
- Anorexia nervosa, bulimia, and obesity can lead to malnutrition.

SECTION 2

Vocabulary
- drug (p. 155)
- addiction (p. 156)
- nicotine (p. 157)
- alcoholism (p. 158)
- narcotics (p. 159)

Section Notes
- A drug is any chemical substance that causes a physical or emotional change in a person.
- Some drugs can cause addiction or psychological dependence.
- Over-the-counter and prescription drugs are legal drugs, many of which are powerful healing agents. Some also have side effects.
- Alcohol and nicotine are legal drugs for adults. Both are addictive and hazardous to your health.

✓ Skills Check

Math Concepts

PERCENTAGE If your recommended daily value of fat is 72 g and you eat a candy bar that has 12 g of fat, what percentage of your daily value of fat have you eaten? To find the percentage, divide the grams of fat in the candy bar by the daily value. Then multiply by 100.

$$12 \text{ g} \div 72 \text{ g} = 0.17 \times 100\% = 17\%$$

The fat in the candy bar is 17 percent of your total daily value.

Visual Understanding

NUTRITION INFORMATION LABELS All packaged foods in the United States are required to have a nutrition label like the one on page 153. On this label you can see that the serving size is 1 cup and that there are two servings in the container. The number of Calories in a serving and the number of Calories from fat are listed. You will also see the percent daily value for a number of nutrients that can be found in one serving. These percentages are based on 2,000 Calories a day.

Lab and Activity Highlights

To Diet or Not to Diet PG 166

Keep It Clean PG 167

 Datasheets for LabBook (blackline masters for these labs)

SECTION 2

- A drug abuser takes a drug, often illegally, for non-medical reasons.
- Drug abuse oftens begins with the use of tobacco, alcohol, or marijuana—the gateway drugs.
- Getting off drugs requires admitting addiction, deciding to stop, going through withdrawal symptoms, and experiencing cravings for the drug.

SECTION 3

Vocabulary
hygiene *(p. 161)*
aerobic exercise *(p. 162)*
stress *(p. 163)*

Section Notes

- Good hygiene is essential to good health.
- Having a healthy body requires getting regular aerobic exercise and getting enough sleep. Posture is also important to good health.
- Some stress is a normal and necessary part of life. Too much stress can result in poor health.
- It is possible to prevent some accidents by using common sense and following basic rules of safety.
- In an emergency, call for help as soon as possible. Do not attempt any lifesaving procedure for which you are not trained.

VOCABULARY DEFINITIONS, continued

nicotine a chemical stimulant that increases heart rate and blood pressure

alcoholism a disorder in which a person is physically and psychologically dependent on alcohol

narcotics drugs made from opium

SECTION 3

hygiene methods of preserving and protecting your health

aerobic exercise vigorous, sustained exercise of the whole body for 20 minutes or more

stress the physical and mental response to situations that create pressure

 Blackline masters of these Chapter Highlights can be found in the **Study Guide.**

---

internet connect

 GO TO: go.hrw.com

Visit the **HRW** Web site for a variety of learning tools related to this chapter. Just type in the keyword:

KEYWORD: HSTBD7

 GO TO: www.scilinks.org

Visit the **National Science Teachers Association** on-line Web site for Internet resources related to this chapter. Just type in the *sci*LINKS number for more information about the topic:

TOPIC:	sciLINKS NUMBER:
Nutrition	HSTL680
Vitamins	HSTL685
Food Pyramids	HSTL690
Nutritional Disorders	HSTL695
Drug and Alcohol Abuse	HSTL700

Lab and Activity Highlights

LabBank

 Labs You Can Eat, Snack Attack

Inquiry Lab, Consumer Challenge

 Long-Term Projects & Research Ideas, Breakfast, Lunch, and Dinner of Champions

Chapter 7 • Chapter Highlights **169**

Chapter Review
Answers

USING VOCABULARY
1. Saturated fats
2. addiction
3. food pyramid
4. bulimia
5. drug user

UNDERSTANDING CONCEPTS
Multiple Choice
6. c
7. d
8. b
9. b
10. d
11. d
12. c

Short Answer
13. Not all narcotics are illegal. Some narcotics are prescription drugs used to treat severe pain.
14. The three nutrients that provide energy in Calories are carbohydrates, proteins, and fats. Three nutrients that do not provide energy are water, vitamins, and minerals.
15. Beer contain alcohol, which depresses the central nervous system. Alcohol is a drug. Therefore, if you drink beer, you are taking a drug.

Concept Mapping
16. An answer to this exercise can be found at the front of this book.

Chapter Review

USING VOCABULARY

To complete the following sentences, choose the correct term from each pair of terms listed below:

1. __?__ are linked to high blood cholesterol levels. *(Saturated fats* or *Unsaturated fats)*
2. Physical dependence on a drug is called __?__. *(addiction* or *tolerance)*
3. The __?__ divides foods into six groups and gives a recommended number of servings for each. *(nutrition information label* or *food pyramid)*
4. One of the characteristics of __?__ is binge eating. *(bulimia* or *anorexia nervosa)*
5. A person who uses drugs for their intended purpose and with the proper dosage is a __?__. *(drug abuser* or *drug user)*

UNDERSTANDING CONCEPTS

Multiple Choice

6. Which of the following is *not* a function of water in the body?
   a. transport substances
   b. regulate temperature
   c. provide Calories
   d. provide lubrication

7. Side effects of over-the-counter and prescription medicines may include
   a. nausea.
   b. headaches.
   c. drowsiness.
   d. All of the above

8. Which of the following nutrients does *not* provide energy for the body?
   a. carbohydrates
   b. vitamins
   c. fats
   d. proteins

9. What are the effects of nicotine on the body?
   a. liver damage and decrease in physical endurance
   b. loss of appetite and decrease in physical endurance
   c. loss of appetite and brain damage
   d. liver damage and loss of appetite

10. Which of the following statements about drugs is true?
    a. All drugs are illegal.
    b. Smoking just one or two cigarettes is safe for anyone.
    c. Alcohol is not a drug.
    d. Withdrawal symptoms may be painful.

11. Aerobic exercise does *not*
    a. burn calories.
    b. increase the heart rate.
    c. strengthen the heart, lungs, and bones.
    d. make you weak.

12. When talking to a 911 operator, you should *not*
    a. describe the accident.
    b. ask what to do.
    c. hang up before the dispatcher hangs up.
    d. speak slowly and clearly.

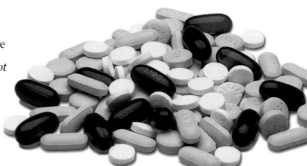

Short Answer

Answer each of the following questions with a few sentences:

13. Are all narcotics illegal? Explain.
14. What are the three types of nutrients that provide energy in Calories? Which three nutrients do not provide energy?
15. If you drink a beer, are you taking a drug? Explain your answer.

Concept Mapping

16. Use the following terms to create a concept map: unsaturated fats, carbohydrates, water, proteins, nutrients, simple sugars, starch, fats, vitamins, minerals, saturated fats.

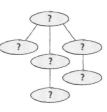

CRITICAL THINKING AND PROBLEM SOLVING

Write one or two sentences to answer the following questions:

17. Many people eat little or no meat. Meat is often the major source of protein in American diets. What other sources of protein can vegetarians choose?
18. You are at a party and a friend offers you a cigarette. He says that one cigarette won't hurt you. Using what you know about tobacco and addiction, explain why his reasoning is false.

MATH IN SCIENCE

19. Assume your diet is 20 percent fat, 30 percent protein, and 50 percent carbohydrates. In order to get 2,000 Calories a day, how many Calories of each nutrient should you be getting?

INTERPRETING GRAPHICS

List the unsafe habits in the following illustrations. For each unsafe habit, tell what the person should be doing instead.

What's wrong here?

20.

21.

 Take a minute to review your answers to the Pre-Reading Questions found at the bottom of page 146. Have your answers changed? If necessary, revise your answers based on what you have learned since you began this chapter.

CRITICAL THINKING AND PROBLEM SOLVING

17. Sample answer: Other foods that contain protein include eggs, dairy products, and fish.
18. Sample answer: The nicotine in cigarettes is highly addictive. One cigarette could easily lead to many more.

MATH IN SCIENCE

19. 600 Cal from protein, 400 Cal from fat, and 1,000 Cal from carbohydrates

INTERPRETING GRAPHICS

20. The girl in this picture is riding without her seat belt. She needs to wear her seat belt.
21. In this picture, a pot handle is hanging over the edge of the stove. The woman should use a pot holder to turn the handle of the pot so that it is not hanging over the edge of the stove.

 Concept Mapping Transparency 28

 Blackline masters of this Chapter Review can be found in the **Study Guide.**

Science, Technology, and Society

Bacteria at Your Service

Background

Bacteria have many uses. For instance, some bacteria are currently being used to help break down toxic substances, such as cyanide.

Certain bacteria store energy in the form of a biodegradable plastic called PHB. Originally, researchers found that the plastic was stiff and brittle. It also had a high melting point, which made it difficult to work with. However, scientists found that by feeding the bacteria a mixture of glucose and organic acids, the bacteria could produce a stronger, more flexible plastic with a lower melting point.

Genetically engineered bacteria have been developed to create more-successful fruit and vegetable crops. This technology could provide fresher and more nutritious food for consumption around the world.

Science, Technology, and Society

Bacteria at Your Service

Wanted: Hard worker to clean up trash. You must have experience curing diseases. The ability to make plastics is a plus. Only microorganisms need apply for this position.

What could possibly be flexible enough to perform so many different activities? Would you believe it's bacteria?

Cleaning Up Our Act

Without bacteria, Earth would be littered with the remains of plants and animals. That's because many bacteria decompose once-living matter. Some bacteria also break down other substances. In fact, they offer solutions to some of our toughest pollution problems.

When an oil spill occurs, it often severely damages the environment. However, scientists have engineered bacteria that actually feed on oil! As the bacteria eat the oil, they break it down into harmless substances. Scientists hope to use the bacteria to clean up large oil spills, but they must first be certain that introducing a large number of these bacteria into the ocean will not harm the environment.

Producing Plastics and Pills

Did you know that some bacteria act like factories? For example, one kind of bacteria makes biodegradable plastic! These amazing bacteria store energy as plastic granules, just as animals store energy as fat and plants store energy as starch. Originally, researchers found that the plastic was stiff and brittle. However, scientists found that when certain substances are added to the bacteria's diet, the plastic they produce is flexible enough to be made into consumer products.

Other bacteria can produce important drugs used to fight disease. In fact, some bacteria even produce antibiotics that can fight infections caused by other bacteria!

Helping Plants Grow

Genetic engineers have designed bacteria that can help make plants pest resistant and cold resistant. Some bacteria even keep foods fresh longer on the grocery store shelves. From the garbage dump to the grocery store to the medicine cabinet, bacteria are at our service!

What Do You Think?

▶ Many scientists are concerned about the potential side effects of genetic engineering. What could be some consequences of altering the genetic makeup of crops? Do some research to find out for yourself. Write a report or organize a debate with your classmates.

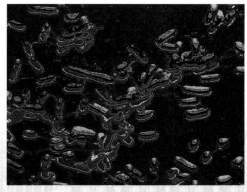

▲ *These bacteria are used to clean up oil spills.*

Answer to What Do You Think?
Answers will vary.

Meatless Munching

What'll it be today, the hamburger special, chicken surprise, or veggie platter? More and more people are opting for the veggie platter. In fact, research indicates that more than 12 million Americans are now vegetarians, and this number appears to be growing.

It's Not Just Salads

When you think of a vegetarian diet, you might think only of vegetables. Of course, vegetables are important to a vegetarian diet, but not all vegetarian diets are alike. Some vegetarian diets are based solely on plant products, while semivegetarian diets may also include dairy products, eggs, fish, and poultry.

Why the Trend?

There are many different reasons for choosing a vegetarian lifestyle. Some people believe that a vegetarian diet is healthy because a decreased consumption of animal products can lower their intake of saturated fat and cholesterol. Other people have ecological reasons for eating a vegetarian diet. For example, producing a serving of meat requires more land, water, and chemicals than producing a serving of grain. Finally, many vegetarians believe that it is unethical to kill animals for meat when plants are available.

Benefits and Risks

Recent statistics suggest that a vegetarian diet may reduce the risk of heart disease, adult-onset diabetes, and some forms of cancer. However, this type of diet takes careful thought. It is not simply a matter of eliminating meat. People may replace the meat with too many dairy products and eggs, which are higher in fat than most meats. Others may substitute high carbohydrate foods (such as pasta) and junk food (such as french fries) for their meat choices instead of increasing their fruit and vegetable intake. This could lead to nutritional deficiencies. The key to consuming the recommended amount of calories and nutrients for a healthy diet is to eat a wide variety of nutritious, low-fat foods. This is a good idea whether you want to decrease your meat intake or not.

▲ *All of the foods shown here are plant products and are commonly used in vegetarian diets.*

Prepare a Healthy Menu

▶ Choose one of the nutrients that meats provide, and do some research to find a vegetarian substitute. What difficulties did you encounter in your search? Would you eat the substitute you found?

HEALTH WATCH
Meatless Munching

Background

To be nutritionally complete, our bodies need 20 different amino acids, 10 of which we manufacture ourselves. The other 10 are called essential amino acids. By eating a variety of foods, a vegetarian can be assured of getting all the protein he or she needs. Indeed, many experts consider dietary variety to be the primary factor in achieving a nutritionally complete diet and maintaining good health, for both meat eaters and vegetarians.

Teaching Strategy

Explain to the class that a protein consists of a long chain of amino acids. Proteins are used in the body to regulate hormones, to repair damaged tissue, and to help fight infections. Assign each student one of the 20 common amino acids to investigate. Tell students to find out what foods the amino acid can be found in and what function it serves in the body. Then have students write the information in outline form on a small index card, along with illustrations of the foods the amino acid is found in. Use the index cards to make a large poster for the class.

Answers to Prepare a Healthy Menu

Some nutrients found in meats include vitamin B_{12}, iron, zinc, calcium, and protein. Vitamin B_{12} is not found in plants in appreciable amounts, so many vegetarians must eat foods fortified with B_{12} or take supplements.

Exploring, inventing, and investigating are essential to the study of science. However, these activities can also be dangerous. To make sure that your experiments and explorations are safe, you must be aware of a variety of safety guidelines.

You have probably heard of the saying, "It is better to be safe than sorry." This is particularly true in a science classroom where experiments and explorations are being performed. Being uninformed and careless can result in serious injuries. Don't take chances with your own safety or with anyone else's.

Following are important guidelines for staying safe in the science classroom. Your teacher may also have safety guidelines and tips that are specific to your classroom and laboratory. Take the time to be safe.

Safety Rules!

Start Out Right

Always get your teacher's permission before attempting any laboratory exploration. Read the procedures carefully, and pay particular attention to safety information and caution statements. If you are unsure about what a safety symbol means, look it up or ask your teacher. You cannot be too careful when it comes to safety. If an accident does occur, inform your teacher immediately, regardless of how minor you think the accident is.

Safety Symbols

All of the experiments and investigations in this book and their related worksheets include important safety symbols to alert you to particular safety concerns. Become familiar with these symbols so that when you see them, you will know what they mean and what to do. It is important that you read this entire safety section to learn about specific dangers in the laboratory.

If you are instructed to note the odor of a substance, wave the fumes toward your nose with your hand. Never put your nose close to the source.

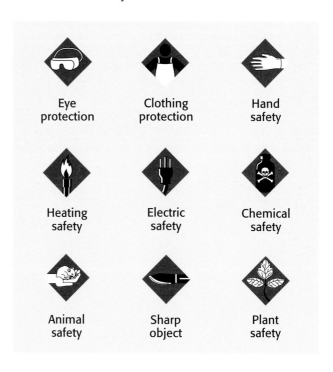

Eye protection • Clothing protection • Hand safety

Heating safety • Electric safety • Chemical safety

Animal safety • Sharp object • Plant safety

Eye Safety

Wear safety goggles when working around chemicals, acids, bases, or any type of flame or heating device. Wear safety goggles any time there is even the slightest chance that harm could come to your eyes. If any substance gets into your eyes, notify your teacher immediately, and flush your eyes with running water for at least 15 minutes. Treat any unknown chemical as if it were a dangerous chemical. Never look directly into the sun. Doing so could cause permanent blindness.

Avoid wearing contact lenses in a laboratory situation. Even if you are wearing safety goggles, chemicals can get between the contact lenses and your eyes. If your doctor requires that you wear contact lenses instead of glasses, wear eye-cup safety goggles in the lab.

Safety Equipment

Know the locations of the nearest fire alarms and any other safety equipment, such as fire blankets and eyewash fountains, as identified by your teacher, and know the procedures for using them.

Be extra careful when using any glassware. When adding a heavy object to a graduated cylinder, tilt the cylinder so the object slides slowly to the bottom.

Neatness

Keep your work area free of all unnecessary books and papers. Tie back long hair, and secure loose sleeves or other loose articles of clothing, such as ties and bows. Remove dangling jewelry. Don't wear open-toed shoes or sandals in the laboratory. Never eat, drink, or apply cosmetics in a laboratory setting. Food, drink, and cosmetics can easily become contaminated with dangerous materials.

Certain hair products (such as aerosol hair spray) are flammable and should not be worn while working near an open flame. Avoid wearing hair spray or hair gel on lab days.

Sharp/Pointed Objects

Use knives and other sharp instruments with extreme care. Never cut objects while holding them in your hands. Place objects on a suitable work surface for cutting.

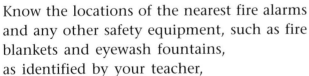

Heat

Wear safety goggles when using a heating device or a flame. Whenever possible, use an electric hot plate as a heat source instead of an open flame. When heating materials in a test tube, always angle the test tube away from yourself and others. In order to avoid burns, wear heat-resistant gloves whenever instructed to do so.

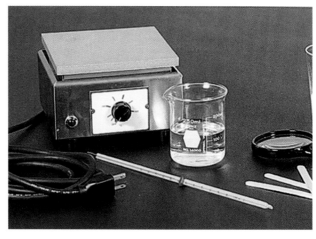

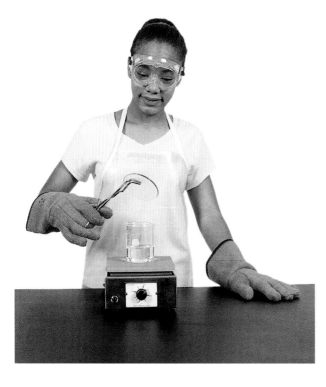

Chemicals

Wear safety goggles when handling any potentially dangerous chemicals, acids, or bases. If a chemical is unknown, handle it as you would a dangerous chemical. Wear an apron and safety gloves when working with acids or bases or whenever you are told to do so. If a spill gets on your skin or clothing, rinse it off immediately with water for at least 5 minutes while calling to your teacher.

Never mix chemicals unless your teacher tells you to do so. Never taste, touch, or smell chemicals unless you are specifically directed to do so. Before working with a flammable liquid or gas, check for the presence of any source of flame, spark, or heat.

Electricity

Be careful with electrical cords. When using a microscope with a lamp, do not place the cord where it could trip someone. Do not let cords hang over a table edge in a way that could cause equipment to fall if the cord is accidentally pulled. Do not use equipment with damaged cords. Be sure your hands are dry and that the electrical equipment is in the "off" position before plugging it in. Turn off and unplug electrical equipment when you are finished.

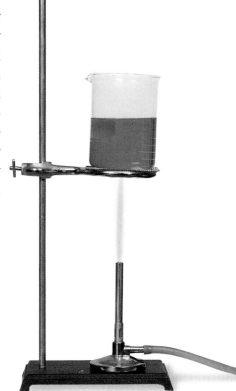

Animal Safety

Always obtain your teacher's permission before bringing any animal into the school building. Handle animals only as your teacher directs. Always treat animals carefully and with respect. Wash your hands thoroughly after handling any animal.

Plant Safety

Do not eat any part of a plant or plant seed used in the laboratory. Wash hands thoroughly after handling any part of a plant. When in nature, do not pick any wild plants unless your teacher instructs you to do so.

Glassware

Examine all glassware before use. Be sure that glassware is clean and free of chips and cracks. Report damaged glassware to your teacher. Glass containers used for heating should be made of heat-resistant glass.

Safety First! LabBook

Seeing is Believing
Teacher's Notes

Time Required
One 45-minute class period, and 5 to 10 minutes every other day for 2 weeks

Lab Ratings

TEACHER PREP 🧪
STUDENT SET-UP 🧪
CONCEPT LEVEL 🧪
CLEAN UP 🧪

MATERIALS
The materials listed on the student page are enough for 1–2 students. This lab may be done with several different types of marking methods. The fingernail is very hard and not very porous. Marking the nail permanently is a challenge. A permanent marker, such as a laundry-marking pen, may need to be refreshed only once a day. Fingernail polish may be an acceptable alternative. Acrylic paint may also be used.

Safety Caution
Remind students to review all safety cautions and icons before beginning this lab activity.

Lab Notes
Few topics are as important to students as acquiring knowledge and understanding of their own body. As they develop, students can't help but observe the ways they are changing physically. One part of the body that grows quickly and requires a great deal of their attention is their fingernails. In this lab, students are able to witness the growth of their own body.

178 Chapter 1 • LabBook

Seeing Is Believing

Fingernails are part of your body's integumentary system, which includes the skin that covers your entire body. Nails are a modification of the outer layer of the skin, and they grow continuously throughout your life. In this activity, you will measure the rate at which fingernails grow.

Materials
- permanent marker
- metric ruler
- graph paper (optional)

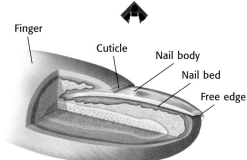

Procedure

1. Trace around each of your hands. Then fill in some of the details, such as the fingernails. Choose a finger on your drawing, and label the parts of the fingernail, as shown at right. Notice that the nail bed is the area where the nail is attached to the finger. The illustration at right shows how far inside your finger your fingernail begins.

2. Find the center of the nail bed on your right index finger (the finger next to your thumb). Make a mark with the permanent marker on the center of your nail bed, as shown at right. **Caution:** Do not get the permanent marker ink on your clothing.

3. Measure from the mark to the base of your nail. Record this measurement on your hand drawing. Label this measurement "Day 1."

4. Repeat steps 2–3 for your left index finger. Then switch roles with your lab partner.

5. Let your fingernails grow for 2 days. Normal, daily activity will not wash away the stain completely, but you may need to freshen the mark periodically throughout this lab.

6. Measure the distance from the mark on your nail to the base of your nail. Record this distance on your hand drawing. Label this measurement "Day 3."

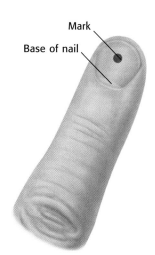

 Datasheets for LabBook

Kathy LaRoe
East Valley Middle School
East Helena, Montana

7. Continue measuring and recording the growth of your nails every other day for 2 weeks. Refresh the mark as necessary. You may continue to file or trim your nails as usual.

8. After you have completed your measurements, prepare a graph similar to the one below.

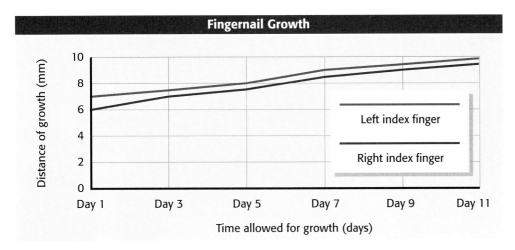

Analysis

9. Did one hand have a faster-growing nail? Write two possible explanations for this observation.

10. Who has the fastest-growing nails among your classmates? Who has the slowest-growing nails? What is the difference in the total nail growth between these two students?

11. Among your classmates, do the nail-growth rates for males and females differ? Is there a relationship between nail growth and other physical characteristics, such as height?

Going Further

Do some research in the library or on the Internet to find answers to the following questions:

- How are nails important to you? What do they help you do? Give at least three examples to support your answers.
- Are your fingernails an indication of your health or state of nutrition?

Lab Notes

Tell students that the graphed data shown in this lab is only an example and will not be the same as their own data. A female adult index fingernail, for example, may be about 12 mm from the cuticle to the beginning of the free edge.

Answers

All answers to the Analysis questions will depend on student observations and measurements.

Going Further

Nails are extremely versatile. Most of us take for granted the many and unique functions they preform. Fingernails, for example, intensify our tactile sensitivity; they help us when we try to pick up small objects; they protect our fingertips; they serve as tools for scratching; they can be used as weapons. Your students may think of many more ways that toenails and fingernails are important to the body.

Science Skills Worksheet 26 "Grasping Graphing"

Chapter 6 • LabBook **179**

Enzymes in Action
Teacher's Notes

Time Required
One 45-minute class period

Lab Ratings

- TEACHER PREP 🧪🧪
- STUDENT SET-UP 🧪🧪
- CONCEPT LEVEL 🧪🧪🧪
- CLEAN UP 🧪🧪

MATERIALS

The materials listed on the student page are enough for 1 student or 1 group of 2 to 4 students. If you do not have enough mortar and pestles or small plates, you may have students use any small container to mash the liver with a fork. Beef liver is readily obtained at the grocery store.

Safety Caution

Remind students to review all safety cautions and icons before beginning this lab activity. Use a dilute solution of hydrogen peroxide. Hydrogen peroxide can be harmful to skin and clothing. Caution students to be careful not to spill or splatter the solution while pouring. If hydrogen peroxide comes into contact with skin, wash the skin immediately with plenty of running water.

180 Chapter 3 • LabBook

Enzymes in Action

You know how important enzymes are in the process of digestion. This lab will help you see enzymes at work. Hydrogen peroxide is continuously produced by your cells. If it is not quickly broken down, hydrogen peroxide will kill your cells. Luckily, your cells contain an enzyme that converts hydrogen peroxide into two non-poisonous substances. This enzyme is also present in the cells of beef liver. In this lab, you will observe the action of this enzyme on hydrogen peroxide.

Procedure

1. In your ScienceLog, draw a data table similar to the one below. Be sure to leave enough space to write your observations.

Data Table		
Size and condition of liver	Experimental liquid	Observations
1 cm cube beef liver	2 mL water	
1 cm cube beef liver	2 mL hydrogen peroxide	
1 cm cube beef liver (mashed)	2 mL hydrogen peroxide	

Materials

- 1 cm cubes of beef liver (3)
- tweezers
- small plate
- 10 mL graduated cylinder
- water
- 3 test tubes
- test-tube rack
- 4 mL of fresh hydrogen peroxide
- mortar and pestle (or fork and watch glass)
- spatula
- protective gloves

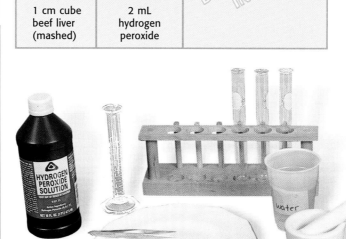

 Datasheets for LabBook

James Chin
Frank A. Day Middle School
Newtonville, Massachusetts

2. Get three equal-sized pieces of beef liver from your teacher, and use your forceps to place them on your plate.

3. Pour 2 mL of water into a test tube labeled "Water and liver."

4. Using the tweezers, carefully place one piece of liver in the test tube. Record your observations in your data table.

5. Pour 2 mL of hydrogen peroxide into a second test tube labeled "Liver and hydrogen peroxide."
Caution: Do not splash hydrogen peroxide on your skin. If you do get hydrogen peroxide on your skin, rinse the affected area with running water immediately, and tell your teacher.

6. Using the tweezers, carefully place one piece of liver in the test tube. Record your observations of the second test tube in your data table.

7. Pour another 2 mL of hydrogen peroxide into a third test tube labeled "Ground liver and hydrogen peroxide."

8. Using a mortar and pestle (or fork and watch glass), carefully grind the third piece of liver.

9. Using the spatula, scrape the ground liver into the third test tube. Record your observations of the third test tube in your data table.

Analysis

10. What was the purpose of putting the first piece of liver in water? Why was this a necessary step?

11. Describe the difference you observed between the liver and the ground liver when each was placed in the hydrogen peroxide. How can you account for this difference?

Going Further

Do plant cells contain enzymes that break down hydrogen peroxide? Try this experiment using potato cubes instead of liver to find out.

Answers

10. The piece of liver in water is the control. It is necessary to have a control so that differences in reactions to other substances can be observed.

11. Students should observe that the cube of liver produced a quick, foaming reaction to the hydrogen peroxide as the solution came into contact with the cells and destroyed them. The mashed liver should also produce a very quick reaction; however, students should observe that the effervescence subsides faster than with the cube. This is because the mashed liver has a greater surface area than the cube, exposing more liver enzymes to the hydrogen peroxide. There should be no reaction in water.

Going Further

Students will discover that plant cells do contain enzymes that will break down hydrogen peroxide. Have them experiment with different types of plants and vegetables so they can observe the different rates of reaction.

My, How You've Grown!
Teacher's Notes

Time Required
One 45-minute class period

Lab Ratings

- TEACHER PREP 🧪
- STUDENT SET-UP 🧪
- CONCEPT LEVEL 🧪🧪
- CLEAN UP 🧪

Randy Christian
Stovall Junior High School
Houston, Texas

182 Chapter 5 • LabBook

My, How You've Grown!

In humans, the process of development that takes place between fertilization and birth lasts about 266 days. In 4 weeks, the new individual grows from a single fertilized cell to an embryo whose heart is beating and pumping blood. All of the organ systems and body parts are completely formed by the end of the seventh month. During the last 2 months before birth, the baby grows and its organ systems mature. At birth, the average mass of a baby is about 33,000 times as much as that of an embryo at 2 weeks of development! In this activity you will discover just how fast a fetus grows.

Materials
- graph paper
- colored pencils

TRY at HOME

Procedure

1. Using graph paper, make two graphs—one titled "Length" and one titled "Mass"—in your ScienceLog. On the length graph, use intervals of 25 mm on the y-axis. Extend the y-axis to 500 mm. On the mass graph, use intervals of 100 g on the y-axis. Extend this y-axis to 3,300 g. Use 2-week intervals for time on the x-axes for both graphs. Both x-axes should extend to 40 weeks.

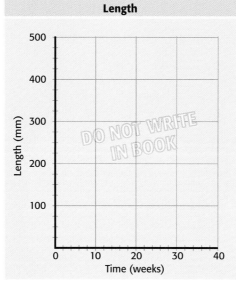
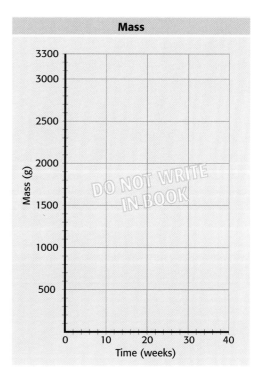

2. Examine the data table at right. Plot the data in the table on your graphs. Use a colored pencil to draw the curved line that joins the points on each graph.

Analysis

3. Describe the change in mass of a developing fetus. How can you explain this change?

4. Describe the change in length of a developing fetus. How does the change in mass compare to the change in length?

Going Further

Using the information in your graphs, estimate how tall a child would be at age 3 if he or she continued to grow at the same average rate a fetus grows.

Increase of Mass and Length of Average Human Fetus

Time (wks)	Mass (g)	Length (mm)
2	0.1	1.5
3	0.3	2.3
4	0.5	5.0
5	0.6	10.0
6	0.8	15.0
8	1.0	30.0
13	15.0	90.0
17	115.0	140.0
21	300.0	250.0
26	950.0	320.0
30	1,500.0	400.0
35	2,300.0	450.0
40	3,300.0	500.0

Answers

2. Students' graphs should look like those below.
3. The change in mass of a developing fetus is steadily increasing, approximately tripling each month of the first and second trimesters. This is the period of rapid cell division.
4. The change in length steadily increases, doubling and even tripling each month in the first two trimesters. In the third trimester, the rate of lengthening slows.

Going Further

The child would be 2.45 m (8.04 ft) tall!

Datasheets for LabBook

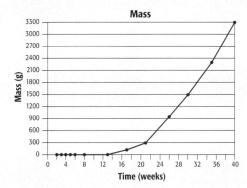
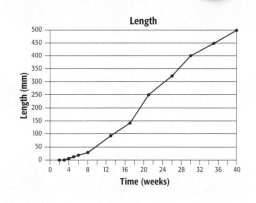

Chapter 3 • LabBook

Passing the Cold
Teacher's Notes

Time Required
One 45-minute class period

Lab Ratings

- TEACHER PREP 🧪🧪
- STUDENT SET-UP 🧪🧪
- CONCEPT LEVEL 🧪🧪
- CLEAN UP 🧪🧪

MATERIALS

Prepare a phenolphthalein indicator solution ahead of time. Dilute the indicator solution in water. Add 10 mL of the indicator to 40 mL of water. This is enough solution for one student. Mix a 1.5 percent NaOH solution. Mix 15 g of NaOH to 1 L of water. All but one student will receive 50 mL of this solution.

Safety Caution

Remind students to review all safety cautions and icons before beginning this lab activity. Also remind students that although they are working with a very low concentration of an alkaline solution, they should work safely with all materials. All spills should be cleaned up immediately. Skin exposed to solutions should be washed immediately with plenty of running water. Remind students that they must never mix unknown solutions without supervision and teacher approval.

184 Chapter 6 • LabBook

Using Scientific Methods

Passing the Cold

DISCOVERY LAB

There are more than 100 viruses that cause the symptoms of the common cold. Any of the viruses can be passed from person to person—through the air or through direct contact. In this activity you will track the progress of an outbreak in your class.

Ask a Question

1. How are cold viruses passed from person to person? How can the progress of an outbreak be modeled?

Conduct an Experiment

2. Obtain an empty cup or beaker, an eyedropper, and 50 mL of one of the solutions from your teacher. Only one of you will have the "cold virus" solution. You will see a change in your solution when you have become infected.

3. Your teacher will divide the class into two equal groups. If there is an extra student, that person will record data on the chalkboard. Otherwise, the teacher will act as the recorder.

4. Each group will form a straight line and face each other.

5. Each time your teacher says "Mix," fill your eyedropper with your solution, and place 10 drops of your solution in the beaker of the person in the line opposite you without touching your eyedropper to the liquid.

6. Gently stir the liquid in your cup with your eyedropper. Do not put your eyedropper in anyone else's solution.

7. If your solution changes color, raise your hand so that the recorder can record the number of students who have been "infected."

8. Your teacher will instruct one line to move one person to the right. The person at the end of the line without a partner should go to the other end of the line.

9. Repeat steps 5–8 nine more times for a total of 10 trials.

Materials
- 200 mL beaker or a cup of similar size
- eyedropper
- 50 mL of an unknown solution
- protective gloves

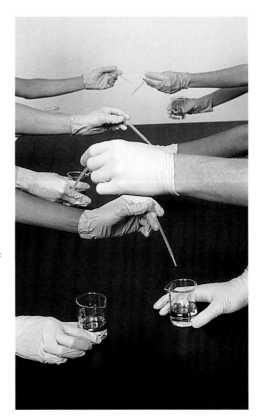

 Datasheets for LabBook

Edith McAlanis
Socorro Middle School
El Paso, Texas

Collect Data

10. Return to your desk, and create a data table in your ScienceLog similar to the one below. The column with the title "Total number of people" will remain the same in every row. Enter the data on the board into your data table.

11. Find the percentage of infected people for the last column by dividing the number of infected people by the total number of people and multiplying by 100 in each line.

Results

Trial	Number of infected people	Total number of people	Percentage of infected people
1			
2			
3			
4			
5			
6			
7			
8			
9			
10			

DO NOT WRITE IN BOOK

Analyze the Results

12. Did you become infected? If so, during which trial did you become infected?

13. Did everyone eventually become infected? If so, how many trials were necessary to infect everyone?

14. Explain at least one reason why this simulation may underestimate the number of people who might have been infected in real life.

15. Use your results to create a line graph showing the change in the infection percentage per trial.

Going Further

Research in the library or on the Internet to find out some of the factors that contribute to the spread of a cold virus. What is the best and easiest way to reduce your chances of catching a cold? Explain your answer.

Going Further

Colds are usually spread by close contact with a person who is actively infected. Colds may be more prevalent in the winter because we tend to stay indoors where we are in closer contact with each other. The virus is carried on the microdroplets in a cough or sneeze and by unwashed hands. The best and easiest way to reduce your chances of catching a cold is to try to avoid crowded places, wash your hands thoroughly and frequently when people around you have colds, and avoid putting your hands on your face near your eyes. Eat a healthy diet and get plenty of rest and exercise to stay in good general health.

Preparation Notes

Phenolphthalein is a base indicator and will turn pink in the presence of NaOH. One student (or two, if your class is large) will be given 50 mL of the indicator. This student will represent the original "infected" individual. (It is more fun if no one knows who the original infected student is at first.) All other students will be given the NaOH solution.

Prepare a results chart similar to the student table on the board. You will need to record results while the students are performing the experiment.

When switching the students between trials, the student on the end will need to move to the other end of the line so that all students will have a new partner for each trial. In case of an odd number of students, you will need to participate, or you may use one student volunteer to record results on the board.

Answers

11. The percentage of infected people can be found by dividing column 1 by column 2 and multiplying by 100. For example, if 2 people are infected in trial 1 and there are 16 people in all, then: 2 ÷ 16 × 100 = 11 percent.

12.–13. If there are 10 students (two rows of five facing each other) it will take six trials for everyone to become infected.

14. In real life, colds are spread by more than one means. Coughing, sneezing, touching with unwashed hands are all ways to spread a cold. More than one person in a classroom may have a cold.

Concept Mapping: A Way to Bring Ideas Together

What Is a Concept Map?

Have you ever tried to tell someone about a book or a chapter you've just read and found that you can remember only a few isolated words and ideas? Or maybe you've memorized facts for a test and then weeks later discovered you're not even sure what topics those facts covered.

In both cases, you may have understood the ideas or concepts by themselves but not in relation to one another. If you could somehow link the ideas together, you would probably understand them better and remember them longer. This is something a concept map can help you do. A concept map is a way to see how ideas or concepts fit together. It can help you see the "big picture."

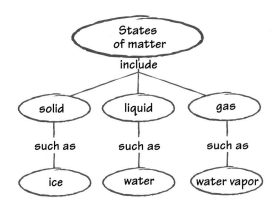
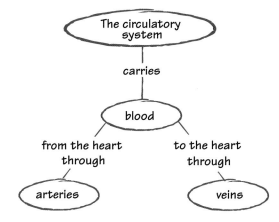

How to Make a Concept Map

❶ Make a list of the main ideas or concepts.

It might help to write each concept on its own slip of paper. This will make it easier to rearrange the concepts as many times as necessary to make sense of how the concepts are connected. After you've made a few concept maps this way, you can go directly from writing your list to actually making the map.

❷ Arrange the concepts in order from the most general to the most specific.

Put the most general concept at the top and circle it. Ask yourself, "How does this concept relate to the remaining concepts?" As you see the relationships, arrange the concepts in order from general to specific.

❸ Connect the related concepts with lines.

❹ On each line, write an action word or short phrase that shows how the concepts are related.

Look at the concept maps on this page, and then see if you can make one for the following terms:

plants, water, photosynthesis, carbon dioxide, sun's energy

One possible answer is provided at right, but don't look at it until you try the concept map yourself.

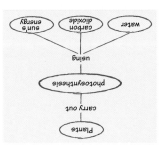

186 Appendix

SI Measurement

The International System of Units, or SI, is the standard system of measurement used by many scientists. Using the same standards of measurement makes it easier for scientists to communicate with one another.

SI works by combining prefixes and base units. Each base unit can be used with different prefixes to define smaller and larger quantities. The table below lists common SI prefixes.

SI Prefixes			
Prefix	**Abbreviation**	**Factor**	**Example**
kilo-	k	1,000	kilogram, 1 kg = 1,000 g
hecto-	h	100	hectoliter, 1 hL = 100 L
deka-	da	10	dekameter, 1 dam = 10 m
		1	meter, liter
deci-	d	0.1	decigram, 1 dg = 0.1 g
centi-	c	0.01	centimeter, 1 cm = 0.01 m
milli-	m	0.001	milliliter, 1 mL = 0.001 L
micro-	µ	0.000 001	micrometer, 1 µm = 0.000 001 m

SI Conversion Table		
SI units	**From SI to English**	**From English to SI**
Length		
kilometer (km) = 1,000 m	1 km = 0.621 mi	1 mi = 1.609 km
meter (m) = 100 cm	1 m = 3.281 ft	1 ft = 0.305 m
centimeter (cm) = 0.01 m	1 cm = 0.394 in.	1 in. = 2.540 cm
millimeter (mm) = 0.001 m	1 mm = 0.039 in.	
micrometer (µm) = 0.000 001 m		
nanometer (nm) = 0.000 000 001 m		
Area		
square kilometer (km^2) = 100 hectares	1 km^2 = 0.386 mi^2	1 mi^2 = 2.590 km^2
hectare (ha) = 10,000 m^2	1 ha = 2.471 acres	1 acre = 0.405 ha
square meter (m^2) = 10,000 cm^2	1 m^2 = 10.765 ft^2	1 ft^2 = 0.093 m^2
square centimeter (cm^2) = 100 mm^2	1 cm^2 = 0.155 $in.^2$	1 $in.^2$ = 6.452 cm^2
Volume		
liter (L) = 1,000 mL = 1 dm^3	1 L = 1.057 fl qt	1 fl qt = 0.946 L
milliliter (mL) = 0.001 L = 1 cm^3	1 mL = 0.034 fl oz	1 fl oz = 29.575 mL
microliter (µL) = 0.000 001 L		
Mass		
kilogram (kg) = 1,000 g	1 kg = 2.205 lb	1 lb = 0.454 kg
gram (g) = 1,000 mg	1 g = 0.035 oz	1 oz = 28.349 g
milligram (mg) = 0.001 g		
microgram (µg) = 0.000 001 g		

Temperature Scales

Temperature can be expressed using three different scales: Fahrenheit, Celsius, and Kelvin. The SI unit for temperature is the kelvin (K).

Although 0 K is much colder than 0°C, a change of 1 K is equal to a change of 1°C.

Three Temperature Scales

	Fahrenheit	Celsius	Kelvin
Water boils	212°	100°	373
Body temperature	98.6°	37°	310
Room temperature	68°	20°	293
Water freezes	32°	0°	273

Temperature Conversions Table

To convert	Use this equation:	Example
Celsius to Fahrenheit °C ⟶ °F	$°F = \left(\frac{9}{5} \times °C\right) + 32$	Convert 45°C to °F. $°F = \left(\frac{9}{5} \times 45°C\right) + 32 = 113°F$
Fahrenheit to Celsius °F ⟶ °C	$°C = \frac{5}{9} \times (°F - 32)$	Convert 68°F to °C. $°C = \frac{5}{9} \times (68°F - 32) = 20°C$
Celsius to Kelvin °C ⟶ K	$K = °C + 273$	Convert 45°C to K. $K = 45°C + 273 = 318 \text{ K}$
Kelvin to Celsius K ⟶ °C	$°C = K - 273$	Convert 32 K to °C. $°C = 32 \text{ K} - 273 = -241°C$

Appendix

Measuring Skills

Using a Graduated Cylinder

When using a graduated cylinder to measure volume, keep the following procedures in mind:

❶ Make sure the cylinder is on a flat, level surface.

❷ Move your head so that your eye is level with the surface of the liquid.

❸ Read the mark closest to the liquid level. On glass graduated cylinders, read the mark closest to the center of the curve in the liquid's surface.

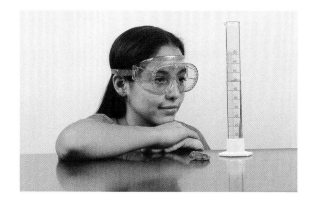

Using a Meterstick or Metric Ruler

When using a meterstick or metric ruler to measure length, keep the following procedures in mind:

❶ Place the ruler firmly against the object you are measuring.

❷ Align one edge of the object exactly with the zero end of the ruler.

❸ Look at the other edge of the object to see which of the marks on the ruler is closest to that edge. **Note:** Each small slash between the centimeters represents a millimeter, which is one-tenth of a centimeter.

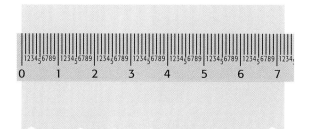

Using a Triple-Beam Balance

When using a triple-beam balance to measure mass, keep the following procedures in mind:

❶ Make sure the balance is on a level surface.

❷ Place all of the countermasses at zero. Adjust the balancing knob until the pointer rests at zero.

❸ Place the object you wish to measure on the pan. **Caution:** Do not place hot objects or chemicals directly on the balance pan.

❹ Move the largest countermass along the beam to the right until it is at the last notch that does not tip the balance. Follow the same procedure with the next-largest countermass. Then move the smallest countermass until the pointer rests at zero.

❺ Add the readings from the three beams together to determine the mass of the object.

❻ When determining the mass of crystals or powders, use a piece of filter paper. First find the mass of the paper. Then add the crystals or powder to the paper and re-measure. The actual mass of the crystals or powder is the total mass minus the mass of the paper. When finding the mass of liquids, first find the mass of the empty container. Then find the mass of the liquid and container together. The mass of the liquid is the total mass minus the mass of the container.

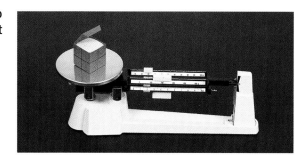

Appendix

Scientific Method

The series of steps that scientists use to answer questions and solve problems is often called the **scientific method.** The scientific method is not a rigid procedure. Scientists may use all of the steps or just some of the steps of the scientific method. They may even repeat some of the steps. The goal of the scientific method is to come up with reliable answers and solutions.

Six Steps of the Scientific Method

1 Ask a Question Good questions come from careful **observations.** You make observations by using your senses to gather information. Sometimes you may use instruments, such as microscopes and telescopes, to extend the range of your senses. As you observe the natural world, you will discover that you have many more questions than answers. These questions drive the scientific method.

Questions beginning with *what, why, how,* and *when* are very important in focusing an investigation, and they often lead to a hypothesis. (You will learn what a hypothesis is in the next step.) Here is an example of a question that could lead to further investigation.

Question: How does acid rain affect plant growth?

2 Form a Hypothesis After you come up with a question, you need to turn the question into a **hypothesis.** A hypothesis is a clear statement of what you expect the answer to your question to be. Your hypothesis will represent your best "educated guess" based on your observations and what you already know. A good hypothesis is testable. If observations and information cannot be gathered or if an experiment cannot be designed to test your hypothesis, it is untestable, and the investigation can go no further.

Here is a hypothesis that could be formed from the question, "How does acid rain affect plant growth?"

Hypothesis: Acid rain causes plants to grow more slowly.

Notice that the hypothesis provides some specifics that lead to methods of testing. The hypothesis can also lead to predictions. A **prediction** is what you think will be the outcome of your experiment or data collection. Predictions are usually stated in an "if . . . then" format. For example, **if** meat is kept at room temperature, **then** it will spoil faster than meat kept in the refrigerator. More than one prediction can be made for a single hypothesis. Here is a sample prediction for the hypothesis that acid rain causes plants to grow more slowly.

Prediction: If a plant is watered with only acid rain (which has a pH of 4), then the plant will grow at half its normal rate.

3 **Test the Hypothesis** After you have formed a hypothesis and made a prediction, you should test your hypothesis. There are different ways to do this. Perhaps the most familiar way is to conduct a **controlled experiment.** A controlled experiment tests only one factor at a time. A controlled experiment has a **control group** and one or more **experimental groups.** All the factors for the control and experimental groups are the same except for one factor, which is called the **variable.** By changing only one factor, you can see the results of just that one change.

Sometimes, the nature of an investigation makes a controlled experiment impossible. For example, dinosaurs have been extinct for millions of years, and the Earth's core is surrounded by thousands of meters of rock. It would be difficult, if not impossible, to conduct controlled experiments on such things. Under such circumstances, a hypothesis may be tested by making detailed observations. Taking measurements is one way of making observations.

4 **Analyze the Results** After you have completed your experiments, made your observations, and collected your data, you must analyze all the information you have gathered. Tables and graphs are often used in this step to organize the data.

5 **Draw Conclusions** Based on the analysis of your data, you should conclude whether or not your results support your hypothesis. If your hypothesis is supported, you (or others) might want to repeat the observations or experiments to verify your results. If your hypothesis is not supported by the data, you may have to check your procedure for errors. You may even have to reject your hypothesis and make a new one. If you cannot draw a conclusion from your results, you may have to try the investigation again or carry out further observations or experiments.

6 **Communicate Results** After any scientific investigation, you should report your results. By doing a written or oral report, you let others know what you have learned. They may want to repeat your investigation to see if they get the same results. Your report may even lead to another question, which in turn may lead to another investigation.

Scientific Method in Action

The scientific method is not a "straight line" of steps. It contains loops in which several steps may be repeated over and over again, while others may not be necessary. For example, sometimes scientists will find that testing one hypothesis raises new questions and new hypotheses to be tested. And sometimes, testing the hypothesis leads directly to a conclusion. Furthermore, the steps in the scientific method are not always used in the same order. Follow the steps in the diagram below, and see how many different directions the scientific method can take you.

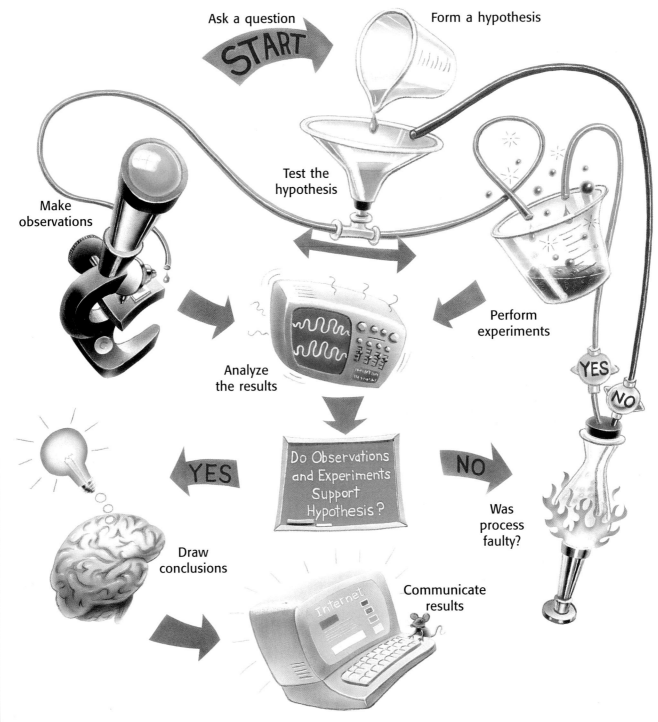

Making Charts and Graphs

Circle Graphs

A circle graph, or pie chart, shows how each group of data relates to all of the data. Each part of the circle represents a category of the data. The entire circle represents all of the data. For example, a biologist studying a hardwood forest in Wisconsin found that there were five different types of trees. The data table at right summarizes the biologist's findings.

Wisconsin Hardwood Trees	
Type of tree	Number found
Oak	600
Maple	750
Beech	300
Birch	1,200
Hickory	150
Total	3,000

How to Make a Circle Graph

❶ In order to make a circle graph of this data, first find the percentage of each type of tree. To do this, divide the number of individual trees by the total number of trees and multiply by 100.

$$\frac{600 \text{ oak}}{3{,}000 \text{ trees}} \times 100 = 20\%$$

$$\frac{750 \text{ maple}}{3{,}000 \text{ trees}} \times 100 = 25\%$$

$$\frac{300 \text{ beech}}{3{,}000 \text{ trees}} \times 100 = 10\%$$

$$\frac{1{,}200 \text{ birch}}{3{,}000 \text{ trees}} \times 100 = 40\%$$

$$\frac{150 \text{ hickory}}{3{,}000 \text{ trees}} \times 100 = 5\%$$

❷ Now determine the size of the pie shapes that make up the chart. Do this by multiplying each percentage by 360°. Remember that a circle contains 360°.

$20\% \times 360° = 72°$ $25\% \times 360° = 90°$
$10\% \times 360° = 36°$ $40\% \times 360° = 144°$
$5\% \times 360° = 18°$

❸ Then check that the sum of the percentages is 100 and the sum of the degrees is 360.

$20\% + 25\% + 10\% + 40\% + 5\% = 100\%$
$72° + 90° + 36° + 144° + 18° = 360°$

❹ Use a compass to draw a circle and mark its center.

❺ Then use a protractor to draw angles of 72°, 90°, 36°, 144°, and 18° in the circle.

❻ Finally, label each part of the graph, and choose an appropriate title.

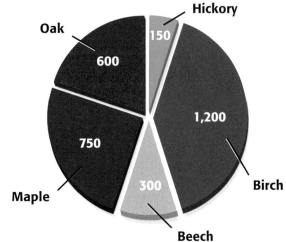

A Community of Wisconsin Hardwood Trees

Appendix

Line Graphs

Population of Appleton, 1900–2000	
Year	Population
1900	1,800
1920	2,500
1940	3,200
1960	3,900
1980	4,600
2000	5,300

Line graphs are most often used to demonstrate continuous change. For example, Mr. Smith's science class analyzed the population records for their hometown, Appleton, between 1900 and 2000. Examine the data at left.

Because the year and the population change, they are the *variables*. The population is determined by, or dependent on, the year. Therefore, the population is called the **dependent variable**, and the year is called the **independent variable**. Each set of data is called a **data pair**. To prepare a line graph, data pairs must first be organized in a table like the one at left.

How to Make a Line Graph

1. Place the independent variable along the horizontal (*x*) axis. Place the dependent variable along the vertical (*y*) axis.

2. Label the *x*-axis "Year" and the *y*-axis "Population." Look at your largest and smallest values for the population. Determine a scale for the *y*-axis that will provide enough space to show these values. You must use the same scale for the entire length of the axis. Find an appropriate scale for the *x*-axis too.

3. Choose reasonable starting points for each axis.

4. Plot the data pairs as accurately as possible.

5. Choose a title that accurately represents the data.

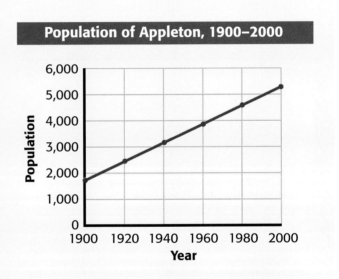

How to Determine Slope

Slope is the ratio of the change in the *y*-axis to the change in the *x*-axis, or "rise over run."

1. Choose two points on the line graph. For example, the population of Appleton in 2000 was 5,300 people. Therefore, you can define point *a* as (2000, 5,300). In 1900, the population was 1,800 people. Define point *b* as (1900, 1,800).

2. Find the change in the *y*-axis.
(*y* at point *a*) − (*y* at point *b*)
5,300 people − 1,800 people = 3,500 people

3. Find the change in the *x*-axis.
(*x* at point *a*) − (*x* at point *b*)
2000 − 1900 = 100 years

4. Calculate the slope of the graph by dividing the change in *y* by the change in *x*.

$$\text{slope} = \frac{\text{change in } y}{\text{change in } x}$$

$$\text{slope} = \frac{3{,}500 \text{ people}}{100 \text{ years}}$$

$$\text{slope} = 35 \text{ people per year}$$

In this example, the population in Appleton increased by a fixed amount each year. The graph of this data is a straight line. Therefore, the relationship is **linear**. When the graph of a set of data is not a straight line, the relationship is **nonlinear**.

Using Algebra to Determine Slope

The equation in step 4 may also be arranged to be:

$y = kx$

where y represents the change in the y-axis, k represents the slope, and x represents the change in the x-axis.

$$\text{slope} = \frac{\text{change in } y}{\text{change in } x}$$

$$k = \frac{y}{x}$$

$$k \times x = \frac{y \times x}{x}$$

$$kx = y$$

Bar Graphs

Bar graphs are used to demonstrate change that is not continuous. These graphs can be used to indicate trends when the data are taken over a long period of time. A meteorologist gathered the precipitation records at right for Hartford, Connecticut, for April 1–15, 1996, and used a bar graph to represent the data.

Precipitation in Hartford, Connecticut April 1–15, 1996

Date	Precipitation (cm)	Date	Precipitation (cm)
April 1	0.5	April 9	0.25
April 2	1.25	April 10	0.0
April 3	0.0	April 11	1.0
April 4	0.0	April 12	0.0
April 5	0.0	April 13	0.25
April 6	0.0	April 14	0.0
April 7	0.0	April 15	6.50
April 8	1.75		

How to Make a Bar Graph

1. Use an appropriate scale and a reasonable starting point for each axis.
2. Label the axes, and plot the data.
3. Choose a title that accurately represents the data.

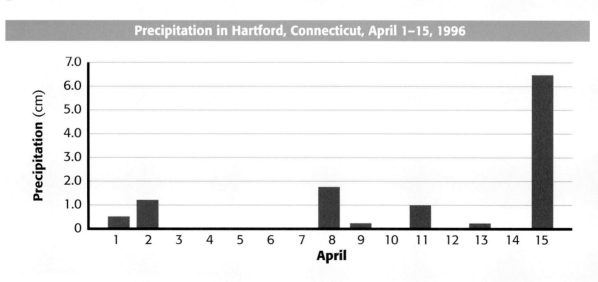

Appendix **195**

Periodic Table of the Elements

Each square on the table includes an element's name, chemical symbol, atomic number, and atomic mass.

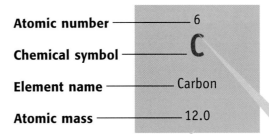

- Atomic number — 6
- Chemical symbol — C
- Element name — Carbon
- Atomic mass — 12.0

The background color indicates the type of element. Carbon is a nonmetal.

The color of the chemical symbol indicates the physical state at room temperature. Carbon is a solid.

Background
- Metals
- Metalloids
- Nonmetals

Chemical Symbol
- Solid
- Liquid
- Gas

Period 1

1
H
Hydrogen
1.0

	Group 1	Group 2	Group 3	Group 4	Group 5	Group 6	Group 7	Group 8	Group 9
Period 2	3 Li Lithium 6.9	4 Be Beryllium 9.0							
Period 3	11 Na Sodium 23.0	12 Mg Magnesium 24.3							
Period 4	19 K Potassium 39.1	20 Ca Calcium 40.1	21 Sc Scandium 45.0	22 Ti Titanium 47.9	23 V Vanadium 50.9	24 Cr Chromium 52.0	25 Mn Manganese 54.9	26 Fe Iron 55.8	27 Co Cobalt 58.9
Period 5	37 Rb Rubidium 85.5	38 Sr Strontium 87.6	39 Y Yttrium 88.9	40 Zr Zirconium 91.2	41 Nb Niobium 92.9	42 Mo Molybdenum 95.9	43 Tc Technetium (97.9)	44 Ru Ruthenium 101.1	45 Rh Rhodium 102.9
Period 6	55 Cs Cesium 132.9	56 Ba Barium 137.3	57 La Lanthanum 138.9	72 Hf Hafnium 178.5	73 Ta Tantalum 180.9	74 W Tungsten 183.8	75 Re Rhenium 186.2	76 Os Osmium 190.2	77 Ir Iridium 192.2
Period 7	87 Fr Francium (223.0)	88 Ra Radium (226.0)	89 Ac Actinium (227.0)	104 Rf Rutherfordium (261.1)	105 Db Dubnium (262.1)	106 Sg Seaborgium (263.1)	107 Bh Bohrium (262.1)	108 Hs Hassium (265)	109 Mt Meitnerium (266)

A row of elements is called a period.

A column of elements is called a group or family.

Lanthanides

| 58 Ce Cerium 140.1 | 59 Pr Praseodymium 140.9 | 60 Nd Neodymium 144.2 | 61 Pm Promethium (144.9) | 62 Sm Samarium 150.4 |

Actinides

| 90 Th Thorium 232.0 | 91 Pa Protactinium 231.0 | 92 U Uranium 238.0 | 93 Np Neptunium (237.0) | 94 Pu Plutonium 244.1 |

These elements are placed below the table to allow the table to be narrower.

This zigzag line reminds you where the metals, nonmetals, and metalloids are.

			Group 13	Group 14	Group 15	Group 16	Group 17	Group 18
								2 **He** Helium 4.0
			5 **B** Boron 10.8	6 **C** Carbon 12.0	7 **N** Nitrogen 14.0	8 **O** Oxygen 16.0	9 **F** Fluorine 19.0	10 **Ne** Neon 20.2
			13 **Al** Aluminum 27.0	14 **Si** Silicon 28.1	15 **P** Phosphorus 31.0	16 **S** Sulfur 32.1	17 **Cl** Chlorine 35.5	18 **Ar** Argon 39.9
Group 10	Group 11	Group 12						
28 **Ni** Nickel 58.7	29 **Cu** Copper 63.5	30 **Zn** Zinc 65.4	31 **Ga** Gallium 69.7	32 **Ge** Germanium 72.6	33 **As** Arsenic 74.9	34 **Se** Selenium 79.0	35 **Br** Bromine 79.9	36 **Kr** Krypton 83.8
46 **Pd** Palladium 106.4	47 **Ag** Silver 107.9	48 **Cd** Cadmium 112.4	49 **In** Indium 114.8	50 **Sn** Tin 118.7	51 **Sb** Antimony 121.8	52 **Te** Tellurium 127.6	53 **I** Iodine 126.9	54 **Xe** Xenon 131.3
78 **Pt** Platinum 195.1	79 **Au** Gold 197.0	80 **Hg** Mercury 200.6	81 **Tl** Thallium 204.4	82 **Pb** Lead 207.2	83 **Bi** Bismuth 209.0	84 **Po** Polonium (209.0)	85 **At** Astatine (210.0)	86 **Rn** Radon (222.0)
110 **Uun*** Ununnilium (271)	111 **Uuu*** Unununium (272)	112 **Uub*** Ununbium (277)		114 **Uuq*** Ununquadium (285)		116 **Uuh*** Ununhexium (289)		118 **Uuo*** Ununoctium (293)

A number in parenthesis is the mass number of the most stable form of that element.

63 **Eu** Europium 152.0	64 **Gd** Gadolinium 157.3	65 **Tb** Terbium 158.9	66 **Dy** Dysprosium 162.5	67 **Ho** Holmium 164.9	68 **Er** Erbium 167.3	69 **Tm** Thulium 168.9	70 **Yb** Ytterbium 173.0	71 **Lu** Lutetium 175.0
95 **Am** Americium (243.1)	96 **Cm** Curium (247.1)	97 **Bk** Berkelium (247.1)	98 **Cf** Californium (251.1)	99 **Es** Einsteinium (252.1)	100 **Fm** Fermium (257.1)	101 **Md** Mendelevium (258.1)	102 **No** Nobelium (259.1)	103 **Lr** Lawrencium (262.1)

*The official names and symbols for the elements greater than 109 will eventually be approved by a committee of scientists.

The Six Kingdoms

Kingdom Archaebacteria

The organisms in this kingdom are single-celled prokaryotes.

Archaebacteria		
Group	**Examples**	**Characteristics**
Methanogens	*Methanococcus*	found in soil, swamps, the digestive tract of mammals; produce methane gas; can't live in oxygen
Thermophiles	*Sulpholobus*	found in extremely hot environments; require sulphur, can't live in oxygen
Halophiles	*Halococcus*	found in environments with very high salt content, such as the Dead Sea; nearly all can live in oxygen

Kingdom Eubacteria

There are more than 4,000 named species in this kingdom of single-celled prokaryotes.

Eubacteria		
Group	**Examples**	**Characteristics**
Bacilli	*Escherichia coli*	rod-shaped; free-living, symbiotic, or parasitic; some can fix nitrogen; some cause disease
Cocci	*Streptococcus*	spherical-shaped, disease-causing; can form spores to resist unfavorable environments
Spirilla	*Treponema*	spiral-shaped; responsible for several serious illnesses, such as syphilis and Lyme disease

Kingdom Protista

The organisms in this kingdom are eukaryotes. There are single-celled and multicellular representatives.

Protists		
Group	**Examples**	**Characteristics**
Sacodines	*Amoeba*	radiolarians; single-celled consumers
Ciliates	*Paramecium*	single-celled consumers
Flagellates	*Trypanosoma*	single-celled parasites
Sporozoans	*Plasmodium*	single-celled parasites
Euglenas	*Euglena*	single-celled; photosynthesize
Diatoms	*Pinnularia*	most are single-celled; photosynthesize
Dinoflagellates	*Gymnodinium*	single-celled; some photosynthesize
Algae	*Volvox*, coral algae	4 phyla; single- or many-celled; photosynthesize
Slime molds	*Physarum*	single- or many-celled; consumers or decomposers
Water molds	powdery mildew	single- or many-celled, parasites or decomposers

Kingdom Fungi

There are single-celled and multicellular eukaryotes in this kingdom. There are four major groups of fungi.

Fungi		
Group	Examples	Characteristics
Threadlike fungi	bread mold	spherical; decomposers
Sac fungi	yeast, morels	saclike; parasites and decomposers
Club fungi	mushrooms, rusts, smuts	club-shaped; parasites and decomposers
Lichens	British soldier	symbiotic with algae

Kingdom Plantae

The organisms in this kingdom are multicellular eukaryotes. They have specialized organ systems for different life processes. They are classified in divisions instead of phyla.

Plants		
Group	Examples	Characteristics
Bryophytes	mosses, liverworts	reproduce by spores
Club mosses	*Lycopodium*, ground pine	reproduce by spores
Horsetails	rushes	reproduce by spores
Ferns	spleenworts, sensitive fern	reproduce by spores
Conifers	pines, spruces, firs	reproduce by seeds; cones
Cycads	*Zamia*	reproduce by seeds
Gnetophytes	*Welwitschia*	reproduce by seeds
Ginkgoes	*Ginkgo*	reproduce by seeds
Angiosperms	all flowering plants	reproduce by seeds; flowers

Kingdom Animalia

This kingdom contains multicellular eukaryotes. They have specialized tissues and complex organ systems.

Animals		
Group	Examples	Characteristics
Sponges	glass sponges	no symmetry or segmentation; aquatic
Cnidarians	jellyfish, coral	radial symmetry; aquatic
Flatworms	planaria, tapeworms, flukes	bilateral symmetry; organ systems
Roundworms	*Trichina*, hookworms	bilateral symmetry; organ systems
Annelids	earthworms, leeches	bilateral symmetry; organ systems
Mollusks	snails, octopuses	bilateral symmetry; organ systems
Echinoderms	sea stars, sand dollars	radial symmetry; organ systems
Arthropods	insects, spiders, lobsters	bilateral symmetry; organ systems
Chordates	fish, amphibians, reptiles, birds, mammals	bilateral symmetry; complex organ systems

Using the Microscope

Parts of the Compound Light Microscope

- The **ocular lens** magnifies the image 10×.
- The **low-power objective** magnifies the image 10×.
- The **high-power objective** magnifies the image either 40× or 43×.
- The **revolving nosepiece** holds the objectives and can be turned to change from one magnification to the other.
- The **body tube** maintains the correct distance between the ocular lens and objectives.
- The **coarse-adjustment knob** moves the body tube up and down to allow focusing of the image.
- The **fine-adjustment knob** moves the body tube slightly to bring the image into sharper focus.
- The **stage** supports a slide.
- **Stage clips** hold the slide in place for viewing.
- The **diaphragm** controls the amount of light coming through the stage.
- The **light source** provides a **light** for viewing the slide.
- The **arm** supports the body tube.
- The **base** supports the microscope.

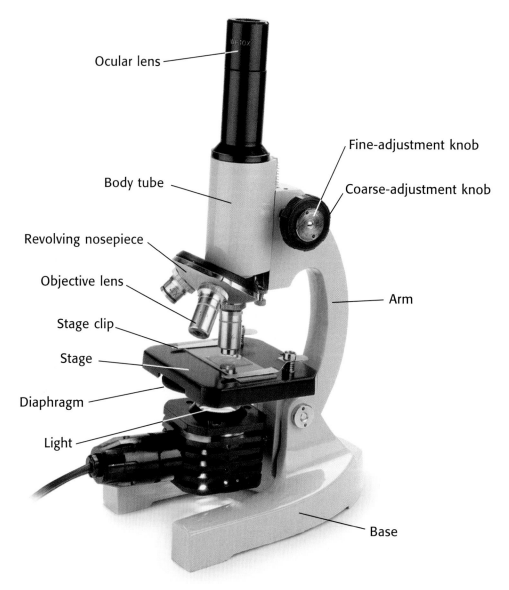

200 Appendix

Proper Use of the Compound Light Microscope

❶ Carry the microscope to your lab table using both hands. Place one hand beneath the base, and use the other hand to hold the arm of the microscope. Hold the microscope close to your body while moving it to your lab table.

❷ Place the microscope on the lab table at least 5 cm from the edge of the table.

❸ Check to see what type of light source is used by your microscope. If the microscope has a lamp, plug it in, making sure that the cord is out of the way. If the microscope has a mirror, adjust it to reflect light through the hole in the stage.
Caution: If your microscope has a mirror, do not use direct sunlight as a light source. Direct sunlight can damage your eyes.

❹ Always begin work with the low-power objective in line with the body tube. Adjust the revolving nosepiece.

❺ Place a prepared slide over the hole in the stage. Secure the slide with the stage clips.

❻ Look through the ocular lens. Move the diaphragm to adjust the amount of light coming through the stage.

❼ Look at the stage from eye level. Slowly turn the coarse adjustment to lower the objective until it almost touches the slide. Do not allow the objective to touch the slide.

❽ Look through the ocular lens. Turn the coarse adjustment to raise the low-power objective until the image is in focus. Always focus by raising the objective away from the slide. *Never focus the objective downward.* Use the fine adjustment to sharpen the focus. Keep both eyes open while viewing a slide.

❾ Make sure that the image is exactly in the center of your field of vision. Then switch to the high-power objective. Focus the image, using only the fine adjustment. *Never use the coarse adjustment at high power.*

❿ When you are finished using the microscope, remove the slide. Clean the ocular lens and objective lenses with lens paper. Return the microscope to its storage area. Remember, you should use both hands to carry the microscope.

Making a Wet Mount

❶ Use lens paper to clean a glass slide and a coverslip.

❷ Place the specimen you wish to observe in the center of the slide.

❸ Using a medicine dropper, place one drop of water on the specimen.

❹ Hold the coverslip at the edge of the water and at a 45° angle to the slide. Make sure that the water runs along the edge of the coverslip.

❺ Lower the coverslip slowly to avoid trapping air bubbles.

❻ Water might evaporate from the slide as you work. Add more water to keep the specimen fresh. Place the tip of the medicine dropper next to the edge of the coverslip. Add a drop of water. (You can also use this method to add stain or solutions to a wet mount.) Remove excess water from the slide by using the corner of a paper towel as a blotter. Do not lift the coverslip to add or remove water.

Glossary

A

addiction a physical dependence on a drug (156)

aerobic exercise vigorous, sustained exercise of the whole body for 20 minutes or more (162)

alcoholism a disorder in which a person is physically and psychologically dependent on alcohol (158)

allergy an inappropriate immune-system reaction to a harmless antigen (135)

alveoli (al VEE uh LIE) tiny sacs that form the bronchiole branches of the lungs (41)

amnion a thin, fluid-filled membrane surrounding a placental mammal's fetus (111)

anorexia nervosa a disorder characterized by self-starvation and an intense fear of gaining weight (154)

antibiotic a substance used to kill or slow the growth of bacteria or other microorganisms (129)

antibody a special protein that can recognize specific pathogens (131)

antigen pieces of a pathogen that generate an immune response from immune-system cells (132)

anus the opening at the end of the digestive tract through which feces pass to the outside (60)

arteries blood vessels that carry blood away from the heart (33)

atrium an upper chamber of the heart (32)

autoimmune disease a disease in which the immune system attacks the cells of the body it is meant to protect (136)

axon a long cell fiber in the nervous system that transfers intercellular messages (77)

B

B cell an immune-system cell that matures in bones and makes antibodies (131)

bile a green liquid made by the liver and stored in the gallbladder; used in fat digestion (59)

blood a connective tissue made up of platelets, white blood cells, red blood cells, and plasma (30)

blood pressure the amount of force exerted by blood on the inside walls of a blood vessel (35)

brain the mass of nerve tissue that is the main organ of the nervous system (79)

bronchi (BRAHNG kie) the two tubes that connect the lungs with the trachea (41)

budding a type of asexual reproduction in which a small part of the parent's body develops into an independent organism (102)

bulimia a disorder characterized by binge eating followed by induced vomiting to rid the body of food (154)

C

Calorie a unit that expresses the amount of energy found in food (148)

cancer a condition in which certain body cells begin dividing at an uncontrolled rate (136)

capillaries the smallest blood vessels (33)

carbohydrate a biochemical composed of one or more simple sugars bonded together that is used to provide and store energy (149)

cardiac muscle the type of muscle found in the heart (12)

cardiovascular system a collection of organs that transport blood to and from your body's cells; the organs in this system include the heart, the arteries, and the veins (30)

cartilage a flexible tissue that gives support and protection but is not rigid like bone (10)

cellular respiration the process of producing ATP in the cell from oxygen and glucose; releases carbon dioxide and water (42)

central nervous system a collection of organs that processes all incoming and outgoing messages from the nerves; the organs in this system include the brain and the spinal cord (76)

cerebellum (SER uh BEL uhm) the part of the brain that keeps track of the body's position (80)

cerebrum the part of the brain that detects touch, sight, sound, odor, taste, pain, heat, and cold and controls all voluntary acts, including thought (79)

chemical digestion the process in which large molecules are broken down into simpler molecules or chemical building blocks (55)

cochlea (KAHK lee uh) an ear organ that converts sound waves into electrical impulses (86)

compact bone the type of bone tissue that does not have open spaces (9)

cones photoreceptors that can detect bright light and help you see colors (84)

connective tissue one of the four main types of tissue in the body; functions include support, protection, insulation, and nourishment (5)

D

dendrite a short, branched extension of a neuron where the neuron receives impulses from other cells (77)

depressant a drug that slows the actions of the central nervous system (155)

dermis the layer of skin below the epidermis (17)

diaphragm (DIE uh FRAM) the sheet of muscle underneath the lungs of mammals that helps draw air into the lungs (42)

digestive system a collection of organs that break down food so that it can be used by the body; the organs in this system include the stomach, the pancreas, the liver, the gallbladder, the small intestine, and the large intestine (54)

drug any chemical substance that causes a physical or emotional change in a person (155)

drug abuser a person who takes drugs for a purpose other than to relieve a medical condition (159)

E

egg a sex cell produced by a female (103)

embryo an organism in the earliest stage of development (110)

enamel the outermost layer of a tooth; the hardest material in the body (56)

endocrine system a collection of glands that control body-fluid balance, growth, and sexual development (88)

enzyme a protein that makes it possible for certain chemical reactions to occur quickly (55)

epidermis the outermost layer of the skin (17)

epididymis (EP uh DID i mis) the area of the testes where sperm are stored before they enter the vas deferens (106)

epithelial tissue one of the four main types of tissue in the body; the tissue that covers and protects underlying tissue (4)

esophagus (i SAWF uh guhs) a long, straight tube that connects the throat to the stomach (56)

excretion the process of removing wastes from the body; term used only when substances must pass through a membrane in order to leave the body (62)

extensor a muscle that straightens part of the body (13)

external fertilization the fertilization of eggs by sperm that occurs outside the body of the female (104)

F

fallopian tube the tube that leads from an ovary to the uterus (107)

farsighted describes someone who has better vision for distant objects than for near ones (85)

fat energy-storing nutrients that help the body store some vitamins (149)

feedback controls a system that turns endocrine glands on or off (90)

fetus an embryo during the later stages of development within the uterus (112)

flexor a muscle that bends part of the body (13)

fragmentation a type of reproduction in which an organism breaks into two or more parts, each of which may grow into a separate individual (102)

G

gallbladder a small, baglike organ that stores bile (59)

gland a group of cells that make special chemicals for the body (88)

H

hair follicle a small organ in the dermis layer of the skin that produces hair (18)

hallucinogen (huh LOO si nuh juhn) a drug that distorts the senses, causes changes in mood and thought processes, and causes hallucinations (158)

helper T cell an immune-system cell that activates killer T cells and B cells (132)

hemoglobin (HEE moh GLOH bin) the protein in red blood cells that attaches to oxygen so that oxygen can be carried through the body (30)

homeostasis (HOH mee OH STAY sis) the maintenance of a stable internal environment (4)

hormone a chemical messenger that carries information from one part of an organism to the other; in mammals, hormones are made by the endocrine glands (88)

hygiene methods of preserving and protecting your health (161)

I

immune system a collection of cells, tissues, and organs that fight disease-causing agents (131)

immunity resistance to a disease (128)

implantation the process in which an embryo imbeds itself in the lining of the uterus (110)

impulse an electrical message that passes along a neuron (77)

infectious disease a disease caused by a pathogen (126)

infertile the state of being unable to have children (109)

integumentary system (in TEG yoo MEN tuhr ee) a collection of organs that helps the body maintain a stable and healthy internal environment; the organs in this system include skin, hair, and nails (16)

internal fertilization the fertilization of an egg by sperm that occurs inside the body of a female (104)

iris the colored part of the eye (85)

J

joint the place where two or more bones connect (10)

K

kidney a bean-shaped organ that removes many harmful substances from the blood (63)

killer T cell an immune-system cell that kills body cells infected with pathogens (132)

L

large intestine a large organ that reabsorbs water from the digestive tract and stores, compacts, and eliminates indigestible material from the body (60)

larynx (LER ingks) the area of the throat that contains the vocal cords (41)

lens a curved, transparent object that forms an image by refracting light (85)

ligament a strong band of tissue that connects bones to bones (11)

liver a large, reddish brown organ that produces bile and stores nutrients; the liver has more than 200 functions in the body (59)

lung a saclike organ that takes oxygen from the air and delivers it to the blood (41)

lymph the fluid and particles absorbed into lymph capillaries (38)

lymph nodes small, bean-shaped organs that contain small fibers that work like nets to remove particles from the lymph (39)

lymphatic system a collection of organs that collect extracellular fluid and return it to the blood; the organs in this system include the lymph nodes and the lymphatic vessels (38)

M

macrophage (MAK roh FAYJ) an immune-system cell that engulfs pathogens (131)

malnutrition a disorder resulting from not consuming the right combination of nutrients (154)

marsupial a mammal that gives birth to live, partially developed young that continue to develop inside the mother's pouch or skin fold (105)

mechanical digestion the breaking, crushing, and mashing of food (55)

medulla (mi DOOL uh) the part of the brain that connects to the spinal cord and controls many involuntary processes in the body (80)

meiosis (mie OH sis) cell division that produces sex cells (103)

melanin a darkening chemical in the skin that determines skin color (16)

memory B cell an immune-system cell that "remembers" how to make a specialized antibody for a particular pathogen (134)

menstruation the monthly discharge of blood and tissue from the uterus (107)

mineral an element that is essential for good health (150)

monotreme a mammal that lays eggs (105)

motor neuron a neuron that sends impulses from the brain and spinal cord to other systems (78)

muscle tissue one of the four main types of tissue in the body; contains cells that contract and relax to produce movement (5)

muscular system a collection of organs whose primary function is movement; organs in this system include the muscles and the connective tissue that attaches them to bones (12)

N

narcotics drugs made from opium (159)

nearsighted describes someone who has better vision for near objects than for distant ones (85)

nephron a microscopic filter in the kidney that removes a variety of harmful substances from the blood (63)

nerve an axon bundled together with blood vessels and connective tissue (78)

nervous system a collection of organs that gather and interpret information about the body's internal and external environment and respond to that information; the organs in this system include the brain, nerves, and spinal cord (76)

nervous tissue one of the four main types of tissue in the body; the tissue that sends electrical signals through the body (4)

neuron a specialized cell that transfers messages throughout the body in the form of fast-moving electrical energy (77)

nicotine a chemical stimulant found in tobacco that increases heart rate and blood pressure (157)

noninfectious disease a disease that cannot spread from one person to another (126)

nutrient a substance that must be consumed or taken in by an organism to promote normal growth, maintenance, and repair (55, 148)

O

obesity a disorder characterized by an extremely high percentage of body fat (154)

optic nerve a nerve that transfers electrical impulses from the eye to the brain (84)

organ a combination of two or more tissues that work together to perform a specific function in the body (5)

organ system a group of organs that works together to perform body functions (5)

ovulation the process in which an egg is ejected through the ovary wall (107)

P

pancreas a fish-shaped organ between the stomach and small intestine that produces enzymes for chemical digestion (58)

pasteurization (PAS tuhr i ZAY shuhn) a method of heating food and beverages to kill bacteria (128)

pathogen an agent that causes a disease (126)

penis the male reproductive organ that transfers semen into the female's body during sexual intercourse (106)

peripheral nervous system the collection of communication pathways, or nerves, whose primary function is to transfer information from all areas of the body and the outside environment to the central nervous system and from the central nervous system to the rest of the body (76)

peristalsis (PER uh STAHL sis) a rhythmic muscle contraction in the digestive tract (56)

pharynx (FER ingks) the upper portion of the throat (41)

photoreceptors specialized neurons in the retina that detect light (84)

placenta a special organ of exchange that provides a developing fetus with nutrients and oxygen (111)

placental mammal a mammal that nourishes its unborn offspring with a placenta inside the uterus and gives birth to well-developed young (105)

plasma the fluid part of blood (30)

platelet a cell fragment that helps clot blood (31)

prescription a note written by a doctor to allow a patient to buy a medicine (156)

protein a biochemical that is composed of amino acids; its functions include regulating chemical reactions, transporting and storing materials, and providing support (149)

puberty the time of life when the sex organs become mature (106)

pulmonary circulation the circulation of blood between the heart and lungs (34)

pupil the opening to the inside of the eye (85)

R

receptor a specialized cell, sometimes a dendrite, that detects changes inside or outside the body (78)

red blood cell a cell that carries oxygen from the lungs to all cells of the body and carries carbon dioxide back to the lungs to be exhaled (30)

reflex a quick, involuntary response to a stimulus (82)

respiration the exchange of gases between living cells and their environment; includes breathing and cellular respiration (40); see cellular respiration

respiratory system a collection of organs whose primary function is to take in oxygen and expel carbon dioxide; the organs of this system include the lungs, the throat, and the passageways that lead to the lungs (40)

retina a layer of light-sensitive cells in the back of the eye (84)

rods photoreceptors that detect very dim light (84)

S

salivary glands organs located around the mouth that produce a liquid that begins chemical digestion (56)

saturated fat a type of fat found in meats, dairy products, coconut oil, and palm oil; known to raise blood cholesterol levels (150)

scrotum a skin-covered sac that hangs from the male body and contains the testes (106)

semen a mixture of sperm and fluids (106)

seminiferous tubules (SEM uh NIF uhr uhs TOO BYOOLZ) the coiled tubes inside the testes where sperm cells are produced (106)

sensory neuron a special neuron that gathers information about what is happening in and around the body and sends this information on to the central nervous system (78)

sexual reproduction reproduction in which two sex cells join to form a zygote; sexual reproduction produces offspring that share characteristics of both parents (103)

sexually transmitted disease a disease that can pass from an infected person to an uninfected person during sexual contact (109)

skeletal muscle the type of muscle that moves bones and helps protect inner organs (12)

skeletal system a collection of organs whose primary function is to support and protect the body; the organs in this system include bones, cartilage, ligaments, and tendons (8)

small intestine a muscular tube about 2.5 cm in diameter and up to 6 m long; the site of most chemical digestion (58)

smooth muscle the type of muscle found in the blood vessels and the digestive tract (12)

sperm a sex cell produced by a male (103)

spleen an organ that filters blood and produces lymphocytes (39)

spongy bone a type of bone tissue that has many open spaces and contains marrow (9)

stimulant a drug that speeds up the action of the central nervous system (155)

stomach a muscular, baglike organ of the digestive tract that is attached to the lower end of the esophagus (57)

stress a physical and mental response to situations that create pressure (163)

sweat glands small organs in the dermis layer of the skin that release sweat (16)

systemic circulation the circulation of blood between the heart and the body (excluding the lungs) (34)

T

T cell an immune-system cell that matures in the thymus (131)

tendon a tough connective tissue that connects skeletal muscles to bones (13)

testes the organs in the male reproductive system that make sperm and testosterone (106)

thymus a lymph organ that produces lymphocytes (39)

tissue a group of similar cells that work together to perform a specific job in the body (4)

tonsils small masses of soft tissue located at the back of the nasal cavity, on the inside of the throat, and at the back of the tongue (39)

trachea (TRAY kee uh) the air passageway from the larynx to the lungs (41)

U

umbilical cord the cord that connects an embryo to a placenta (111)

unsaturated fat a type of fat that usually comes from plant sources and helps reduce blood cholesterol levels (150)

ureter a slender tube that carries urine from each kidney to the urinary bladder (63)

urethra in males, a slender tube that carries urine and semen through the penis to the outside; in females, a slender tube that carries urine to the outside (63, 106)

urinary bladder a baglike organ that stores urine until it can be eliminated through the urethra (63)

urinary system a collection of organs that remove waste from the blood; this system includes the kidneys, ureters, urethra, and the urinary bladder (62)

urine a concentrated mixture of waste materials that forms in the nephrons of the kidney (63)

uterus the organ in the female reproductive system where a zygote grows and develops (107)

V

vaccine a substance that helps the body develop an immunity to a pathogen (128)

vagina the passageway in the female reproductive system that receives sperm during sexual intercourse (107)

vas deferens (vas DEF uh RENZ) a tube in males where sperm is mixed with fluids to make semen (106)

veins blood vessels that direct blood to the heart (33)

ventricle a lower chamber of the heart (32)

villi fingerlike projections on the inside wall of the small intestine (58)

vitamin an organic compound that controls many body functions, including cell growth and hormone production (151)

W

white blood cell a blood cell that protects the body against pathogens (31)

Z

zygote a fertilized egg (103)

Index

Boldface numbers refer to an illustration on that page.

A

ABO blood group, 36, **36**
accidents
 alcohol-related, 158, **158**
 calling for help, 165, **165**
 first aid for victims of, 165, **165**
 prevention of, 164, **164**
acupuncture, 144
addiction, 156, 160
adolescence, 114, **114**
adrenal gland, 88, **89**
adrenaline, 88
adulthood, **114,** 115
aerobic exercise, 162, **162**
aging process, 115
AIDS (acquired immune deficiency syndrome), 109, 137, **137,** 159
air, at high elevation, 41
alcohol, use of, 158–160, **159**
allergy, 135, **135,** 144
alveoli, 41, **41**
amebic dysentery, **126**
amino acids, 55, **55,** 149
amnion, 111, **111**
amphetamines, 159
anabolic steroids, 15
analgesics, 155
Animalia (kingdom), 199
animals, human pathogens spread by, 127
anorexia nervosa, 154
antibiotic-resistant microorganisms, 145
antibiotics, 155
 produced by animals, 145
 produced by bacteria, 172
 treatment of bacterial disease with, 129
antibodies, 31, 36, **36,** 131, **131,** 133–134
antidiuretic hormone, 64, 72
antigens, 36, **36**
antihistamines, 155
anus, 54, **54,** 60, **60**
artery, 33, **33–34**
arthritis, 10
 rheumatoid, 136, **136**
asexual reproduction, 102, **102**
aspirin, 155
asthma, 135
atherosclerosis, 37, **37**
athlete, water requirement of, 64, 72, 150, **150**
autoimmune disease, 136, **136**
axon, 77, **77**

B

B cells, 131–132, **131,** 134
bacteria
 antibiotic resistance in, 145
 decomposers, 172
 pathogens, 65, 126, **126**
 products useful to humans, 172
bacterial infection, 126, **126**
 treatment of, 129
 of urinary system, 65
balanced diet, 148–151, **148**
ball-and-socket joint, **10**
bar graphs, 195, **195**
barbiturates, 159
bicarbonate, 58
bile, 59
biodegradable, 172
biotin, 151
blood, **5,** 30–31, **31**
 amount of water in, 64, 72
 removal of waste products from, 62, **62**
blood cells, formation of, 8–9, 30–31
blood clotting, 31, **31**
blood flow, 14, 35
 exercise and, 35
 through heart, 32, **32**
 through body, 34, **34**
blood pressure, 35, **35,** 80, 154, **158**
blood type, 36, **36**
blood vessels, **17,** 33, **33**
blood-sugar level, **90,** 91
body temperature
 fever, 134, **134**
 regulation of, 16, 18
bone
 compact, 9, **9**
 composition of, 9, **9**
 functions of, 8, **8**
 growth of, 10
 spongy, 9, **9**
brain, 79, **79,** 80
breathing, 40–42, **42**
bronchioles, 41, **41**
bronchitis, 157
bronchus, **40,** 41, **41**
budding, 102, **102**
bulimia, 154

C

caffeine, 64
calcium, 150
Calorie, 148
cancer, 19, 136, **136,** 149, 173
 breast, 109
 colon, 61
 destruction of cancer cell, 136, **136**
 lung, 43, 157
 mouth and throat, 157
 prostate, 109
 skin, 16, 19, **19**
canine teeth, 56, **56**
capillary, 33, **33–34**
carbohydrates
 complex, 149
 in diet, 148–149, **149**
 digestion of, 55, 58
 simple, 149
carbon dioxide
 production in cellular respiration, 42, **42**
 transport in blood, **34,** 42, **42**
cardiovascular system, **6,** 30–37, **30,** 43
careers, naturopathic physician, 144
cartilage, **9–10,** 10
cell membrane, 145
cellular respiration, 40, 42
central nervous system, 76, **76,** 79–81
cerebellum, 79–80, **80**
cerebrum, 79, **79–80**
chemical digestion, 55–57, **55**
childhood, 114, **114**
chloride, 150
cholesterol, 37, 59, 150, 173
chromosomes, 103
chyme, 57–58
cigarette smoking, 37, 43, **43,** 113, 126, **135,** 157, **157,** 159–160
cilia, **157**
circle graph, 193, **193**
circulation
 pulmonary, 34, **34**
 systemic, 34, **34**
collagen, **5,** 17
colon cancer, 61
common cold, 129
compact bone, 9
complete protein, 149
complex carbohydrates, 149
compound light microscope, 200–201, **200**
concave lens, **85**
concept mapping, 186
cones (photoreceptors), 84
connective tissue, 4, **5,** 17
constipation, 61
contact lenses, 85
control group, 191
controlled experiment, 191
conversion tables, SI, 187
convex lens, **85**
cornea, **84**
cowpox, 128
CPR course, 165, **165**
cramping, 14

Index **207**

D

decomposers, bacteria 227, 172
dehydration, 61
dendrite, 77, **77**
depressants, 155
dermis, 17, **17**
development
    of humans, 110–115, **110–114**
    of seeds,
diabetes mellitus, 136, 154, 173
diaphragm, **40,** 42
diarrhea, 61, **61**
diastolic pressure, 35
diet
    balanced, 148–151, **148**
    healthy, 152–153
    vegetarian, 173
digestion, **5,** 54–61
    chemical, 55–57, **55**
    mechanical, 55–57, **57**
digestive enzymes, 55, **55,** 58
digestive system, **7, 54,** 54–61
digestive tract, 54, **54**
disease, 126–129
    autoimmune, 136, **136**
    causes of, 126, **126**
    infectious, 126
    noninfectious, 126
diuretics, 64
dolphin, 87
drug abuse, 159–160
    getting off drugs, 160
    signs of, 160
drugs, 155
    dependence, addiction, and
        tolerance, 156
    effects of, 155–156
    gateway, 159, **159**
    myths about, 160
    products containing, **155**
    side effects of, 156
    types of, 156–159
    use of, 159–160

E

ear, 86, **86**
eating disorder, 154
echidna, 105
eggs, 103, **103,** 107
    of amphibians, 98, **104**
    of humans, 103, 107–108, **110**
    of monotremes, 105, **105**
electrolytes, muscle function and, 14
embryo, 110–111, **110–111**
enamel, tooth, 56, **56**
endocrine system, **7,** 88–91, **89**
energy, from food, 148, **149**
environment, human responses to, 83–87
enzymes
    digestive, 55, **55,** 58
    in small intestine, 58
    in stomach, 57
epidermis, of skin, 17, **17**

epididymis, 106, **106**
epinephrine, 88
epithelial tissue, 4, **4–5,** 17
esophagus, 41, 54, **54,** 56, **57,** 60
essential amino acids, 149
estrogen, 107
excretion, 62
exercise, 162–163
    aerobic, 14, **14,** 162, **162**
    blood flow and, 35
    effect on muscular system, 14
    heart and, 35
    lack of, 126
    replacing water lost in, 64, 72, 150, **150**
    resistance, 14, **14**
experiment, 191
eyes
    color of, 85, **85**
    defenses against pathogens, 130
    human, **84,** 85
    lens of, **84,** 85

F

fallopian tube, 107, **107**
farsightedness, 85, **85**
fat
    in diet, 126, 148–150, **149**
    digestion of, 55, 58–59
    saturated, 150, 173
    unsaturated, 150
feces, 60, 73
feedback control, 90, **90**
fertilization, 103, **103,** 110, **110**
    external, 104, **104**
    internal, 104
    twins and, 108, **108–113**
fetal development, 112–113, **112**
fetus, 112, **112–113**
fever, 134, **134**
fiber, 60
fight-or-flight response, 88, **88**
first aid, 165, **165**
fluorine, 150
folic acid, 151
food and pathogens, 127
food allergy, 144
food label, 153
food pyramid, 152, 154
food safety, 127
fragmentation, 102, **102**
frogs, 104, **104**
fulcrum, 11, **11**
fungus, pathogenic, 126, **126**

G

gallbladder, 54, **54,** 59, **59**
gastric ulcer, 61, **61**
gateway drugs, 159, **159**
gland, 88
glasses, 85, **85**
glucose, blood, **88, 90,** 91
graduated cylinder, 189, **189**
growth, of humans, 110–115, **114**

growth hormone, 91
gums (mouth), 56, **56,** 157

H

hair, **17,** 18
    color of, 18
    functions of, 18
    structure of, 18, **18**
hair follicle, **17,** 18
hallucinogens, 158
handwashing, 161
healthy habits, 161–165
    diet, 152–153
    survey of, 147
hearing, 86, **86**
heart
    atrium, 32, **32, 34**
    blood flow through, 32, **32**
    exercise and, 29
    failure of, 37
    of humans, 32, **32,** 34
    valves, 32, **32**
    ventricle, 32, **32, 34**
heart disease, 149–150, 154, 157, 173
helper T cells, 132, 134, **136–137,** 137
hemispheres
    cerebral, 79, **79**
hemoglobin, 30
hemophilia, 126
herbal medicines, 157, **157**
heroin, 159
high-fat diet, 126
hinge joint, **10**
HIV, 137, **137**
homeopathy, 144
homeostasis, 4
hormones, 19, 88–91, **89–91**
    control of fluid balance by, 64
host, 73
humans
    body defenses against pathogens, 130–134
    chromosomes of, 103
    digestive system of, 54–61
    diseases of, 126–129
    growth and development of, 101, 110–115, **110–115**
    reproduction in, 106–109
    responses to environment, 83–87, **83–87**
    urinary system of, **62,** 62–65
hydra, 102, **102**
hygiene, 161
hypertension, 37
hypothesis, 190–192

I

immune system, 131–134
    cells of, 131, **131**
    challenges to, 135–137
    response to pathogens, 132–133
immunity, 128

impulse, 77, **77**
incisors, 56, **56**
incomplete protein, 149
infancy, 113–114, **114**
infectious disease, 126
infertile, 109
injury prevention, 164, **164**
insulin, **90**, 91, **91**
integumentary system, **6**, 16–19, **16–19**
involuntary muscle, 12
involuntary process, 79
iodine, 150
iris (eye), 85
iron, 150

J

Jenner, Edward, 128
joint, 10–11, **10**

K

Kelvin scale, 188
kidney, **62**, 63, **63**
   diseases of, 65, **65**
   filtration of blood, 63, **63**
kidney machine, 65, **65**
kidney stones, 65, **65**
killer T cells, 132, 136, **136**
knee-jerk reflex, 82

L

labor (childbirth), 113
large intestine, 54, **54**, 60, **60**
larynx, **40**, 41
lens
   concave, **85**
   convex, **85**
   of eye, **84–85**, 85
   of microscope, **200**
lever, 11, **11**
   first class, 11, **11**
   second class, 11, **11**
   third class, 11, **11**
ligament, 11
light
   response of eyes to, 84–85, **84**
liver, 54, **54**, 59, **59**
load, 11, **11**
lung cancer, 157
lungs, **6**, 40–43, **40–43**
   pulmonary circulation, 34, **34**
lupus, 136
Lyme disease, 127
lymphatic system, **7**, 38–39, **38–39**

M

macrophages, 131–134
magnesium, 150
malnutrition, 154
mammals
   reproduction in, 105, **105**
marijuana, 158–159, **159**
marrow, 9, 30–31

mechanical digestion, 55–57, **57**
medulla, 79–80, **80**
meiosis, 103
melanin, 16, **16**, 18
memory B cells, 134
menstruation, 107, 114
middle ear, 86, **86**
milk, pasteurized, 128
minerals, 148, 150
molars, 56, **56**
mole (skin), 19, **19**
monotremes, 105
motor neuron, 78, **78**, 81
motor vehicle accident, **158**, 164
mouth, 54, **54**, 56
   cancer, 157
   defenses against pathogens, 130
   dry, 63, 72
movement
   bones and, 8
   muscular system and, 13, **13**
muscle
   injury to, 15, **15**
   involuntary, 12
   neural control of, 78, **78**
   of skin, **17**, 18
   spasm, 14
   strain, 15, **15**
   types of, 12, **12**
   voluntary, 12
muscle tissue, 4, **5**
muscular system, **6**, 12–15

N

nails, 18, **18**
narcotics, 159
nearsightedness, 85, **85**
needle sharing, 159
nephron, 63, **63**
nerve, 76, 78, **78**
   auditory, **86**
   optic, 84, **84**
nerve endings
   in skin, 16, **17**
nervous system, **7**, 76–82, **76**
   of humans, 76–82
nervous tissue, 4, **4–5**
neuron, 77
   motor, 78, **78**, 81
   sensory, 78, 81
   structure of, 77, **77**
nicotine, 157, 159, **159**
noninfectious disease, 126
nose, 40, **40**, 87, **87**, 130
nutrients, 148
   absorption from small intestine, 58
   storage in liver, 59
nutrition, 148–151
   healthy eating habits, 152–153
   naturopathic medicine and, 144
nutrition facts label, 153
nutritional disorders, 154

O

obesity, 154
oil glands, **17**, 19
oil spill, 172
oils, 150
olfactory cells, 87
opossum, **105**
organ, 5, **5**
organ system, 5, **6–7**
outdoor safety, 164
ovary, human, **89**, 107, **107**
over-the-counter drugs, 156
ovulation, 107
oxygen
   in air at high elevation, 41
   cellular respiration and, 42, **42**
   transport in blood, 30, **30**, **34**, 42, **42**

P

palm oil, 150
pancreas, 54, **54**, 58, **59**, **89–90**, 91
pancreatic juice, 58
pantothenic acid, 151
parathyroid gland, **89**
Pasteur, Louis, 128
pasteurization, 128, **128**
pathogens, 31, **126–126**
   body defenses against, 130–134
   preventing spread of, 128
   spread of, 127, **127**
PCP, 158
peer pressure, 159
pelvis, **8**
penguins, 104, **104**
penis, 106, **106**
percentages, 153
peripheral nervous system, 76–78, **76**
peristalsis, 56
pharynx, **40**, 41
phosphorus, 150
photoreceptors, 84, **84**
pituitary gland, **89**, 91
placenta, 111, **111**, 113
plasma, 30
platelets, 30–31, **31**, 131
pneumonia, 43
pollen, 43, 135, **135**
pollutants
   similar to female hormones, 109
posture, 161, **161**
potassium, 72, 150
pregnancy, 110–113
   ectopic, 108
premolars, 56, **56**
prescription drugs, 156, **156**
protein
   complete, 149
   in diet, 148–149, **149**
   digestion of, 55, **55**, 58
   in immune response, 133
   incomplete, 149
   pathogenic, 126, **126**

puberty, 106, 114
pulse, 33
pupil (eye), 84–85, **84**

R

rabies, **126**
ragweed pollen, **135**
receptors, 78, 83, **83**
rectum, 54, **54,** 60, **60**
red blood cells, 8–9, 30, **33**
    blood type and, 36, **36**
    breakdown of, 39
reflex, 82, **82**
refrigeration, 127
reproduction
    asexual, 102, **102**
    in humans, 106–109
    in mammals, 105, **105**
    in marsupials, 104–105, **105**
    in monotremes, 105, **105**
    in placental mammals, 105
    sexual, 103, **103**
reproductive system (human)
    diseases of, 109
    female, **6,** 107, **107**
    male, **6,** 106, **106**
respiration, 40. *See also* cellular respiration
respiratory system, **6,** 40–43, **40–42**
    air flow through, 40–42, **40, 42**
    cilia in, **157**
    disorders of, 43, **43**
    parts of, 40–41, **40–41**
retina, 84, **84**
rheumatoid arthritis, 136, **136**
rib cage, **8,** 42
ringworm, 127
rods (photoreceptors), 84

S

safety
    drug safety tips, 156
    at home, 164, **164**
    outdoors, 164
    on road, 164
saliva, 56, 64, 72, 145
salivary glands, 54, **54,** 56
Salmonella, 127
saturated fats, 150, 173
scolex, 73
scrotum, 106, **106**
sea star, 102, **102**
seat belts, 164
sedatives, 155
semen, 106, **106**
seminiferous tubules, 106, **106**
semivegetarian diet, 173
senses, 83–87
sensory neuron, 78, 81
sensory receptors, 83, **83**
sewage treatment plant, 60
sex cells, 103, **103**
sexual reproduction, 103, **103**

sexually transmitted disease, 109
SI units, 187
simple carbohydrates, 149
skeletal muscle, 8, 12–13, **12–13,** 38
skeletal system, 6, 8–11, **8–11**
    storage function, 8
    support function, 8
skin
    as a protective barrier, 19, 130, **130**
    color of, 16, **16**
    functions of, 16
    infections on, 19
    layers of, 17, **17**
    receptors in, 83, **83**
    wound healing, 19, **19**
skin cancer, 16, 19, **19**
sleep, 162, **162**
sliding joint, **10**
small intestine, 54, **54,** 57–58, **57–58**
    enzymes in, 55, **55**
    tapeworm in, 73
smallpox, 128
smell, 87, **87**
smooth muscle, 12, **12**
sodium, 72, 150
sound wave, 86, **86**
sperm, 103, **103,** 106, **108,** 110
sphincter, 60
spinal cord, 78, 81, **81**
spleen, 39, **39**
steroids, 15
stimulants, 155
stomach, 5, **5,** 54, **54,** 57, **57**
    enzymes in, 55, **55**
    ulcers of, 61, **61**
stomach acid, **5,** 57
strep throat, **126**
stress, 163, **163**
stroke, 37
sweat, 16, **17,** 64, 72
sweat glands, 16, **17**
systolic pressure, 35

T

T cells, 131, 134
    helper, 132, 134, **136–137,** 137
    killer, 132, 136, **136**
tapeworms, 73
taste, 87, **87**
taste bud, 83, 87, **87**
teeth, **54**
    care of, 161
    tooth structure, 56, **56**
    types of, 56, **56**
tendon, 13, **13**
    injury to, 15
testes, **89,** 106, **106**
testosterone, 15, 106
thirst, 64, 72
throat, 54, **54,** 157
thymus gland, 39, **39, 89**

thyroid gland, **89,** 91
tissue, 4–5, **4–5**
tobacco, 157, **157**
tolerance of drug, 156
tongue, 56, 87, **87**
tonsils, 39, **39**
trachea, **40,** 41, **41**
transfusion, 36
twins, 108, **108**

U

ulcer, gastric, 61, **61**
umbilical cord, 111, **111,** 113, **113**
unsaturated fats, 150
ureter, **62,** 63, **63**
urethra, **62,** 63, 106, **106**
urinary bladder, **62–63,** 63
urinary system, 6, 62–65, **62**
urination, 63
urine, 63
uterus, 107, **107,** 110, **110**

V

vaccine, 128
vagina, 107, **107**
valve
    heart, 32, **32**
    venous, 33
vas deferens, 106, **106**
vein
    circulatory system, 33, **33–34**
vertebrae, 81, **81**
vertebral column, **8**
villi, 58, **58**
viruses
    treatment of viral diseases, 129
vision, 84–85, **84–85**
vitamins, 84, 148, 151
vocal cords, 41
voice box, 41
voluntary muscle, 12

W

water
    balance in human body, 64
    contaminated with pathogens, 127
    human requirement for, 148, 150, **150**
    reabsorption in small intestine, 60
white blood cells, 30–31, **31,** 39
windpipe. *See* trachea
withdrawal symptoms, 156, 159–160
wound healing, 19, **19,** 145

Z

zebra, 104, **104**
zinc, 150
zygote, 103

Credits

Abbreviations used: (t) top, (c) center, (b) bottom, (l) left, (r) right, (bkgd) background

ILLUSTRATIONS

All illustrations, unless otherwise noted below by Holt, Rinehart and Winston.

Table of Contents vi(tl), Kip Carter; vii(tr), Keith Kasnot; vii(br), Morgan-Cain Associates.

Scope and Sequence: T11, Paul DiMare, T13, Dan Stuckenschneider/Uhl Studios, Inc.

Chapter One Page 4 (c,cl), Morgan-Cain & Associates; 5 (cl), Morgan-Cain & Associates; 5 (c), Morgan-Cain & Associates; 5 (tr), Morgan-Cain & Associates; 6, Christy Krames; 7, Christy Krames; 9, Keith Kasnot; 10, John Huxtable/Black Creative; 11, John Huxtable/Black Creative; 13 (br), Christy Krames; 17 (br), Morgan-Cain & Associates; 17 (cr), Marty Roper/Planet Rep; 19 (t), Morgan-Cain & Associates; 22 (br), John Huxtable/Black Creative; 24 (br), Christy Krames; 25 (tr), Morgan-Cain & Associates.

Chapter Two Page 30 (tr), Christy Krames; 31 (b), Keith Kasnot; 32, Kip Carter; 33 (c), Kip Carter; 34 (b), Kip Carter; 36 (tl), Jared Schneidman Design; 36 (br), Marty Roper/Planet Rep; 38 (br), Kip Carter; 39 (tr), Christy Krames; 40 (bl), Christy Krames; 41 (b), Christy Krames; 42 (br), Christy Krames; 42 (bc,bl), Kip Carter; 46 (br), Christy Krames; 48 (tr), Kip Carter; 49 (cr), Kip Carter.

Chapter Three Page 54 (bc), Christy Krames; 55 (b), Brian Evans; 56 (tl,bl), Keith Kasnot; 57 (cl), Christy Krames; 57 (cr), Brian Evans; 58 (tl), Marty Roper/Planet Rep; 58 (cl), Christy Krames; 58 (c), Brian Evans; 59, Christy Krames; 60 (tl), Christy Krames; 62 (cl), Christy Krames; 63 (cr), Keith Kasnot; 68 (cr), Brian Evans; 68 (tr), Keith Kasnot; 69 (tr), Christy Krames; 70 (br), Keith Kasnot; 70 (tr), Brian Evans.

Chapter Four Page 76 (bl), Christy Krames; 77 (b), Scott Thorn Barrows/The Neis Group; 78 (b), Scott Thorn Barrows/The Neis Group; 79 (bc), Brian Evans; 80 (b), Brian Evans; 81 (tr), Christy Krames; 83 (b), Morgan-Cain & Associates; 84 (bc), Keith Kasnot; 84 (bl), Carlyn Iverson; 85 (bl), Keith Kasnot; 86 (b), Christy Krames; 87 (tr), Keith Kasnot; 88 (b), Dan McGeehan/Koralick Associates; 89 (b), Christy Krames; 90, Christy Krames; 94 (br), Keith Kasnot; 95 (cl), Dan McGeehan/Koralick Associates; 96 (tr), Christy Krames; 97 (tr), Christy Krames.

Chapter Five Page 103 (bl), Rob Schuster/Hankins and Tegenborg; 106 (bl), Keith Kasnot; 107 (tr), Keith Kasnot; 108 (br), Rob Schuster/Hankins and Tegenborg; 110 (cl), David Fischer; 111 (cr), Christy Krames; 121 (cr), Sidney Jablonski; 122 (bl), Morgan-Cain & Associates.

Chapter Six Page 130 (br), Scott Thorn Barrows/The Neis Group; 131 (bl), Scott Thorn Barrows/The Neis Group; 132-133, Blake Thornton/Rita Marie; 134 (l), Stephen Durke/Washington Artists; 140 (br), Blake Thornton/Rita Marie; 142 (bc), Scott Thorn Barrows/The Neis Group; 143 (tr), Sidney Jablonski.

LabBook Page 178, Morgan-Cain Associates;

Chapter Seven Page 159 (br), Marty Roper/Planet Rep; 164 (c), Uhl Studios, Inc.; 169 (tl), Marty Roper/Planet Rep.

Appendix 188 (t), Terry Guyer; 192 (b), Mark Mille/Sharon Langley; 196-197, Kristy Sprott.

PHOTOGRAPHY

Cover: VCG/FPG International

Title page: (cr), Frans Lanting/Minden Pictures; (tc), Kim Taylor/Bruce Coleman, Inc.

Feature Borders: Unless otherwise noted below, all images copyright ©2001 PhotoDisc/HRW. "Across the Sciences" 72, 122, all images by HRW; "Careers" 144, sand bkgd and Saturn, Corbis Images; DNA, Morgan Cain & Associates; scuba gear, ©1997 Radlund & Associates for Artville; "Eureka" 27, 99 ©2001 PhotoDisc/HRW; "Health Watch" 51, 73, 145, 173, dumbbell, Sam Dudgeon/HRW Photo; aloe vera, EKG, Victoria Smith/HRW Photo; basketball, ©1997 Radlund & Associates for Artville; shoes, bubbles, Greg Geisler; "Scientific Debate" 95, Sam Dudgeon/HRW Photo; "Weird Science" 50, mite, David Burder/Stone; atom balls, J/B Woolsey Associates; walking stick, turtle, EclectiCollection.

Table of Contents: iv(tl), Lennart Nilsson/Albert Bonniers Forlag AB, A CHILD IS BORN; iv(cl), Tektoff-RM/CNRI/Science Photo Library/Photo Researchers, Inc.; iv(bl), Chris Hamilton; v(tr), K. H. Kjeldsen/Science Photo Library/Photo Researchers, Inc.; v(tcr), John Huxtable/Black Creative; v(bl), SuperStock; v(bc), Sam Dudgeon/HRW Photo; vi(cl), Prof. P. Motta/Department of Anatomy/University "La Sapienza" Rome/Science Photo Library/Photo Researchers, Inc.; vi(b), Sam Dudgeon/HRW Photo; vii(tr), Sam Dudgeon/HRW Photo.

Scope and Sequence: T8(l), Lee F. Snyder/Photo Researchers, Inc.; T8(r), Stephen Dalton/Photo Researchers, Inc.; T10, E. R. Degginger/Color-Pic, Inc., T12(l), Rob Matheson/The Stock Market

Master Materials List: T26(bl), Sam Dudgeon/HRW Photo; T26(c, br), Image ©2001 PhotoDisc; T27(t, cl, cr), Image ©2001 PhotoDisc

Chapter One: 2-3, AFP/Corbis; 3, HRW Photo; 4-5(b), David Madison/Stone; 10(tl), Peter Dazeley/Stone; 10(bc, br, bl), SP/FOCA/HRW Photo; 12(bc), Bob Torrez/Stone; 12(bl), Dr. E.R. Degginger; 12(br), Manfred Kage/Peter Arnold, Inc.; 12(cl), G.W. Willis/BPS/Stone; 14(bl), Chris Hamilton; 15 Shelby Thorner/David Madison; 18(c), Dr. Robert Becker/Custom Medical Stock Photo; 19(cr), Dr. P. Marazzi/Science Photo Library/Photo Researchers, Inc.; 23 Peter Dazeley/Stone; 26(tr), Dan McCoy/ Rainbow; 27(cl), Gamma-Liaison; 27(tr), Huntsville Times.

Chapter Two: 28-29, Bruce Iverson; 29, HRW Photo; 30(bl), Dr. Dennis Kunkel/Phototake NYC; 31, Don Fawcett/Photo Researchers, Inc.; 33(cl, tr), O. Meckes/Nicole Ottawa/Photo Researchers, Inc.; 33(cr), David Phillips/Science Source/Photo Researchers, Inc.; 35(tr), Custom Medical Stock Photo; 35(br), James Wilson/Woodfin Camp & Associates; 37(tr), Ken Wagner/Phototake NYC; 43(cl, cr), Matt Meadows/Peter Arnold, Inc.; 46(c), Dr. Dennis Kunket/Phototake, Inc.; 47(tr), Don Fawcett/Photo Researchers, Inc.; 49(bl), Dr. Dennis Kunkel/Phototake NYC; 50(tr), Index Stock Photography; 51(tr), Russell Dian/HRW Photo; 51(bl), Jim Gripe/Pivot Media.

Chapter Three: 52-53, Bod Daemmrich; 53, HRW Photo; 61(br), Prof. P. Motta/Dept. of Anatomy/University "La Sapienza" Rome/Science Photo Library/Photo Researchers, Inc.; 61(tr), The Stock Market; 64(br), Image Bank; 65(tr), Stephen J. Krasemann/ DRK Photo; 65(cr), E.K. Martin and Associates; 66, Sam Dudgeon/HRW Photo; 73 J. H. Robinson/Photo Researchers, Inc.

Chapter Four: 74-75, Omikron/Photo Researchers, Inc.; 75, HRW Photo; 85 Bruno Joachim/Liaison; 87 Louis Psihoyos/Matrix; 91 Will & Deni McIntyre/Photo Researchers, Inc.; 99 Journal of Nuclear Medicine.

Chapter Five: 100-101, Lennart Nilsson; 101, HRW Photo; 102(bl), Visuals Unlimited/Cabisco; 102(br), Innerspace Visions; 104(tl), Michael Fogden/Animals Animals; 104(bl), Photo Researchers, Inc.; 104(cr), Guy Mannering/Bruce Coleman; 105 Dr. E. R. Degginger/Bruce Coleman; 108(tl), Chip Henderson/Stone; 110(b), Lennart Nilsson; 111 Petit Format/Nestle/Science Source/Photo Researchers, Inc.; 112 Lennart Nilsson; 113(tr), Lennart Nilsson/Albert Bonniers Forlag AB, A CHILD IS BORN; 113(cr), Keith/Custom Medical Stock Photo; 115 NASA; 117, Victoria Smith/HRW Photo; 118 Guy Mannering/Bruce Coleman, Inc.; 120 Lennart Nilsson/Albert Bonniers Forlag AB, BEING BORN; 123(inset), Vince Viverito, Jr./Richard Wolf Medical Instruments Corp., Vernon Hills, IL; 123(cr), Tom McCarthy/ Rainbow.

Chapter Six: 124-125, Oliver Meckes/Photo Researchers; 125, HRW Photo; 126(br), CNRI/Science Photo Library/Photo Researchers, Inc.; 126(bc), Tektoff-RM/CNRI/Science Photo Library/Photo Researchers, Inc.; 126(bl), Manfred Kage/Peter Arnold; 127 Kent Wood/Photo Researchers, Inc.; 135(tr), Visuals Unlimited/George Musil; 135(bc), Image Copyright ©<2000>PhotoDisc, Inc.; 135(cr), K. H. Kjeldsen/Science Photo Library/Photo Researchers, Inc.; 135(cl), SuperStock; 136(tl), Clinical Radiology Dept., Salisbury District Hospital/Science Photo Library/Photo Researchers, Inc.; 136(bl, br), Dr. A. Liepins/Science Photo Library/Photo Researchers, Inc.; 137 Lennart Nilsson; 141 Dr. A. Liepins/Science Photo Library/Photo Researchers, Inc.; 144(all), Chris Mooney/HRW Photo; 145 E. R. Degginger/Bruce Coleman.

Chapter Seven: 146-147, Arthur Tilley/FPG International; 147, HRW Photo; 152 John Kelly/Stone; 157(bl), E. Dirksen/Photo Researchers, Inc.; 157(tr), ©<2000>Stephen Foster; 157(br), Dr. Andrew P. Evans/Indiana University; 158 Spencer Grant/Photo Researchers, Inc.; 162 Rob Van Petten/Image Bank; 163 Wally McNamee/Corbis; 166, Peter Van Steen/HRW Photo172 Manfred Kage/Peter Arnold; 173 SuperStock.

LabBook: "LabBook Header": "L", Corbis Images, "a", Letraset-Phototone, "b" and "B", HRW, "o" and "k", Images Copyright ©<2000>PhotoDisc, Inc. 175(cl), Michelle Bridwell/HRW Photo; 175(br), Image Copyright ©<2000>Photodisc, Inc.; 176(bl), Stephanie Morris/HRW Photo; 177(tr), Jana Birchum/HRW Photo.

Appendix: 200 CENCO.

Sam Dudgeon/HRW Photos: all Systems of the Body background photos, p. viii-1, 65 (bl), 6, 8, 13(tr, tc), 14(c), 16, 21, 22(c), 44, 54-55(t), 66, 67, 80, 82, 89, 90, 92, 93, 94, 119, 161, 169(cr), 174, 175(bc), 176(br, tl), 176(tr), 177(tl), 180, 181, 184, 189(br).

Peter Van Steen/HRW Photos: p. 11(all), 18(l, br), 72(cl), 114(all), 128, 130, 135(br), 140, 148, 149(all), 150, 153, 155, 156, 157(cr), 162(br), 163(br), 165(all), 168, 170, 171(tr, cr), 177(b), 189(tr).

John Langford/HRW Photos: p. 129, 175(tr).

Acknowledgements continued from page iv.

Alyson Mike
Science Teacher
East Valley Middle School
East Helena, Montana

Donna Norwood
Science Teacher and Dept. Chair
Monroe Middle School
Charlotte, North Carolina

James B. Pulley
Former Science Teacher
Liberty High School
Liberty, Missouri

Terry J. Rakes
Science Teacher
Elmwood Junior High School
Rogers, Arkansas

Elizabeth Rustad
Science Teacher
Crane Middle School
Yuma, Arizona

Debra A. Sampson
Science Teacher
Booker T. Washington Middle School
Elgin, Texas

Charles Schindler
Curriculum Advisor
San Bernadino City Unified Schools
San Bernadino, California

Bert J. Sherwood
Science Teacher
Socorro Middle School
El Paso, Texas

Patricia McFarlane Soto
Science Teacher and Dept. Chair
G. W. Carver Middle School
Miami, Florida

David M. Sparks
Science Teacher
Redwater Junior High School
Redwater, Texas

Elizabeth Truax
Science Teacher
Lewiston-Porter Central School
Lewiston, New York

Ivora Washington
Science Teacher and Dept. Chair
Hyattsville Middle School
Washington, D.C.

Elsie N. Waynes
Science Teacher and Dept. Chair
R. H. Terrell Junior High School
Washington, D.C.

Nancy Wesorick
Science and Math Teacher
Sunset Middle School
Longmont, Colorado

Alexis S. Wright
Middle School Science Coordinator
Rye Country Day School
Rye, New York

John Zambo
Science Teacher
E. Ustach Middle School
Modesto, California

Gordon Zibelman
Science Teacher
Drexel Hill Middle School
Drexell Hill, Pennsylvania

Self-Check Answers

Chapter 1—Body Organization and Structure

Page 14: Curl-ups use flexor muscles; push-ups use extensor muscles.

Page 17: Blood vessels belong to the cardiovascular system.

Chapter 2—Circulation and Respiration

Page 34: The hollow tube shape of arteries and veins allows blood to reach all parts of the body. Valves in the veins prevent blood from flowing backward.

Page 38: Like blood vessels, lymph capillaries receive fluid from the spaces surrounding cells. The fluid absorbed by lymph capillaries flows into lymph vessels. These vessels drain into large neck veins instead of into an organ, such as the heart. Lymph does not deliver oxygen and nutrients.

Chapter 3—The Digestive and Urinary Systems

Page 59: Bile is involved in the physical digestion because emulsification does not change the chemical composition of the fat molecules; it only increases the surface area of each fat droplet.

Chapter 4—Communication and Control

Page 81: 1. cerebrum 2. cerebellum 3. to protect the spinal cord

Chapter 5—Reproduction and Development

Page 103: In asexual reproduction, one animal produces offspring that are genetically identical to itself. In sexual reproduction, the genes of two individuals are mixed when sex cells join to form a zygote. This zygote develops into a unique individual.

Page 111: The uterus provides the nutrients and protection that the embryo needs to continue growing. The uterus is also the only place the placenta will form.

Chapter 6—Body Defenses and Disease

Page 127: If Jackie did not wash the counter after cutting up the meat, bacteria could grow on the counter where the meat was. This bacteria could contaminate her brother's sandwich.

Chapter 7—Staying Healthy

Page 153: You have eaten two servings from the bread, cereal, rice, pasta group; one serving from the fruit group, and one from the milk, yogurt and cheese group.

Page 156: Regular use of some drugs may cause tolerance or addiction. Withdrawal symptoms occur when the body does not receive a drug that it is addicted to.

Page 159: 1. A hallucination is a vision or sound that is not real. 2. Heroin is highly addictive and potentially deadly. If shared needles are used to inject it, users risk getting diseases like hepatitis or AIDS.